HAZARDOUS MATERIALS AND HAZARDOUS WASTE MANAGEMENT: A TECHNICAL GUIDE

Gayle Woodside

A Wiley-Interscience Publication

JOHN WILEY & SONS, INC.

New York • Chichester • Brisbane • Toronto • Singapore

Library of Congress Cataloging in Publication Data:

Woodside, Gayle, 1951–
  Hazardous materials and hazardous waste management :
a technical guide / by Gayle Woodside.
      p. cm.
  Includes bibliographical references and index.
  ISBN 0-471-54676-3 (alk. paper)
  1. Hazardous wastes—management. 2. Hazardous
substances—Management. I. Title.
TD1030.W66 1993
604.7—dc20   92-38582

Printed in the United States of America

10 9 8 7 6 5 4 3 2 1

To my father, John A. Woodside

PREFACE

This book began from a series of papers that I wrote in 1989 and 1990 on topics pertaining to hazardous materials and hazardous waste management. I presented these papers at various technical conferences around the country, and I noticed that there always seemed to be large crowds of people trying to learn more about this complex topic. While attending one of these conferences, I met a university professor who stated that there was a dearth of reference and text books on the subject of hazardous materials and hazardous waste. Several of my co-workers had made similar statements. I gathered ideas from professors and co-workers about the type and complexity of material that would prove useful to them, and I began creating the book *Hazardous Materials and Hazardous Waste Management: A Technical Guide.*

The book covers aspects of managing hazardous materials and hazardous waste that an engineer, scientist, or other professional who works in the field will encounter when working in industry or government. Included are regulatory references for the various topics, as well as field applications. Tables have been included whenever possible to give the reader information in a summary form, and figures have been added for clarity. I structured the book by discrete topic, with inclusive information in each chapter, so that a professor can teach any of the sections or chapters in any order, and can leave out a chapter or section not germane to his or her syllabus. Additionally, I hope that professionals working in the field will find the information structured in a way that is easy to use.

The intent of writing on the subjects of hazardous materials *and* hazardous waste (and not just on one topic or the other) was to give students and professionals a sense of the interconnection between hazardous materials and hazardous waste regulations. Traditionally, reference or text books have addressed only one topic, such as industrial hygiene or hazardous waste management. In today's world, this is not enough. Many times, the professional who works in the field is asked to be conversant in all aspects of hazardous materials and hazardous waste management, including exposure assessment, chemical safety, waste analysis and classification, emergency planning, training, labeling, manifesting, and many other aspects. Thus, I have addressed topics that I feel will be beneficial to the student who is just entering this field as well as to the professional who has been part of shaping regulations and management practices over the years.

I would like to acknowledge Mark C. Stuckey for his contributions to the chapters pertaining to treatment technologies and pollution prevention and Teresa Gerber for her help with many of the drawings in the book. Additionally, I would like to thank the following people for their help with technical information: John Woodside, John Prusak, Kelvin Langlois, Dianna Kocurek, Lial Tischler, Sheila Payne, Chris Bauer, Greg Davis, David Dalke, Bonnie Blam, John Reynolds, Jeff Erb, Mary Winton, Mike Slapik, Jim Smart, Barbara Salomon, Maria Martinez, Evelyn Ortiz, Dwayne Cox, Bill LaBonville, Paul Depres, Jim Doersam, Bob Tassan, Stuart Hurwitz, and Ken Takvam. A special thanks goes to the people who made this publication possible: Dan Sayre, Millie Torres, Tracey Thornblade, and Lisa Gargano. And, of course, I would like to thank my husband and friend, Bruce Almy, for all his support and encouragement.

TABLE OF CONTENTS

LIST OF FIGURES

LIST OF TABLES

SECTION I

HAZARDOUS MATERIALS
AND HAZARDOUS WASTE:
AN OVERVIEW

1

HAZARDOUS MATERIALS AND HAZARDOUS WASTE: A REGULATORY OVERVIEW

Regulations pertaining to environmental and workplace hazards have been in place for over three decades. The National Environmental Policy Act of 1969 became the basis for environmental protection in the United States, and all environmental laws enacted thereafter—including those pertaining to air quality, water quality, and hazardous waste—were an outgrowth of the policies and goals set forth in this Act. The Occupational Safety and Health Act of 1970 and subsequent amendments have been the basis for regulation of hazardous materials in the workplace, as well as other aspects of industrial safety and health. Likewise, the Atomic Energy Act of 1954 and subsequent Acts set forth the basis for regulation of radioactive source materials and waste. This chapter provides an overview of the regulatory process and a summary of key regulations that govern hazardous materials and hazardous waste.

THE PROCESS OF REGULATING

Rulemaking

The process of regulating hazardous materials and hazardous waste starts with Acts of Congress in the form of Bills, which must be passed by the House of Representatives and Senate and signed into law by the President. This process is lengthy and complicated and typically requires hearings, followed by amendments or modifications to the original piece of legislation. In order to become law, the Bill must receive majority approval by both chambers of the

Legislature. Thus, many Bills get tabled or otherwise stalled during the process, and are never enacted into law. An Act can take several years to pass, with modifications and amendments introduced throughout the process.

Acts typically embody a concept or goal of the Legislature and set forth a timeframe in which these goals are to be achieved. Acts also empower the Executive Branch of the government, through an agency such as the Environmental Protection Agency (EPA), Occupational Safety and Health Administration (OSHA), Nuclear Regulatory Commission (NRC), or Department of Transportation (DOT), to create regulations that take the concept and refine it into specific rules that are meant to accomplish the goals set forth in the Act.

In some instances, the executive agency lacks scientific or other information required to properly craft regulations, and will commission studies before proposing specific standards or requirements. These studies then become the technical basis of the regulation or future agency policies. In the area of hazardous materials and hazardous waste, the studies may be used to set requirements for permissible exposure limits for a material, treatment standards for a waste, or allowable emissions for a particular pollutant.

When the executive agency begins drafting a regulation, the rulemaking procedure must follow the terms defined in the Administrative Procedure Act. These include providing adequate notice, allowing public comment, publishing substantive rules at specified times, and giving interested parties the right to petition for issuance, amendment, or repeal.

The vehicle used for providing notice of proposed rulemaking is the *Federal Register*.[1] Once the notice is published, the agency is required to give interested persons an opportunity to participate through submission of written data, views, or arguments about the proposed rule. The opportunity for oral presentation is normally part of the process, but is not required under the statute. This public comment period on proposed legislation is usually 60 to 90 days. The basic steps generally followed in the rulemaking process for hazardous materials and hazardous waste are:

- Advanced Notice of Proposed Rulemaking (ANPR)—With this published notice, the regulating agency signals its intent to begin studying key aspects of a regulatory objective. The agency solicits from the public input on how to best achieve its regulatory goal.
- Proposed Rule—A draft regulation, or proposed rule, is published for public comment. Any significant input from the public received during the ANPR comment period, along with comments of agency rationale for including or excluding specific aspects from the rule, is published in the

---

[1]The *Federal Register* system of publication was originally established by the Federal Register Act. Its function was expanded and amended by the Administrative Procedure Act. The *Federal Register* is published daily and contains federal agency regulations, proposed rules, meetings and proceedings, and other documents of the Executive Branch.

preamble. Written comments once again are solicited from the public, and there may be public hearings about the rule held at strategic locations in regions throughout the nation.

The overall intent of the rule normally is not changed through public comments, although definition of terms, applicability, and administrative aspects of the rule such as recordkeeping requirements and reporting deadlines may get modified. In some cases, however, the development and publication of the final rule may be postponed for a significant period because of adverse public opinion about the rule, a change in rulemaking priorities at the agency, or other reasons.

- Final Rule—This rule, once published in the *Federal Register*, can become legally enforceable 30 days after publication, with some exceptions allowing for earlier implementation. It is not uncommon, however, for the agency to set the effective date at six months to one year after publication of the rule. This is meant to give the regulated community time to understand the rule and come into compliance. Final rules are published in the *Code of Federal Regulations*.[2]

Permitting

Once regulations are finalized, the executive agency may establish a permits program to ensure compliance with the new regulations by the regulated community. An industrial facility might need a permit if the facility uses, produces or generates, treats, stores, emits or discharges, or disposes of a hazardous material or hazardous waste.

Most of the permits necessary for hazardous materials and hazardous waste management are required by EPA. EPA can delegate its permitting authority to a state if that state has an acceptable implementation plan for the permitting process and regulates with at least as much restriction as EPA. If the state has not received permitting authority from EPA, the state may require a separate permit in addition to the EPA permit for specific industrial activities such as wastewater discharge or hazardous waste treatment, storage, and disposal. In some instances, local governments also may require a separate permit.

OSHA has outlined recordkeeping and other responsibilities in its regulations for facilities that use chemicals, but individual facility permits are not presently required. The agency maintains the right to inspect a facility's operations and records at any time for compliance with OSHA regulations. OSHA, like EPA, can delegate its inspection authority and other regulatory aspects to a state.

[2]Rules and regulations are codified in the *Code of Federal Regulations* (CFR), which is updated annually. The CFR is divided into 50 titles, encompassing broad subject categories such as labor (29CFR), which includes occupational safety and health, energy (10 CFR), which includes radioactive materials and waste, protection of environmental (40 CFR), and transportation (49 CFR). The *Federal Register* serves as a daily update to the CFR.

The permitting process begins when a facility files an application for a specific permit. Examples of permit applications include: Resource Conservation and Recovery Act (RCRA) waste management permit, Toxic Substance Control Act (TSCA) premanufacture notice, National Pollution Discharge Elimination System (NPDES) wastewater discharge permit, EPA air emissions permit, and NRC permit (license) to dispose of radioactive waste. Examples of facilities that typically are required to apply for and receive permits include chemical-producing facilities, manufacturing facilities, hazardous waste treatment, storage, or disposal facilities, industrial wastewater treatment facilities, publicly owned treatment works (POTW), and others.

The application usually requires general information pertaining to the facility such as name, location, standard industrial classification, and type of manufacturing or operational activities engaged in at the facility. Information specific to the type of permit also is required. The completed application can range from a short application of several pages to a complex application with volumes of information included. Table 1.1 gives examples of activities that typically require permitting.

After the permit application is submitted, the agency reviews it for administrative completeness and technical soundness. Where applicable, the agency relies on health studies, risk assessments, demonstrated technology, established standards (regulatory and nonregulatory), and other engineering data, such as modeling, to determine acceptability of the facility's application.

If the specifics of the application are approved by the agency, a draft permit will be issued and public notice will be made to solicit comments on the permit, as required. At this point, the public has an opportunity to ask questions about the specifics of the permit or request a public hearing to oppose granting the permit. If a citizen or other group does not request a public hearing within a specified timeframe—usually 30 days—the permit is granted by the agency.

If a public hearing is requested, opponents of the permit application are given a chance to explain their reasons for opposition. The agency defends the technical basis of the permit application and the applicant provides input on its own behalf. An unbiased hearing examiner from the agency listens to all arguments and makes the final determination as to whether the permit will be granted or whether specific permit provisions require modification.

Once a permit is issued to regulate certain facility activities, the permit provisions become the facility's enforceable standards. Generally, standards set in the permit remain the facility's enforceable standards until the permit expires. Any changes to the standards developed by the agency typically are incorporated into the permit during the permit renewal process, although some new or revised standards may become enforceable earlier.

TABLE 1.1   Selected Examples of Activities that Typically Require Permitting

Activity	Permit Requirements	Comments
Treatment, storage, or disposal of hazardous waste	Notify EPA or state of activity; obtain an EPA identification number Apply for and obtain a RCRA Part B Permit	Facility must meet interim status or permitted standards Hazardous waste can be stored for up to 90 days without a permit; waste can be treated during this period
Off-site discharge of wastewater or stormwater to land or stream from municipal or industrial facility	Apply for and obtain NPDES and/or state permit	Stormwater discharges can be permitted under the general permit system
Wastewater discharge to publicly owned treatment works (POTW)	If required, apply for a permit or authorization from the POTW	Discharge must meet applicable EPA effluent guidelines and pre-treatment standards
Emissions of air pollutants such as criteria pollutants or hazardous air pollutants	Federal permit required for major sources State permit required in some states for other sources	New Clean Air Act Amendments will alter permitting and technology standards
Manufacture or importation of chemicals	Premanufacture notice must be submitted to EPA under TSCA	Health/environmental effects must be submitted for each new chemical
Disposal of radioactive waste	Apply for and obtain NRC license	Requirements for disposal vary depending on class of radioactive waste

Monitoring and Recordkeeping

Almost all hazardous materials and hazardous waste regulations require some type of monitoring or recordkeeping. Additionally, a permit can specify a monitoring or inspection schedule to verify that permit provisions are being met. Depending upon permit requirements, the data or records are submitted to the agency on a periodic basis such as monthly, quarterly, annually, biennially, or when an exception occurs.

Once submitted to an agency, monitoring records and other data become public record. Data that is maintained at the facility is subject to audit by the agency, but generally is not considered public information. Under Title III of the Superfund Amendments and Reauthorization Act (SARA), the citizens of the community have a right to know about the chemicals and hazards in the local area. Additionally, citizens have access to reports detailing the releases of certain chemicals to the environment.

Enforcement

Enforcement of the regulations, like permitting, is also the responsibility of the executive agency. As part of its enforcement program, the agency has the right to audit records and inspect the regulated facility for compliance. An audit can include items such as records, data, processes, operating parameters, material use and handling methods, storage practices, and other facility operations.

When noncompliance issues arise, the agency has the power to begin enforcement proceedings. Generally, the agency can issue citations of noncompliance and, in many cases, can levy administrative penalties. Other enforcement actions, such as civil or criminal liability suits, are referred to the Judicial Branch of the government. Fines vary depending on the nature of the violation or discrepancy and can range from a few hundred dollars to $25,000 per day per violation. Criminal penalties may include personal fines and (for the most grievous violations) jail sentences.

Judicial Branch Involvement

In addition to responsibility for enforcement suits, the Judicial Branch of the government must handle law suits that are filed by the public or the regulated community. Citizens or public organizations can bring a lawsuit against the executive agency if they can substantiate that the regulations do not meet the legislative intent required by the Act. This can include missed deadlines, lack of standards, and the exemption of the regulations for certain practices or industries.

Likewise, the regulated community can bring a lawsuit against the executive agency if it has a strong case indicating that the agency has overstepped its regulatory bounds. This can include promulgating rules not specified by the Act or basing rules on flawed studies. For these cases, the role of the judicial body is to determine the intent of the Act of Congress and the facts regarding the actions and regulations of the executive agency. Based on this review, the agency may have to recraft its regulations or the regulations may be allowed to stand.

Acquisition of Information

Under the Freedom of Information Act, an agency is required to make available to the public substantive rules along with statements of general policy and interpretations of general applicability formulated and adopted by the agency. This is accomplished primarily through the *Federal Register*. An agency also is required to make available, for a reasonable fee, any guidance documents that the agency staff may develop or contract. These documents are generally available through the National Technical Information Service (NTIS), which is part of the Department of Commerce.[3]

Additionally, the public has a right to view permits, enforcement actions, and other agency documents. These documents can be viewed at the agency. Copies can be requested and will be provided for a reasonable fee.

In order to make information easily accessible, data base systems containing regulatory compliance information have been developed by EPA and other agencies. These data base systems include information on releases of oil and hazardous substances, toxic release inventory data, enforcement actions, and TSCA test data. These data bases, like the agency manuals, are generally available through NTIS.

KEY HAZARDOUS MATERIALS AND HAZARDOUS WASTE REGULATIONS

National Environmental Policy Act (NEPA) of 1969

This Act established the Council on Environmental Quality and set lofty policies and goals for environmental protection. Included in the Act, as set forth in 42 U.S.C. §4331, is a charge to the federal government to:

(1) fulfill the responsibilities of each generation as trustee of the environment for succeeding generations;

(2) assure for all Americans safe, healthful, productive, and esthetically and culturally pleasing surroundings; (and)

(3) attain the widest range of beneficial uses of the environment without degradation, risk to health or safety, or other undesirable and unintended consequences.

The policies set forth in this Act gave Congress the initiative for subsequent legislation and research and development in the environmental area.

Occupational Safety and Health (OSH) Act of 1970

The OSH Act and its amendments pertain mainly to workplace safety. OSHA, an agency within the Department of Labor, has promulgated a comprehensive set of regulations that set standards for hazardous materials management, electrical safety, fire and life safety, and other areas of workplace safety.

---

[3]The mailing address for NTIS is 5285 Port Royal Road, Springfield, VA 22161; the telephone number is (703) 487-4650.

In terms of hazardous materials management, OSHA regulates numerous safety and health aspects including flammable and compressed gas storage, material labeling and information communication, personal protective equipment, workplace monitoring, medical surveillance, management of ionizing and nonionizing radiation sources, and training requirements. Additionally, numerous chemicals are regulated individually. Most of the regulations pertaining to hazardous materials are defined in 29 CFR Parts 1900–1910. All industrial facilities that use chemicals are subject to all or part of these regulations.

Hazard Communication Standard (Effective 1986).   A key OSHA rule that has affected facilities that manage hazardous materials and hazardous waste is the Hazard Communication Standard. Regulations driven by the law are defined in 29 CFR §1910.1200. Included are requirements for training employees about the hazards associated with chemicals used in the workplace. The training program must be established in writing and kept at the facility. Additionally, material safety data sheets must be available at the workplace and accessible to employees at all times, and chemical containers, including stationary containers, must be labeled appropriately.

Hazardous Waste Operations and Emergency Response (Effective 1990).   Another key rule promulgated and administered by OSHA (as issued under the authority of the Superfund Amendments and Reauthorization Act of 1986) sets forth the requirements for hazardous waste operations at Superfund and other cleanup sites. Hazardous materials emergency response training and other requirements also are outlined in this same rule, which is documented in 29 CFR §1910.120.

Standard for Process Safety Management of Highly Hazardous Chemicals (Effective 1992).   This rule, published in 29 CFR §1910.119, lists over 100 highly hazardous chemicals and threshold quantities for these chemicals. If the chemical is manufactured, stored, or used in an amount greater than the threshold quantity, the law requires the facility to conduct a process hazard analysis for all associated process equipment. In addition, employees are required to receive training on process hazards associated with the equipment.

Clean Air Act (CAA) of 1963 (Amended 1970, 1977, and 1990)

This Act originated in 1963, with most of the key aspects of the law as we know it today defined by the 1970, 1977, and 1990 CAA Amendments. Regulations based on the CAA Amendments govern emissions of pollutants into the atmosphere from industrial and commercial activity. The regulations, delineated in 40 CFR Parts 1–99, focus on categories of industrial processes as well as specific hazardous and toxic pollutants. Definitions of what constitutes a major source, definitions of air quality standards, and control requirements based on these definitions are detailed in the rules. Methods for sampling various constituents coming from stacks and vents also are set forth in the regulations.

Among other things, the 1970 CAA Amendments required EPA to set National Ambient Air Quality Standards (NAAQS) for six air pollutants which posed a significant threat to human health. Congress set 1975 as the deadline for achieving these goals, but this deadline was not met. In an effort to achieve NAAQS and other national air quality goals, the CAA Amendments of 1977 required that new processes or sources be installed with Best Available Control Technology (BACT). Areas that exceeded (or did not attain) NAAQS were considered nonattainment areas. As such, these areas required additional controls on sources in order to achieve attainment of the standards, and proportional decreases (or offsets) for emissions were required for each emission increase.

National standards also were set for specific pollutant sources, such as dry cleaners and plastics manufacturers, and were termed New Source Performance Standards (NSPS). New sources constructed or modified after NSPS are proposed are required to meet these standards. Additionally, numerous hazardous air pollutants were listed and regulated via individual standards.

The CAA Amendments of 1990 altered significantly the national air pollution control strategy. While all new sources are still required to meet BACT, new sources in nonattainment areas are required to meet the Lowest Achievable Emission Rate (LAER) for the pollutants that exceed NAAQS. Existing sources in nonattainment areas are required to retrofit emission sources with Reasonably Available Control Technology (RACT). Additionally, decreases in emissions from sources at ratios greater than one to one (1:1) will have to be demonstrated before a new source will be permitted in a nonattainment area.

The new CAA Amendments of 1990 also affected significantly the regulation of hazardous air pollutants. Rather than regulating these pollutants on a specific basis, EPA established source categories, such as refineries and ethylene manufacturers, and regulates emissions of hazardous air pollutants by these source categories. Maximum Achievable Control Technology (MACT) is required for these sources. Sources are expected to retrofit abatement equipment to meet MACT within three years of promulgation of a standard. With MACT, it is expected that hazardous air pollutants can be reduced significantly.

Clean Water Act (CWA), Formerly the Federal Water Pollution Control Act of 1972 (Amended 1977 and 1987)

The basic statutory structure for water quality management was set in place with the Federal Water Pollution Control Act in 1972. This Act set up the framework for the establishment of minimum acceptable requirements for water quality and wastewater management. In 1977, the Act was renamed the Clean Water Act and was enhanced to include increased controls on toxic pollutants. In 1987, the Water Quality Act provided further amendments and enhancements to the control of toxic pollutants.

Point source discharges affected by this Act include wastewater discharges and stormwater discharges from industrial and commercial facilities, municipalities and POTWs, private treatment plants, and other sources. Regulations governing these discharges are found in several places in the *Code of Federal Regulations*, depending on the particular focus of regulation. Effluent guidelines are defined in 40 CFR Parts 401–471, general pretreatment standards are found in 40 CFR Part 403, and regulations pertaining to permitting are located in 40 CFR Parts 122–125. Water quality standards are found in 40 CFR Part 131.

The effluent guidelines for wastewater discharges cover a limited, but diverse, number of specific industrial and commercial categories such as hospitals, food processors, metal finishers, semiconductor manufacturers, and chemical manufacturers. Both facilities that discharge directly to water bodies and those that discharge to POTWs are covered by effluent guidelines.

The general pretreatment standards regulate nondomestic discharges to POTWs. These pretreatment standards detail responsibilities for POTWs with greater than 5 million gallons per day flow and which receive nondomestic discharges. The standards also set out requirements for those categorical discharges covered by effluent guidelines, discharges that are considered to have significant impacts on the POTW, and all other nondomestic discharges to POTWs.

Permitting of wastewater and stormwater discharges that go directly to water bodies is administered under the NPDES process. Discharges that require permitting include wastewater discharges covered by effluent guidelines, wastewater discharges from POTWs and from privately owned treatment facilities, and stormwater discharges from urban municipalities, construction sites, and certain industries.

Water quality standards define the quality goals of a water body. They designate the use or uses to be made of the water (i.e., for the protection and propagation of fish or for recreational purposes), and they set criteria necessary to protect these uses. Once established, the standards become the basis for the establishment of water quality-based treatment controls and other strategies for water quality protection.

Toxic Substance Control Act (TSCA) of 1976

The intent of TSCA is to assure that chemicals manufactured or imported into the United States are registered and listed in the TSCA registry. Rules for this Act are defined in 40 CFR Parts 700–799. A premanufacture notice (PNM) must be filed with EPA in order to receive authorization to manufacture or import for commercial purposes a new chemical. (Some exceptions apply.) Through this registration process, environmental fate and health effects studies must be developed and submitted. Other aspects of the regulation require periodic reporting of production/importation quantities and record-keeping and reporting of allegations of health/environmental effects and associated investigations.

Chemical manufacturers are most affected by this law and its regulations. They are required to file the PNM and to study and report health and environmental data about each new chemical. Research and development facilities are generally exempt from filing a PNM if certain requirements are met. Facilities that use chemicals in their processes must ensure that appropriate chemical information accompanies the chemical and that the chemical is registered under TSCA before it is received. These facilities are also subject to recordkeeping and reporting requirements related to health/environmental allegations.

In addition to requirements pertaining to new chemicals, TSCA also regulates polychlorinated biphenyls (PCBs) and asbestos. Included in these regulations are labeling requirements, recordkeeping requirements, and other requirements. The requirements for PCBs are published under 40 CFR Part 761. Asbestos requirements are found in 40 CFR Part 763.

Resource Conservation and Recovery Act (RCRA) of 1976

EPA promulgated rules under RCRA to regulate the management of hazardous solid waste from generation to final disposal, also termed "cradle-to-grave." These rules are defined in 40 CFR Parts 260–270. The term "solid waste" under these regulations includes waste solids, sludges, liquids, and containerized gases.

The regulations include criteria for defining a hazardous waste. The definitions for hazardous waste include hazardous characteristics such as ignitability or toxicity, wastes from nonspecific sources such as certain spent halogenated and nonhalogenated solvents, and wastes from specific sources such as wastewater treatment sludges generated from a specific process. In addition, the rules list several hundred toxic and acutely hazardous chemicals that, if discarded or spilled, become a hazardous waste.

The regulations also set forth standards for hazardous waste generators, transporters, and owners and operators of treatment, storage, and disposal facilities. Included in the standards are facility permitting requirements, transporter identification and tracking requirements, recordkeeping and inspection requirements, requirements for financial bonding, and other requirements.

Hazardous and Solid Waste Amendments (HSWA) of 1984.   These amendments were enacted and subsequent regulations were developed to strengthen the RCRA regulations. These regulations provide technical standards for landfill disposal, leak detection systems, and underground storage of petroleum products and CERCLA hazardous substances. Additionally, the regulations prohibit specified hazardous wastes from land disposal unless certain treatment standards are met. HSWA also regulates continuing releases from solid waste management units.

Medical Waste Tracking Act of 1988.   This Act, aimed specifically at medical waste management, required EPA to set up a demonstration program for characterizing and tracking of medical waste, and for evaluating treatment techniques for this waste. The specifics of the demonstration program and other aspects of medical management regulation are published in 40 CFR Part 259. Information from the demonstration program may be used to draft additional regulations for this type of solid waste.

Pollution Prevention Act of 1990.   This Act has changed the focus of cradle-to-grave management of hazardous waste to "cradle-to-cradle" management. The new cradle-to-cradle concept emphasizes prevention of waste through recycling, source reduction, elimination of toxic materials, and other methods to attain environmentally conscious manufacturing. In essence, the Act required EPA to develop programs to minimize pollution and to develop a list of priority chemicals that will be the target of minimization (and in some cases elimination) programs.

Comprehensive Environmental Response, Compensation, and Liability Act of 1980 (CERCLA or Superfund Act) and Amendments

This Act—termed CERCLA or Superfund, interchangeably—and several other Acts passed under the Superfund umbrella provided for liability, compensation, cleanup, and emergency response for hazardous substances released into the environment. Regulations pertaining to the naming and cleanup of Superfund sites, which are designated in the National Priority List, are found in 40 CFR Part 300. There are presently over 1,000 sites and over 100 federal facilities on the list. The 1986 amendments to the Act provided for $8.5 billion to be allocated for cleanup of these sites over a 5-year period, as well as the requirement for potentially responsible parties to share in cleanup and associated costs.

Another aspect of this Act is spill reporting. The National Response Center must be notified immediately if the material is released to the environment, without a permit, in amounts greater than the reporting quantity. This requirement for reporting allows EPA to track hazardous materials and hazardous waste incidents. Once reported, these incidents become part of the public record.

Emergency Planning and Community Right-to-Know Act (EPCRA) of 1986.   The Emergency Planning and Community Right-to-Know Act is often referred to as "SARA Title III" since the provisions of the Act are incorporated in Title III of the Superfund Amendments and Reauthorization Act (SARA) of 1986. Regulations resulting from this Act are delineated in 40 CFR Parts 355, 370, and 372.

The Act required the establishment of local emergency planning and release notification, which were accomplished under 40 CFR Part 355. The rule requires each state to set up Local Emergency Planning Committees (LEPCs) to collect local or regional data for hazard assessment. The LEPCs were required to develop an emergency response plan by October 1988, which incorporated the hazards found in the local area. This plan was to be provided to the state. Also, release notification requirements are outlined in this part of the regulations. Any off-site release of a listed chemical above a specified quantity is required to be reported to the LEPC and the State Emergency Response Commission (SERC).

Another key aspect of the Act was to provide knowledge to the community about chemicals stored and released from facilities in the area. This is provided by two reports. The first report, known as the Tier I/Tier II report and defined in 40 CFR §§370.40 and 370.41, requires aggregate information pertaining to hazard type (Tier I) or a list of chemicals stored in threshold amounts at the facility, including locations, amounts, and hazards (Tier II). The second report, the toxic release inventory (TRI) report, is defined in 40 CFR §372.85. This report is required by facilities that manufacture or process certain chemicals in quantities over 25,000 pounds or that use these chemicals in quantities over 10,000 pounds. In this report, a facility specifies information pertaining to the chemical. Before 1991, required information included estimates or actual data pertaining to on-site waste treatment and treatment efficiencies, releases to wastewater and air, and on-site and off-site disposal of certain chemicals. As of 1991, the report also must include data pertaining to off-site and on-site waste recycling, energy recovery from disposal of materials, and source reduction activities used at the facility. EPA can use this data to perform technology and other assessments and to recommend best management practices for specific chemicals.

Radon Gas and Indoor Air Quality Research Act of 1986.   Incorporated into the SARA amendments was the provision under Title IV for setting up a research program for radon gas and indoor air quality. The program is intended to identify, characterize, and monitor the sources and levels of indoor air pollution, particularly radon. Control technologies and other mitigation measures to prevent or abate radon gas and other indoor air pollution are to be researched and developed. It is likely that regulations pertaining to indoor air quality will be promulgated after the research is completed.

Hazardous Materials Transportation Act (HMTA) of 1975

This Act made the Department of Transportation responsible for regulating the transportation of hazardous materials. Regulations pertaining to packaging, container handling, labeling, vehicle placarding, and other safety aspects are detailed in 49 CFR Parts 171–180. Also included are requirements for reporting accidents involving hazardous materials and hazardous waste.

Some DOT regulations overlap with regulations from other agencies such as NRC packaging requirements and EPA hazardous waste manifest requirements.

Atomic Energy Act of 1954

This Act was the original Act governing the production and use of source materials, special nuclear materials, and by-product materials for defense and peaceful purposes. The Atomic Energy Commission (AEC) was set up to license the processing and use of these nuclear materials. Additionally, the Act granted to the AEC the responsibility for regulating health and safety (and the environment) at nuclear facilities.

The Energy Reorganization Act of 1974.   This Act divided the AEC into two agencies, the agency that is now known as Department of Energy (DOE) and the Nuclear Regulatory Commission (NRC). The Act made DOE responsible for energy development and defense production activities, while the NRC was given licensing authority for civilian nuclear energy activities and certain defense activities.

Uranium Mill Tailings Radiation Control Act of 1978.   This Act established the basis for regulation of uranium and thorium mill tailings separately from other radioactive materials and wastes. In general, the Act deals with the control and stabilization of these wastes for protection of public health and the environment. Regulations were promulgated with respect to this Act by EPA and NRC under 40 CFR Part 192 and 10 CFR Part 40, respectively. EPA standards govern control and cleanup of residual radioactive materials at inactive uranium processing sites. NRC regulations establish technical criteria for siting and design of disposal facilities for protection of the groundwater.

Low-Level Radioactive Waste Policy Act of 1980 and 1985 Amendments.   This Act defined low-level radioactive wastes and made each state responsible for providing disposal capacity for these wastes generated within its border. It encouraged regional compacts and allowed the compacts ratified by Congress to exclude waste generated outside their borders beginning January 1, 1986.

When it became evident that regions without waste sites would be unable to have facilities operating by the deadline of 1986, the Low-Level Radioactive Waste Policy Act Amendments of 1985 were enacted. These amendments extended the deadline for facility sitings to 1993, while providing a series of specific dates for progress toward new facility construction. The constitutional right to exclude waste from an outside state or region is being challenged in the courts.

The amendments also specified which categories of low-level radioactive waste are a state responsibility, established volume ceilings for individual nuclear reactors and for operating disposal sites, and set forth other requirements.

Standards that regulate the disposal of low-level radioactive waste are found in 10 CFR Part 61. Included are performance objectives, technical requirements for land disposal facilities, and financial assurances. Licensing requirements for disposal sites and administrative requirements such as recordkeeping, reporting, and inspections are addressed.

Packaging and transport of radioactive materials, including low-level radioactive wastes, any regulated in 10 CFR Part 71. These packaging and transport standards address NRC and DOT requirements.

Nuclear Waste Policy Act of 1982.   This Act established a program for the disposal of civilian spent fuel and high-level waste in geologic repositories. The facilities are operated by DOE and licensed by the NRC. High-level waste is defined by NRC in 10 CFR Part 60. Included in these regulations are licensing criteria for geologic repositories. Management and disposal of this type of waste also are regulated by EPA under 40 CFR part 191. Included are standards for public protection, containment requirements, qualitative assurance requirements, and groundwater protection requirements.

CONCLUSION

Governmental statutes enacted for the purpose of regulating hazardous materials and hazardous waste have been in place for over three decades. During this time, numerous aspects of environmental protection and chemical safety have been the focus of regulation. Regulations now cover diverse issues such as testing requirements for new chemicals, chemical and waste labeling, cradle-to-grave tracking of hazardous waste, technology standards for waste treatment and disposal, and hazard communication to workers and the community.

The regulation process allows input from all interested parties including citizen groups, technical groups, industry, and others. The process is designed to incorporate the goals of Congress in a way that balances environmental protection and chemical safety with technical feasibility, enforceability, and economic impact.

Integral to the regulation process are definitions of regulated materials. These definitions can include testing requirements, threshold quantities, and/or lists of chemicals and wastes covered under a particular regulation. Chapter 2 provides information on definitions of hazardous materials and hazardous waste as established in key regulations.

REFERENCES

Code of Federal Regulations, 10 CFR Parts 60, 61, and 71, Nuclear Regulatory Commission, Washington, D.C.

Code of Federal Regulations, 29 CFR Parts 1900–1910, Department of Labor, Occupational Safety and Health Administration, Washington, D.C.

Code of Federal Regulations, 40 CFR Parts 1–99, U.S. Environmental Protection Agency, Washington, D.C.

Code of Federal Regulations, 40 CFR Parts 122–125, 131, and 401–471, U.S. Environmental Protection Agency, Washington, D.C.

Code of Federal Regulations, 40 CFR Parts 191–192, U.S. Environmental Protection Agency, Washington, D.C.

Code of Federal Regulations, 40 CFR Parts 259–270, U.S. Environmental Protection Agency, Washington, D.C.

Code of Federal Regulations, 40 CFR Parts 355, 370, and 372, U.S. Environmental Protection Agency, Washington, D.C.

Code of Federal Regulations, 40 CFR Parts 700–799, U.S. Environmental Protection Agency, Washington, D.C.

Environmental Law Deskbook, Environmental Law Institute, Washington, D.C., 1989.

Environmental Statutes, 1992 ed., Government Institutes, Inc., Rockville, MD, 1992.

5 U.S.C. §§550–559, Administrative Procedure Act.

5 U.S.C. §552, Freedom of Information Act.

15 U.S.C. §§2601–2671, Toxic Substances Control Act (TSCA).

15 U.S.C. §§2601–2629, TSCA Control of Toxic Substances.

15 U.S.C. §2605e, TSCA Regulation of Polychlorinated Biphenyls (PCBs).

15 U.S.C. §§2641–2655c, TSCA Asbestos Hazard Emergency Response.

15 U.S.C. §§2661–2671, TSCA Indoor Radon Abatement.

29 U.S.C. §§651–678, Occupational Safety and Health Act.

29 U.S.C. §§653, 655, 657, Process Safety Management of Highly Hazardous Chemicals.

29 U.S.C. §655g, Hazard Communication Standard.

29 U.S.C. §§655 and 657, Hazardous Waste Operations and Emergency Response Standard.

33 U.S.C. §§1251–1387, Federal Water Pollution Control Act (Clean Water Act).

42 U.S.C. §§2011–2021, 2022–2286i, Atomic Energy Act.

42 U.S.C. §§2021–2021j, Low-Level Radioactive Waste Policy Act.

42 U.S.C. §§4321–4370a. National Environmental Policy Act.

42 U.S.C. §§6901–6992k, Resource Conservation and Recovery Act (RCRA).

42 U.S.C. §§6921–6939b, RCRA Hazardous Waste Management.

42 U.S.C. §§6991–6991i, RCRA Regulation of Underground Storage Tanks.

42 U.S.C. §§6992–6992k, RCRA Demonstration Medical Waste Tracking Program.

42 U.S.C. §§7401–7626, Clean Air Act, as amended.

42 U.S.C. §7401, Sec. 401–405, Radon Gas and Indoor Air Quality Research Act (Title IV of the Superfund Amendments and Reauthorization Act).

42 U.S.C. §§7901–7942, Uranium Mill Tailings Radiation Control Act.

42 U.S.C. §§9601–9675, Comprehensive Environmental Response, Compensation, and Liability Act, as amended.

42 U.S.C. §§10101–10270, Nuclear Waste Policy Act of 1982.

42 U.S.C. §§11001–11050, Emergency Planning and Community Right-to-Know Act (Title III of the Superfund Amendments and Reauthorization Act).

49 U.S.C. §§1801–1813, Hazardous Materials Transportation Act.

BIBLIOGRAPHY

Bierlein, Lawrence (1987), *Red Book on Transportation of Hazardous Materials*, 2nd ed., Van Nostrand Reinhold, New York.

Burns (1987), *Low Level Radioactive Waste Regulation*, Lewis Publishers, Boca Raton, FL.

CMA (1987), "A Manager's Guide to Title III," Chemical Manufacturers Association, Washington, D.C.

_____ (1989), "Overview of the Resource Conservation and Recovery Act Videotape," Chemical Manufacturers Association, Washington, D.C.

_____ (1991), "NPDES Discharge Permitting and Compliance Issues Manual," Chemical Manufacturers Association, Washington, D.C.

Cooper Musselman, Victoria (1989), *Emergency Planning and Community Right-to-Know*, Van Nostrand Reinhold, New York.

DOE (1991), "OSHA Training Requirements for Hazardous Waste Operations," DOE/EH-0227P, DE92004780, Office of Environment, Safety, and Health, Department of Energy, Washington, D.C.

Environmental Resource Center (1989), *How to Comply with the OSHA Hazard Communication Standard*, Van Nostrand Reinhold, New York.

Fire, Frank L., Nancy K. Grant, and David H. Hoover (1989), *SARA Title III*, Van Nostrand Reinhold, New York.

Gershey, Edward L., Robert C. Klein, and Amy Wilkerson (1990), *Low-Level Radioactive Waste: Cradle to Grave*, Van Nostrand Reinhold, New York.

Government Institutes (1991), *Environmental Law Handbook*, 11th ed., Government Institutes, Inc., Rockville, MD.

Keith, Lawrence H., and Douglas B. Walters (1992), *The National Toxicology Program's Chemical Data Compendium*, Vol. 3, "Standards and Regulations," Lewis Publishers, Boca Raton, FL.

NGA (1986), "The Low-Level Radioactive Waste Handbook; A User's Guide to the Low-Level Radioactive Waste Policy Amendments Act of 1985," NGA 444, National Governors' Association, Center for Policy Research, Washington, D.C.

Stensvaag, John-Mark (1991), *Clean Air Act 1990 Amendments: Law and Practice*, John Wiley & Sons, New York.

Stensvaag, John-Mark (1989), *Hazardous Waste Law and Practice*, Vol. 2, John Wiley & Sons, New York.

U.S. EPA (1989), "CERCLA Compliance with Other Laws Manual: Parts I and II," OSWER Directives 9234.1-01 and 9234.1-02, Office of Solid Waste and Emergency Response, U.S. Environmental Protection Agency, Washington, D.C.

_____ (1991), "Land Disposal Restriction: Summary of Requirements," OSWER 9934.0-1A, Office of Waste Programs Enforcement, Office of Solid Waste and Emergency Response, U.S. Environmental Protection Agency, Washington, D.C.

Wagner, Travis P. (1989), *The Hazardous Waste Q & A*, Van Nostrand Reinhold, New York.

2

DEFINING A HAZARDOUS MATERIAL OR WASTE

Regulators have spent over two decades defining what makes a material or waste "hazardous." Generally, a hazardous chemical or waste is a material that is potentially dangerous to human health or the environment. This chapter reviews some of the more important definitions that are currently being used for regulating hazardous materials and hazardous waste.

OSHA CHEMICAL DEFINITIONS

OSHA defines hazardous chemicals in terms of health hazards and physical hazards. Presented in Table 2.1 are characteristics related to health and physical hazards that can make a chemical hazardous under 29 CFR §1910.1200.

Health Hazards

OSHA has found that health hazards are difficult to define. Thus, OSHA has issued guidelines for health hazard definitions under 29 CFR §1910.1200, Appendices A and B.

Generally, a health hazard is assessed as either chronic or acute. A chronic health hazard occurs as a result of long-term exposure. An acute health hazard occurs rapidly as a result of a short-term exposure. In addition to chronic or acute hazard classifications, there are specific health hazards classified in the regulations. These include: carcinogens, toxics, corrosives, irritants, sensitizers, and health hazards affecting human organs.

TABLE 2.1   Characteristics that Can Make a Chemical Hazardous Under OSHA Regulations

Health Hazards	Physical Hazards
• Carcinogens	• Combustible liquids
• Toxic chemicals or highly toxic chemicals	• Water reactives
• Reproductive toxins	• Flammables
• Irritants	• Organic peroxides
• Hepatoxins (liver)	• Explosives
• Corrosive chemicals	• Oxidizers
• Neurotoxins (nervous system)	• Pyrophorics
• Sensitizers	• Compressed gases
• Nephrotoxins (kidney)	
• Agents that damage the blood, lungs, eyes, or skin	

A chemical is considered to be a carcinogen, or cancer-causing agent, under the OSHA regulations if it has been evaluated by the International Agency for Research on Cancer (IARC)[1] and is listed in IARC's latest edition of *Monographs* as a carcinogen or potential carcinogen. Additionally, if it is listed as a carcinogen or potential carcinogen in the latest edition of the *Annual Report on Carcinogens* published by the National Toxicology Program,[2] OSHA regulates the chemical as a carcinogen. OSHA also has the right to list and regulate other chemicals as carcinogens as the agency deems appropriate.

Toxic chemicals are defined as either highly toxic or toxic. Highly toxic chemicals are defined as follows:

- The chemical has a median lethal dose (LD_{50}) of 50 milligrams or less per kilogram of body weight when administered orally to albino rats weighing between 200 and 300 grams each.
- The chemical has a median lethal dose (LD_{50}) of 200 milligrams or less per kilogram of body weight when administered by continuous contact for 24 hours (or less if death occurs before 24 hours) with the bare skin of albino rabbits weighing between 2 and 3 kilograms each.
- The chemical has a median lethal concentration (LC_{50}) in air of 200 parts per million by volume or less of gas or vapor, or 2 milligrams per liter or less of mist, fume or dust, when administered by continuous inhalation for 1 hour (or less if death occurs before 1 hour) to albino rats weighing between 200 and 300 grams each.

[1]The International Agency for Research on Cancer is part of the World Health Organization.
[2]The National Toxicology Program is part of the National Center for Toxicological Research, Department of Health and Human Services.

A chemical is toxic if it meets any of the following criteria:

- The chemical has a median lethal dose (LD_{50}) of more than 50 milligrams per kilogram but not more than 500 milligrams per kilogram of body weight when administered orally to albino rats weighing between 200 and 300 grams each.
- The chemical has a median lethal dose (LD_{50}) of more than 200 milligrams per kilogram but not more than 1,000 milligrams per kilogram of body weight when administered by continuous contact for 24 hours (or less if death occurs within 24 hours) with the bare skin of albino rabbits weighing between 2 and 3 kilograms each.
- The chemical has a median lethal concentration (LC_{50}) in air of more than 200 parts per million but not more than 2,000 parts per million by volume of gas or vapor, or more than 2 milligrams per liter but not more than 20 milligrams per liter of mist, fume, or dust when administered by continuous inhalation for 1 hour (or less if death occurs within 1 hour) to albino rats weighing between 200 and 300 grams each.

Corrosive chemicals cause a visible destruction of living tissue. The regulations under DOT in 49 CFR Part 173 Appendix A delineate a test method acceptable to OSHA for defining the term corrosive. The test exposes albino rats to a chemical for 4 hours. If, after the exposure period, the chemical has destroyed or changed irreversibly the structure of the tissue at the site of contact, then the chemical is considered corrosive.

Chemicals that are irritants cause a reversible inflammatory effect on skin or eyes. Tests for determining whether the chemical is an irritant are defined in 16 CFR §1500.41. Tests for eye irritants are defined in 16 CFR §1500.42.

A chemical is defined as a sensitizer if a large number of exposed people or animals develop an allergic reaction in normal tissue after repeated exposure to the chemical. The effects typically are reversible once the exposure ceases.

Chemicals also are considered a health hazard if exposure causes damage to any of the human organs. The damage can take many forms. Examples include liver enlargement, kidney disease or malfunction, excessive nervousness or decrease in motor functions, decrease in lung capacity, damage to cornea, and other organ damage.

Selected examples of chemicals that create the health hazards discussed in this section are presented in Table 2.2. Since many chemicals are associated with more than one health hazard, information pertaining to each chemical should be reviewed thoroughly before the chemical is used in the workplace.

Physical Hazards

A chemical is defined as physical hazard if there is scientific evidence that it is a combustible liquid, flammable, explosive, pyrophoric, or unstable (reactive). Additionally, a chemical is deemed hazardous if it is a compressed gas, an organic peroxide, or an oxidizer.

A combustible liquid is any liquid having a flashpoint at or above 100°F, but below 200°F. A flammable material can be any of the following:

- An aerosol that, when tested by the method described in 16 CFR §1500.45, yields a flame projection exceeding 18 inches at full valve opening, or a flashback at any degree of valve opening.
- A gas that, at ambient temperature and pressure, forms a flammable mixture with air at a concentration of 13% by volume or less.

TABLE 2.2   Selected Examples of Chemicals that Create Health Hazards

Health Hazard	Chemicals that Create the Hazard
Carcinogen	Aldrin, formaldehyde, ethylene dichloride, methylene chloride, dioxin
Toxic	Xylene, phenol, propylene oxide
Highly toxic	Hydrogen cyanide, methyl parathion, acetonitrile, allyl alcohol, sulfur dioxide, pentachlorophenol
Reproductive toxin	Methyl cellosolve, lead
Corrosive	Sulfuric acid, sodium hydroxide, hydrofluoric acid
Irritant	Ammonium solutions, stannic chloride, calcium hypochlorite, magnesium dust
Sensitizer	Epichlorohydrin, fiberglass dusts
Hepatotoxin	Vinyl chloride, malathion, dioxane, acetonitrile, carbon tetrachloride, phenol, ethylenediamine
Neurotoxin	Hydrogen cyanide, endrin, mercury, cresol, methylene chloride, carbon disulfide, xylene
Nephrotoxin	Ethylenediamine, chlorobenzene, dioxane, hexachloronaphthalene, acetonitrile, allyl alcohol, phenol, uranium
Agents that damage:	
Blood	Nitrotoluene, benzene, cyanide, carbon monoxide
Lungs	Asbestos, silica, tars, dusts
Eyes or skin	Sodium hydroxide, ethylbenzene, perchloroethane, allyl alcohol, nitroethane, ethanolamine, sulfuric acid, liquid oxygen, phenol, propylene oxide, ethyl butyl ketone

Sources: Information from NIOSH (1987), NFPA (1991), and 29 CFR §1910.1200 Appendix A.

- A gas that, at ambient temperature and pressure, forms a range of flammable mixtures with air wider than 12% by volume, regardless of the lower limit.
- A liquid having a flashpoint below 100°F.
- A solid, other than a blasting agent or explosive, that is likely to cause fire through friction, absorption of moisture, spontaneous chemical change, or retained heat from manufacturing or processing, or that can be ignited readily and when ignited burns so vigorously and persistently as to create a serious hazard.
- A solid that, when tested by the method described in 16 CFR §1500.44, ignites and burns with a self-sustained flame at a rate greater than one-tenth of an inch per second along its major axis.

Other physical hazard definitions include:

- Explosive—An explosive is defined as a chemical that causes a sudden, almost instantaneous, release of pressure, gas, and heat when subjected to sudden shock, pressure, or high temperature.
- Pyrophoric—A pyrophoric is a material that will ignite spontaneously in air at a temperature of 130°F or below.
- Water reactive—A chemical is water reactive if it reacts with water to release a gas that is either flammable or presents a health hazard.
- Oxidizer—An oxidizer is a chemical that initiates or promotes combustion in other materials, thereby causing fire either of itself or through the release of oxygen or other gases.
- Organic peroxides—In addition to being oxidizers, organic peroxides are materials that are extremely unstable. In their pure state, or when produced or transported, they will vigorously polymerize, decompose, condense, or become self-reactive when exposed to shock, heat, or friction.

Selected examples of chemicals that create physical hazards are presented in Table 2.3.

Compressed Gases

Compressed gases are considered hazardous materials under OSHA. A compressed gas is defined as:

- A gas or mixture of gases having, in a container, an absolute pressure exceeding 40 pounds per square inch (psi) at 70°F.
- A gas or mixture of gases having, in a container, an absolute pressure exceeding 104 psi at 130°F regardless of the pressure at 70°F.
- A liquid having a vapor pressure exceeding 40 psi at 100°F as determined by ASTM D 323–72.

TABLE 2.3   Selected Examples of Chemicals that Create Physical Hazards

Physical Hazard	Chemicals that Create the Hazard
Combustible liquids	Fuel oil, crude oil, other heavy oils
Flammables	Gasoline, isopropyl alcohol, acetone, spray cans that use butane propellants
Explosives	Dynamite, nitroglycerine, ammunition
Phyrophorics	Yellow phosphorus, white phosphorus, superheated toluene, silane gas, lithium hydride
Water reactives	Potassium, phosphorus pentasulfide, sodium hydride
Organic peroxides	Methyl ethyl ketone peroxide, dibenzoyl peroxide, dibutyl peroxide
Oxidizers	Sodium nitrate, magnesium nitrate, bromine, sodium permanganate, calcium hypochlorite, chronic acid

Source: Information from NFPA (1991) and 29 CFR §1910.1200.

Examples of compressed gases include oxygen, helium, and acetylene. Hazards associated with these materials include abnormal pressure release and possible rupture if the container is punctured. Liquified gases, a subset of compressed gases, include butane, propane, and vinyl chloride. In addition to the potential rupture hazard, these three materials are also flammable. Some liquified gases are also classified as cryogenic. Cryogenics are substances that change from a gas to a liquid at or below $-200\,°C$. Examples include liquid nitrogen and liquid oxygen. These materials can cause frost bite or burns to the skin if released.

Highly Hazardous Chemicals

OSHA has listed over 100 highly hazardous chemicals that are regulated if manufactured, stored, or used in quantities greater than threshold amounts. These chemicals are toxic, flammable, explosive, or reactive. Examples include ammonia, hydrogen chloride, furan, and methyl vinyl ketone.

Other OSHA-Regulated Chemicals

In addition to regulating chemicals that fall into the categories reviewed previously, OSHA has set regulatory standards for permissible exposure limits to workers for a number of chemicals. Examples include lead, formaldehyde, benzene, vinyl chloride, and others. In establishing limits for allowable worker exposure, OSHA uses data gathered by the National Institute of Occupational Safety and Health (NIOSH).

HAZARDOUS WASTE DEFINITIONS UNDER RCRA

Under RCRA regulations promulgated by EPA, a waste is hazardous if it is a solid waste and meets one of several definitions. The term "solid waste" might be slightly misleading since it might be construed as wastes that are in the solid phase. In fact, solid wastes are liquids, solids, and containerized gases. Definitions of RCRA hazardous wastes are delineated in 40 CFR 261 Subpart C.

A waste is a "characteristic" hazardous waste—as defined in 40 CFR §§261.21–261.24—if it has any of the following characteristics:

- Ignitability—A waste is hazardous if it is a liquid, other than an aqueous solution containing less than 24% alcohol by volume, and has a flashpoint of less than 140°F. A waste is also hazardous if it is not a liquid and is capable, under standard temperature and pressure, of causing fire through friction, absorption of moisture, or spontaneous chemical changes.Further, it is a hazardous waste if it is an ignitable compressed gas or an oxidizer.

- Corrosivity—A waste is hazardous if it is aqueous and its pH is less than or equal to 2 or greater than or equal to 12.5. Additionally, a waste is considered corrosive if it is a liquid and corrodes steel at a rate greater than 0.250 inches per year at 130°F.

- Reactivity—A waste is hazardous if it is normally unstable and readily undergoes violent change, or if it reacts violently or creates toxic fumes when mixed with water. Additionally, a waste is hazardous if it is a cyanide or sulfide bearing waste that can generate toxic gases or fumes when exposed to pH conditions between 2 and 12.5.

- Toxicity—A waste is hazardous if toxic concentrations of compounds enumerated in 40 CFR §261.24 can be leached into water, using EPA's toxicity characteristic leaching procedure (TCLP). If a waste is less than 0.5% filterable solids, it is considered hazardous if the liquid concentration exceeds the TCLP standard. These standards are set for numerous compounds, including metals, pesticides, and organics.

A waste can be deemed hazardous if it is a waste that is specifically listed in the RCRA regulations under 40 CFR Part 261 Subpart D as being hazardous. Included are wastes from specific sources and nonspecific sources. Examples of listed hazardous wastes from specific sources include specific waste from a process such as distillation bottoms from aniline production, untreated process wastewater from the production of toxaphene, and others. Wastes from non-specific sources include spent solvents and general process wastes such as residues, heavy ends, sludges, and other wastes from processes such as electroplating operations, metal heat treating operations, and other manufacturing operations. An additional listing gives the names of several hundred specific chemicals that, when discarded, become hazardous waste. Examples include acrylonitrile, benzene, and toluene.

Any mixture of a listed hazardous waste and a nonhazardous waste renders the entire mixture hazardous. Likewise, a waste derived from a hazardous waste is considered hazardous. These rules are being reviewed for change, and another method for determining whether or not these types of wastes are hazardous may be used in the future.

DEFINITIONS FOR RADIOACTIVE WASTES

Major Categories of Radioactive Wastes

Definitions of radioactive wastes are published by NRC. There are basically three major types of radioactive wastes: high-level radioactive wastes, transuranic radioactive wastes, and low-level radioactive wastes.

High-level radioactive wastes (HLW) are those reprocessing wastes derived from nuclear reactors including irradiated reactor fuel, liquid wastes resulting from the operation of the first cycle solvent extraction system or equivalent, concentrated wastes resulting from subsequent extraction cycles or equivalent, and solids derived from conversion of high-level radioactive liquids. Weapons by-products also can contain high-level radioactivity. The wastes require permanent isolation in a geologic repository or equivalent at the time of disposal, as defined under 10 CFR Part 60.

Transuranic (TRU) wastes include wastes containing elements with atomic numbers greater than uranium (92), such as plutonium and curium, and that contain more than 100 nanocuries per gram (nCi/g) alpha-emitting transuranic isotopes with half-lives greater than 20 years. Some TRU materials are exceptionally long-lived, such as plutonium-239, which has a half-life of 24,400 years. TRU waste is generated from the reprocessing of plutonium-bearing fuel and irradiated targets and from operations required to prepare the recovered plutonium for weapons use. The waste includes TRU metal scrap, glassware, process equipment, soil, laboratory wastes, filters, and wastes contaminated with TRU materials (U.S. Congress 1991). NRC approves the disposal of these wastes, which includes disposal in a geologic repository or equivalent for most cases.

Low-level radioactive wastes (LLW) are defined as wastes containing radioactivity that is neither high-level radioactive nor transuranic. In general, low-level radioactive waste, regulated under 10 CFR Part 61, is divided into three classes: A, B, and C, with Class C containing the highest concentration of radionuclides. Class C requires stabilization and barriers to protect intruders. Class B requires stabilization, but not barriers, to protect intruders. Class A, with the lowest concentration of radionuclides of all the classes, does not require stabilization or barriers. Examples of low-level radioactive wastes include solidified liquids, filters and resins, and lab trash from nuclear reactors, hospitals, research institutions, and industry (NRC 1989).

Other Radioactive Wastes

In addition to the major categories of radioactive wastes, there are definitions and regulations for other types of radioactive waste. These include uranium and thorium mill tailings, waste derived from naturally occurring and accelerator-produced radioactive materials, and radioactive materials or waste mixed with hazardous waste (DOE 1989).

Uranium and thorium mill tailings are regulated separately from low-level radioactive wastes. Technical criteria are set by NRC for siting and design of disposal facilities for these wastes under 10 CFR Part 40. EPA governs the control and cleanup of residual radioactive materials from inactive uranium processing sites, as well as the management of uranium and thorium by-product materials at active sites. These regulations are found under 40 CFR Part 192.

Naturally occurring and accelerator-produced radioactive materials (NARM) are defined as any radioactive material not classified as source, special nuclear, or by-product material (DOE 1989). A waste from this material is regulated as if it were a low-level radioactive waste.

A "mixed waste" is defined as a radioactive waste that is mixed with hazardous waste. Radioactive wastes can be classified as mixed by using engineering knowledge of the hazardous characteristics of the waste. If the determination of hazardous or nonhazardous must be made through testing, a surrogate waste stream that is devoid of radioactive material, but which is equivalent otherwise, can be used.

Once a determination is made that the waste stream is a mixed waste, the waste is regulated under both EPA and NRC rules. EPA has set forth treatment standards for several types of mixed waste. Incineration is recommended for organic and ignitable low-level mixed waste. For transuranic and high-level mixed wastes, EPA has established vitrification as an acceptable treatment technology prior to long-term storage.

OTHER REGULATORY DEFINITIONS

Priority Pollutants

Nearly 200 chemicals are regulated by EPA as priority pollutants under the Clean Water Act. These include heavy metals, toxics, and other chemicals deemed harmful to human health and aquatic life. Effluent guidelines and other wastewater discharge standards are set for these chemicals in order to ensure that all waters of the United States remain fishable and swimmable.

Additionally, the aggregate concentration of toxic chemicals in effluent can make it unacceptable for discharge, even if the individual constituents are within standards or guidelines. An EPA-approved bio-assay test is used for determining total effluent toxicity. The basis of the toxicity determination is the survival rate of certain aquatic species in whole or diluted effluent.

CERCLA Reportable Quantities

Under CERCLA, EPA regulates release reporting of chemicals in amounts that it has determined to be hazardous to human health or the environment. Chemicals regulated in this fashion include metals, solvents, pesticides, and toxics. The reporting quantities range from 1 pound for very hazardous chemicals such as arsenic acid to up to 5,000 pounds for less hazardous chemicals such as phosphoric acid. If these chemicals are released in amounts greater than the reportable quantity, without a permit, the National Response Center must be notified.

Hazardous Air Pollutants

There are numerous chemicals regulated under the Clean Air Act. Chemicals deemed pivotal to achieving National Ambient Air Quality Standards (NAAQS) are defined and include nitrogen dioxide, sulfur oxides, carbon monoxide, ozone, particulate matter, and lead. Additionally, the regulations specify a separate list of pollutants that are regulated under the National Emission Standards for Hazardous Air Pollutants (NESHAP) program. The NESHAP list includes chemicals such as mercury, asbestos, radon emissions from the disposal of uranium mill tailings, vinyl chloride, and benzene emissions from benzene storage vessels. Other hazardous air pollutants are listed in the regulations and include substances such as acrylonitrile, toluene, and phenol.

SARA TITLE III Chemicals

Under SARA Title III, EPA has listed over 300 chemicals that are considered extremely hazardous substances. These include many of the same chemicals regulated under RCRA and CERCLA. Like CERCLA, SARA also has release reporting requirements. The regulations require notification to state and local authorities if a release leaves the facility in an amount greater than the reportable quantity.

Regulation of Asbestos

Asbestos is an example of a hazardous material that is regulated under multiple statutes including the OSH Act, the Clean Air Act, the Clean Water Act, the Toxic Substance Control Act, and the Mine Safety and Health Administration. OSHA regulations, published under 29 CFR §1910.1001 and §1926.58, detail the following:

- Definition of asbestos—The definition includes materials containing asbestos minerals such as chrysotile, crocidolite, amosite, anthophyllite, tremolite, and actinolite.

- PELs—PELS are permissible exposure limits to airborne concentrations of asbestos fibers, including an 8-hour time weighted average and a 30-minute short-term exposure limit.
- Engineering methods—These include methods for controlling the escape of asbestos fibers during abatement procedures. Examples are wet removal of asbestos and maintaining the work area under a negative pressure.
- Work practices—These include items such as acceptable methods for asbestos removal, required personal protection, and monitoring requirements. Medical monitoring requirements of asbestos workers for respiratory disease and pulmonary function also are included in the requirements.

EPA regulates asbestos through its NESHAP program, as required under the Clean Air Act and detailed in 40 CFR 61 Subpart M. The regulations include standards for asbestos mills, standards for demolition and renovation of areas containing friable asbestos, and waste disposal practices. In addition, standards are set for active waste disposal sites as well as inactive waste disposal sites from asbestos mills and manufacturing and fabricating operations.

In addition to air emissions regulations, EPA defines wastewater effluent standards for asbestos manufacturing source categories, as required under the Clean Water Act. These regulations are defined under 40 CFR Part 427. Finally, EPA has promulgated asbestos regulations under the Toxic Substance Control Act, which are detailed under 40 CFR Part 763. These regulations control the commercial and industrial uses of asbestos, asbestos-containing materials in schools, and asbestos abatement projects.

The Mine Safety and Health Administration controls exposure limits to asbestos and respiratory protection for workers in surface and underground mines. These regulations are found under 30 CFR 56 Subpart D and 57 Subpart D.

LIST OF LISTS

Definitions of what constitutes a hazardous material or hazardous waste are numerous and overlap in many statutes. Literally, thousands of chemicals and wastes are regulated in some manner. Presented in Table 2.4 is an overview of lists of chemicals and wastes cited in the regulations under OSHA, NRC, the Department of Transportation (DOT), and EPA. EPA regulations and notations include the Clean Water Act (CWA), the Clean Air Act (CAA), Safe Drinking Water Act (SDWA), Superfund Amendments and Reauthorization Act (SARA), Comprehensive Environmental Response, Compensation, and Liability Act (CERCLA), and the Toxic Substance Control Act (TSCA). The materials in the lists presented in the table are either defined as hazardous or have been given a standard or other regulatory discharge or reporting limit.

TABLE 2.4   Summary of Lists of Regulated Materials

Regulation	Regulated Materials
10 CFR Part 20 Appendix B to §§20.1001–20.2401	Annual limits of intake and derived air concentrations of radionuclides for occupational exposure; effluent concentrations; concentrations for release to sewerage (NRC)
10 CFR §61.55 Table 1	List of long-lived radionuclides used in low-level waste classification (NRC)
10 CFR §61.55 Table 2	List of short-lived radionuclides used in low-level waste classification (NRC)
29 CFR §1910.119 Appendix A	List of highly hazardous chemicals, toxics and reactives (OSHA)
29 CFR Part 1910 Subpart Z, Tables Z-1-A, Z-2, Z-3	Limits for air contaminants (OSHA)
40 CFR §§50.4–50.12	Pollutants with national ambient air quality standards (CAA)
40 CFR §61.01	List of hazardous air pollutants (CAA)
40 CFR §61 Subparts B-FF	List of NESHAP chemicals (CAA)
40 CFR §116.4 Table 116.4A	List of hazardous substances by common name (CWA)
40 CFR §116.4 Table 116.4B	List of hazardous substances by CAS number (CWA)
40 CFR §122 Appendix D Table II	Organic pollutants of concern in NPDES discharges (CWA)
40 CFR §122 Appendix D Table III	Listed metals, cyanides, and phenols of concern in NPDES discharges (CWA)
40 CFR §122 Appendix D Table V	Toxic pollutants and hazardous substances of concern in NPDES discharges (CWA)
40 CFR Part 129	Toxic pollutant effluent standards (CWA)
40 CFR §141 Subpart B	Primary maximum contaminant levels for drinking water (SDWA)
40 CFR §143.3	Secondary maximum contaminant levels for drinking water (SDWA)
40 CFR §261.24	List of maximum concentration of contaminants for the toxicity characteristics—toxicity characteristic leaching procedure (TCLP) limits (RCRA)

TABLE 2.4 *Continued*

Regulation	Regulated Materials
40 CFR §261.31	List of hazardous wastes from nonspecific sources (RCRA)
40 CFR §261.32	List of hazardous wastes from specific sources (RCRA)
40 CFR §261.33(e)	List of acutely hazardous wastes (RCRA)
40 CFR §261.33(f)	List of toxic wastes (RCRA)
40 CFR §268.32	California list wastes with specific prohibitions (RCRA)
40 CFR §§268.33, 268.34, 268.35	Lists of land restricted wastes (RCRA)
40 CFR §268.41	List of treatment standards for land restricted wastes (RCRA)
40 CFR §300 Appendix B	National priority list (CERCLA)
40 CFR §302.4 Table 302.4	List of hazardous substances and reportable quantities (CERCLA)
40 CFR §302.4 Appendix A	Sequential CAS registry number list of CERCLA hazardous substances (CERCLA)
40 CFR §355 Appendix A	List of extremely hazardous substances for emergency planning and notification (SARA)
40 CFR §372.65	List of toxic chemicals for toxic release inventory reporting (SARA)
40 CFR §700, Index (Finders Aid)	Toxic substances CAS number/chemical index (TSCA)
49 CFR §172.101	Hazardous materials table (DOT)
49 CFR §172.101, Appendix	List of hazardous substances and reportable quantities per CERCLA (DOT)

CONCLUSION

Hazardous materials and hazardous waste can be defined in many ways. Physical properties such as flammability, corrosivity, or reactivity can make a material hazardous. Additionally, materials that cause damage to organs, tissue, or health—such as toxicants, carcinogens, and sensitizers—are considered hazardous.

Federal agencies, through promulgation of many regulations, have defined hazardous materials and hazardous waste. Generally, these definitions take the form of listings of chemicals and wastes that are predetermined from testing or other information to be hazardous. Other methods of defining hazardous materials and waste include specifying test methods and limits that must be achieved before a chemical or waste is considered nonhazardous. Constituents that are highly toxic or cause damage to the environment at low concentrations are listed as focus chemicals in multiple regulations. When a single regulation is viewed, definitions for what is considered "hazardous" may seem narrow or noninclusive. However, when regulations pertaining to hazardous materials and hazardous waste are viewed as a whole, the definitions are comprehensive and provide necessary information for safe management of these materials.

The next chapter discusses the major types of hazardous materials and hazardous waste found in manufacturing and nonmanufacturing applications. The chapter presents a summary of data collected and published by EPA detailing waste generation in the United States by waste type and industry type.

REFERENCES

Code of Federal Regulations, 29 CFR §1910.1200 (Including Appendices), U.S. Department of Labor, Occupational Safety and Health Administration, Washington, D.C.

Department of Energy, "Waste Classification—History, Standards, and Requirements for Disposal," DE89 013705, prepared by David C. Kocher, Oak Ridge National Laboratory, Oak Ridge, TN, 1989.

National Fire Protection Association, *Fire Protection Guide for Hazardous Materials*, 10th ed., Quincy, MA, 1991.

National Institute for Occupational Safety and Health, *Pocket Guide to Chemical Hazards*, DHHS (NIOSH) Publication No. 85-114, U.S. Department of Health and Human Services, Public Health Service, Centers for Disease Control, National Institute for Occupational Safety and Health, Washington, D.C., 1987.

Nuclear Regulatory Commission, "Regulating the Disposal of Low-Level Radioactive Waste: A Guide to the Nuclear Regulatory Commission's 10 CFR Part 61," NUREG/BR-0121, Office of Nuclear Material Safety and Safeguards, U.S. Nuclear Regulatory Commission, Washington, D.C., 1989.

U.S. Congress, "Long-Lived Legacy: Managing High Level and Transuranic Waste at the DOE Nuclear Weapons Complex," OTA-BP-O-83, Office of Technology Assessment, Congress of the United States, Government Printing Office, Washington, D.C., 1991.

BIBLIOGRAPHY

ACGIH (1992), *Threshold Limit Values for Chemical Substances and Physical Agents and Biological Exposure Indices*, American Conference of Governmental Industrial Hygienists, Cincinnati, OH.

Alliance of American Insurers (1983), *Handbook of Hazardous Materials, Fire, Safety, Health*, 2nd ed., Alliance of American Insurers, Schaumberg, IL.

API (1988), "Naturally Occurring Radioactive Material (NORM) in Oil and Gas Production Operations: Video Presentation," American Petroleum Institute, Washington, D.C.

Bretherick, Leslie (1990), *Handbook of Reactive Chemical Hazards*, 4th ed., Butterworth-Heinemann, Stoneham, MA.

Clayton, G.D., and F.E. Clayton, eds. (1982), *Patty's Industrial Hygiene and Toxicology*, Vol. 2, "Toxicology," 3rd ed., Wiley-Interscience, New York.

DOE (1988), "Review of EPA, DOE, and NRC Regulations on Establishing Solid Waste Performance Criteria," DE88 015331, ORNL/TM-9322, prepared by A.J. Mattus, T.M. Gilliam, and L.R. Dole, Oak Ridge National Laboratory, Department of Energy, Oak Ridge, TN.

DOL (1990), *OSHA Regulated Hazardous Substances: Health, Toxicity, Economic and Technological Data*, Vols. 1 and 2, Department of Labor, Noyes Data Corporation, Park Ridge, NJ.

Friedman, D., ed. (1988), *Waste Testing and Quality Assurance*, Special Technical Publication 999, American Society of Testing and Materials, Philadelphia.

ILO (1991), *Occupational Exposure Limits for Airborne Toxic Substances*, 3rd ed., International Labour Office, New York.

Keith, Lawrence H., and Douglas B. Walters (1992), *The National Toxicology Program's Chemical Data Compendium*, Vol. 2, "Chemical and Physical Properties," Lewis Publishers, Boca Raton, FL.

Lewis, Richard J., Sr. (1991), *Hazardous Chemicals Desk Reference*, 2nd ed., Van Nostrand Reinhold, New York.

NFPA (1990), "Identification of the Fire Hazards of Materials," NFPA 740, National Fire Protection Association, Qunicy, MA.

Perry, R. (1984), *Perry's Chemical Engineers' Handbook*, 6th ed., McGraw-Hill, New York.

Sax, N. Irving, and Richard J. Lewis, Sr. (1992), *Dangerous Properties of Industrial Materials*, 8th ed., Van Nostrand Reinhold, New York.

Sax, N. Irving, and Richard J. Lewis, Sr. (1987), *Hawley's Condensed Chemical Dictionary*, 11th ed., Van Nostrand Reinhold, New York.

Sittig, Marshall (1991), *Handbook of Toxic and Hazardous Chemicals and Carcinogens*, Noyes Data Corporation, Park Ridge, NJ.

U.S. EPA (1988), *Extremely Hazardous Substances: Superfund Chemical Profiles*, Vols. 1 and 2, U.S. Environmental Protection Agency, Noyes Data Corporation, Park Ridge, NJ.

Weiss, G., ed. (1986), *Hazardous Chemicals Data Book*, 2nd ed., Noyes Data Corporation, Park Ridge, NJ.

Windholz, Martha, ed. (1989), *The Merk Index*, 11th ed., Merck & Co., Rahway, NJ.

Wolman, Yecheskel (1988), *Chemical Information: A Practical Guide to Utilization*, 2nd ed., John Wiley & Sons, New York.

3

MAJOR TYPES OF
HAZARDOUS MATERIALS
AND HAZARDOUS WASTE

Hazardous materials are used in many manufacturing and nonmanufacturing operations, and hazardous waste is typically generated from these operations. This chapter discusses major types of hazardous materials and hazardous waste associated with processes and operations found across the United States. Additionally, information gathered by EPA in a national survey that summarizes types of waste and waste generators is included. The chapter also discusses medical waste generation.

MANUFACTURING OPERATIONS

Chemical-Use Processes and Waste Types

Almost all manufacturing facilities use chemicals in some form in the production of marketable goods. From the manufacture of pure chemicals to the reprocessing of chemicals to produce derivative products to the use of chemicals as a cleaning agent, chemicals are a common part of the industrial sector.

There are too many chemical-use applications to present an all-inclusive list of industrial processes that use hazardous materials; thus, only selected categories of regulated industrial processes and associated hazardous materials are presented in Table 3.1. Some of these processes and associated materials are currently the focus of EPA as candidates for chemical source reduction and waste minimization.

TABLE 3.1   Selected Examples of Manufacturing Operations or Processes that Use Hazardous Materials

Manufacturing Category	Examples of Materials Potentially Used
Chemical reprocessing operations	Nonhalogenated solvents, cupric chloride, pyrophosphate, acids, caustics, others
Coking operations	Ammonia, benzene, phenols, cyanide
Degreasing operations	Perchloroethylene, trichloroethylene, methylene chloride, 1,1,1-trichloroethane, carbon tetrachloride, chlorinated fluorocarbons
Distillation operations	Chlorobenzene, trichloroethylene, perchloroethylene, aniline, cumene, ortho-xylene, naphthalene, others
Electroplating processes	Cyanides, nickel, copper, acids, chrome, cadmium, gold
Ink formulation	Solvents, caustics, chromium- or lead-containing pigments and stabilizers
Leather tanning	Tannic acid, chromium
Painting operations	Methylene chloride, trichloroethylene, toluene, methanol, turpentine
Petroleum processes	Arsenic, cadmium, chromium, lead, halogenated solvents, flammable oils, distillate products
Primary metal processes	Cyanides, salt baths, heavy metals such as chromium and lead
Pulp and paper operations	Chlorine, sodium sulfite, sodium hydroxide, dioxins, furans, phenols
Textile finishing	Solvents, solutions of dyes
Weapons manufacture	Trinitrotoluene (TNT), nitroglycerin, uranium alloys, plutonium
Wood preserving processes	Creosote, pentachlorophenol, other creosote and chlorophenolic formulations, copper, arsenic, chromium

Waste types generated by processes within industry include bottoms or residues, dusts, discarded or off-spec chemicals or by-products, lab packs, slags, sludges or slurries, spent liquors, waste packages, and wastewaters. Selected examples of manufacturing operations that generate these types of waste are detailed in Table 3.2.

TABLE 3.2   Waste Types and Selected Examples of Operations that Generate the Waste Type

Waste Type	Selected Examples of Operations that Generate the Waste Type
Bottoms or residues	Solvent and petroleum distillation operations, chemical purification operations, decant and separation techniques, cleaning operations
Discarded or off-spec chemicals or by-products	Chemical manufacturing operations, bench or pilot scale testing operations, lab analysis activities
Dusts	Grinding operations, brushing operations, packaging operations, asbestos removal and other demolition activities, mining activities
Lab packs	Lab experiments, bench-scale studies, hazardous waste analysis, process analysis
Slags	Metal heat treating operations, metal-processing operations
Sludges or slurries	Wastewater treatment operations, other separation techniques such as filtration or gravity settling
Spent liquors	Metal-finishing operations, wood preserving operations, pulping operations
Waste packages	Nuclear material processing, weapons manufacturing
Wastewaters, inorganic	Chemical products manufacturing, electroplating processes, petroleum processes, primary metal processing, coking processes
Wastewaters, organic	Chemical products manufacturing, cleaning and rinsing operations, other products manufacturing operations

Some of these wastes are listed as hazardous in the RCRA regulations. Others are hazardous as a result of having hazardous characteristics, such as ignitability, corrosivity, reactivity, or toxicity as defined in EPA's toxicity characteristic leaching procedure. Others may be regulated as hazardous by another agency or by the state.

EPA National Survey of Hazardous Waste Management Facilities

EPA has documented two extensive national surveys pertaining to hazardous waste in the manual "National Survey of Hazardous Waste Generators and Treatment, Storage, Disposal, and Recycling Facilities in 1986" (see references). As the title suggests, the two surveys—one for hazardous waste generators and one for hazardous waste treatment, storage, disposal, and recycling facilities—provide comprehensive information about hazardous waste for the baseline year of 1986. Included in the survey is information pertaining to quantities and types of hazardous waste generated, number and types of facilities that generate hazardous waste, and sources from which waste is generated.

According to the survey, the amount of hazardous waste generated in 1986 was approximately 750 million tons, with approximately 700 million tons being classified as RCRA hazardous waste and the remaining 50 million tons being classified as hazardous under state laws or other federal laws. Of the total amount of hazardous waste generated, almost 620 million tons consisted of hazardous inorganic liquids, sludges, or solids. Organic liquids, sludges, and solids totaled approximately 100 million tons, and unknown or other was quantified at approximately 30 million tons. This data is presented in Figures 3.1 and 3.2.

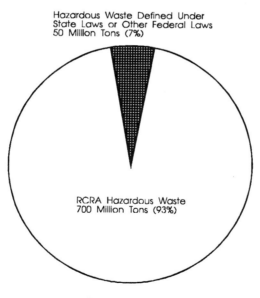

Hazardous Waste Defined Under
State Laws or Other Federal Laws
50 Million Tons (7%)

RCRA Hazardous Waste
700 Million Tons (93%)

Total = 750 Million Tons

Note: Data presented has been rounded.

Source: Data from EPA/530-SW-91-075 (1991).

FIGURE 3.1   Percentages of RCRA Hazardous and Other Hazardous Waste Generated in the U.S. in 1986.

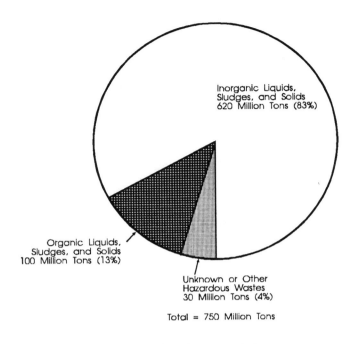

Inorganic Liquids,
Sludges, and Solids
620 Million Tons (83%)

Organic Liquids,
Sludges, and Solids
100 Million Tons (13%)

Unknown or Other
Hazardous Wastes
30 Million Tons (4%)

Total = 750 Million Tons

Note: Data presented has been rounded.

Source: Data from EPA/530-SW-91-075 (1991).

FIGURE 3.2 Percentages of Inorganic, Organic, and Other Hazardous Waste Generated in the U.S. in 1986.

Presented in Figure 3.3 is a summary of amounts of hazardous waste generated in 1986 by industry type. As is shown in the figure, the chemical products industry was overwhelmingly the largest generator, followed by electronics, petroleum and coal products, primary metals, and transportation equipment industries. Examples of subcategories or specific industries defined by EPA that contributed the most to waste generation within each industry category are presented in Table 3.3.

Percentages for major types of waste by RCRA code for the 750 million tons of waste generated in 1986 are presented in Figure 3.4. As can be seen from the figure, the largest amounts of hazardous waste are corrosive wastes and mixtures of ignitable, corrosive, and reactive wastes.

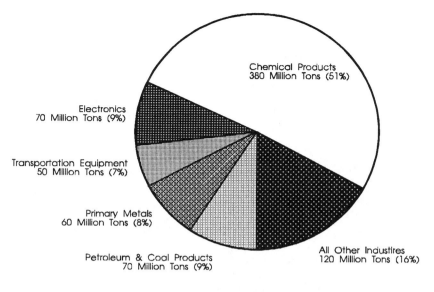

Chemical Products
380 Million Tons (51%)

Electronics
70 Million Tons (9%)

Transportation Equipment
50 Million Tons (7%)

Primary Metals
60 Million Tons (8%)

Petroleum & Coal Products
70 Million Tons (9%)

All Other Industires
120 Million Tons (16%)

Total = 750 Million Tons

Note: Data presented has been rounded.

Source: Data from EPA/530-SW-91-075 (1991).

FIGURE 3.3 Percentages of Hazardous Waste Generated by Specific Industries in the U.S. in 1986.

TABLE 3.3 Examples of Industry Types and Specific Industries that Generated the Most Waste in 1986

Industry Type	Specific Industries that Generated Waste
Chemical products	Organic chemicals, plastics and resins, explosives, cyclic crudes, alkalis and chlorine chemical products, others
Electronics industry	Semiconductors, electronic components, household appliances, TV picture tubes, telephone apparatus, others
Petroleum and coal industry	Petroleum refining, petroleum and coke products, lubricating oils and greases, others
Primary metals industry	Blast furnaces and steel mills, gray iron foundries, aluminum foundries, steel wire, steel pipe and tubes, others
Transportation equipment industry	Aircraft, motor vehicle bodies, motor vehicle parts, aircraft parts, others

Source: Information from EPA/530-SW-91-075 (1991).

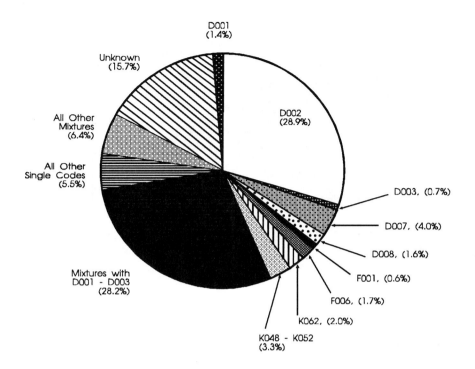

Total = 750 Million Tons

Characteristic Wastes:
    D001 - Ignitable
    D002 - Corrosive
    D003 - Reactive
    D007 - Toxic (Chromium contaminant)
    D008 - Toxic (Lead contaminant)

Hazardous Wastes from Nonspecific Sources:
    F001 - Specific spent halogenated solvents
          used in degreasing
    F006 - Wastewater treatment sludges from
          electroplating operations except
          for certain processes such as zinc
          and aluminum plating

Hazardous Wastes from Specific Sources:
    K062 - Spent pickle liquor generated
          by steel finishing operations
          at facilities within the iron
          and steel industries

    K048 - K052 - Specific wastes from
          petroleum refining including
          dissolved air flotation float,
          slop oil emulsion solids, heat
          exchanger bundle cleaning
          sludge, API separator sludge,
          and leaded tank bottoms

Source: Adapted from EPA/530-SW-91-075 (1991).

FIGURE 3.4 Percentages of Waste Produced by Waste Type in the U.S. in 1986.

Information pertaining to the number of waste generators also is detailed in the EPA survey in terms of industry category and specific industries. There were approximately 12,500 facilities that generated some amount of hazardous waste in 1986. The industries with the largest number of generating facilities included:

- Chemical products industry (approximately 17% of the facilities, totaling 2,133)—The number of generators was largest in the specific industries of paints, organic chemicals, and plastics and resins. Other generators were found in the industries of chemical preparations, printing inks, and others.
- Metal fabrication industry (nearly 14% of the facilities, totaling 1,696)—A large number of generators were found in the specific industries of plating and polishing and fabricated metal products. Other generators came from industries such as metal coating, screw machining products, and others.
- Electronics industry (almost 10% of the facilities, totaling 1,201)—The largest numbers of generators for this industry type were found in the specific industries of electronic components and semiconductors. Other generators were from the industries of electrical industrial apparatus, radio and TV equipment, motors and generators, and others.
- Transportation equipment industry (slightly more than 6% of the facilities, totaling 788)—The largest number of generators from this industry came from the specific industries of aircraft and motor vehicle bodies and motor vehicle parts. Aircraft equipment, aircraft parts, and other specific industries were also generators.
- Primary metals industry (approximately 6% of the facilities, totaling 731)—This industry had the largest number of generators from the specific industries of blast furnace and steel mills. Other specific industries which had generators included gray iron foundries, steel wire, primary metal products, secondary nonferrous metals, and others.

Nonelectrical machinery and electrical, gas, and sanitary services had slightly more than 10% of the total or approximately 675 generators, each. All other industries had 4,580 generators or approximately 37% of the total number of generators.

EPA's survey also has documented sources of waste generation in terms of primary routine and nonroutine industrial activities, secondary routine and nonroutine industrial activities, and unknown or sporadic nonroutine activities. Of these sources, primary activities generated approximately 72% of all hazardous waste in 1986. Secondary and unknown activities generated 18% and 10% of the total waste, respectively. Examples of industrial activities associated with each of these sources and percentages of waste generated by that activity are presented in Table 3.4.

TABLE 3.4   Sources of Waste Generation Documented by EPA

Source Category	Activities that Generate Waste
Primary routine (~530 million tons)	Electroplating processes (13%), hydrogenation (10%), distillation and fractionation (10%), nitration (7%), pickling (4%), spray rinsing (4%), nonspecified production processes (20%), other (32%)
Primary nonroutine sources (~10 million tons)	Cleanout of production processes (98%), discarding of off-spec materials (2%), discarding of out-of-date materials (<1%)
Secondary routine sources (~125 million tons)	Wastewater treatment (62%), incineration (15%), quench cooling (9%), regenerating (6%), other (8%)
Secondary nonroutine sources (~5 million tons)	Cleanup of spill residues (51%) other remedial action (41%), closure of surface impoundments (8%)
Unknown sources (~75 million tons)	Unknown (95%), Accidental spills (3%), other cleanout or closure (2%), other one-time activities (<1%)

Note: The total waste in this table is 745 million tons (instead of 750 million tons) because of rounding.
Source: Information from EPA/530-SW-91-075 (1991).

As can be seen from the table, the major sources of primary waste generation are manufacturing processes, with waste treatment processes contributing as a secondary source. As a result of obtaining data from waste surveys and other reports, EPA has been able to focus its research for pollution prevention technologies and strategies on selected industries and processes that generate the largest amounts of hazardous waste.

Other Sources of Information Pertaining to Hazardous Materials and Hazardous Waste

In addition to periodic surveys, EPA obtains information about hazardous materials use and hazardous waste generation through mandatory reports. These include the Tier I/Tier II report and the toxic release inventory report. In addition, under RCRA, all generators of hazardous waste must prepare and submit annual or biennial reports of types and amounts of waste generated at the facility. EPA requires biennial reports, but some states that have received delegated authority to administer the RCRA program require reports on an annual basis. These reports are available for public review and can be used by EPA or the state for assessing waste reductions from year to year.

NONMANUFACTURING FACILITIES AND ACTIVITIES

Selected examples of nonmanufacturing facilities and activities that use chemicals or hazardous materials and generate waste are listed in Table 3.5. These activities are regulated by several different agencies, some are regulated on the state level, and some are exempt from most of the regulations that apply to industry.

TABLE 3.5 Selected Examples of Hazardous Materials and Wastes Associated with Nonmanufacturing Facilities or Activities

Facility or Activity	Examples of Hazardous Chemicals Used	Examples of Waste Generated
Agricultural activities	Pesticides, fungicides, fertilizers	Potentially toxic runoff from fields and storage areas
Auto servicing or repair facilities	Solvents, paints, battery acid, oils	Used oils, used batteries, spent solvents
Demolition activities	Normally not chemical intensive	Hazardous dusts such as asbestos
Drilling operations	Concentrated brines and muds, considered nonhazardous	Waste excluded under RCRA
Educational research institutions	Lab chemicals, other research chemicals, solvents	Lab packs, discarded chemical mixtures, spent acids and solvents
Electrical/transformer maintenance	Oils, lubricants	Potential for PCBs in old capacitors, used oils
Hospitals	Radiation sources, lab chemicals, disinfectants	Infectious waste, sharps, low-level radioactive waste, lab packs
Municipal water treatment facilities	Chlorine	Chemical residues in tanks or containers
Municipal wastewater treatment facilities	Nonpathological microorganisms, phosphates	Sludges, normally classified as nonhazardous

continued

TABLE 3.5 *Continued*

Facility or Activity	Examples of Hazardous Chemicals Used	Examples of Hazardous Waste Generated
Military equipment repair facilities	Solvents, paints, oils, batteries	Spent solvents, discarded batteries, used oils
Nuclear power plants	Radioactive materials	Radioactive waste, mixed hazardous and radioactive waste, radioactive-contaminated wastewaters
Power plants	High-sulfur coal, natural gas, oil	Sulfur dioxide, carbon monoxide, nitrogen oxides

MEDICAL WASTE

In response to public concern about mismanaged medical waste, Congress enacted the Medical Waste Tracking Act of 1988. As a result, Subtitle J was added to the Resource Conservation and Recovery Act requiring EPA to report on the types, number, and size of generators of medical waste in the United States. Small-quantity generators were to be included. Data compiled by EPA as a result of on-site surveys and research has been compiled into a book entitled *Medical Waste Management and Disposal* (see references). Data from EPA pertaining to amounts of medical waste and generator information is shown in Figures 3.5 and 3.6. As can be seen from the figures, the vast majority of medical waste is generated by hospitals, although they comprise a very small part of the total number of generators.

CONCLUSION

The use of hazardous materials and the generation of hazardous and non-hazardous waste are not limited to industrial and commercial activities, but are also part of agriculture, defense, research, health care, and other non-commercial activities. Applications that use chemicals and generate waste are varied and numerous. Since there is environmental risk associated with releases of hazardous materials to the environment, tracking amounts of chemicals stored and waste generated at a specific location became a focus during the 1980s.

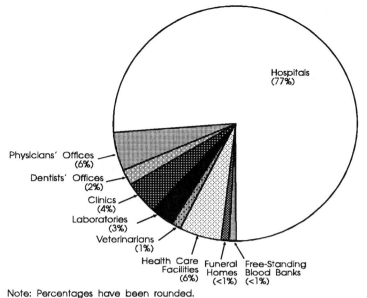

Note: Percentages have been rounded.

Source: Data from EPA (1990).   Total = 465,600 Tons/Year

FIGURE 3.5 Percentages of Medical Waste Produced by Specific Medical Facilities in the U.S.

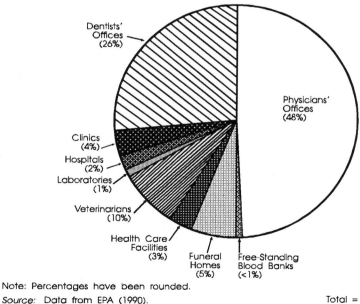

Note: Percentages have been rounded.

Source: Data from EPA (1990).   Total = 377,300 Generators

FIGURE 3.6 Percentages of Medical Waste Generators in the U.S. by Generator Type.

Several reports are now available that give EPA and the public information about chemical use and waste generation. These include the annual or biennial waste generation report, which is required of all generators, the toxic release inventory report, which is required of major chemical users, and the Tier I/Tier II report, which is required of facilities that store hazardous chemicals in large quantities. Additionally, in 1986, EPA surveyed facilities that generate, treat, store, recycle, and dispose of waste and has published data based on that survey. Medical waste also has been tracked through a demonstration program, and EPA has published information about this type of waste.

This section of the book presented an overview of hazardous materials and hazardous waste including key regulations, definitions, and major types of hazardous materials and hazardous waste. The next section discusses the topic of workplace management of hazardous materials and hazardous waste. Included are chapters pertaining to exposure assessment, the use of personal protective equipment, building safety, and regulatory-driven administrative requirements.

REFERENCES

U.S. Environmental Protection Agency, "National Survey of Hazardous Waste Generators and Treatment, Storage, Disposal, and Recycling Facilities in 1986," EPA/530-SW-91-075, Office of Solid Waste and Emergency Response, Washington, D.C., 1991.

U.S. Environmental Protection Agency, *Medical Waste Management and Disposal*, Pollution Technology Review No. 200, Noyes Data Corporation, Park Ridge, NJ, 1991.

U.S. Environmental Protection Agency, "Medical Waste Management in the United States—First Interim Report to Congress," Office of Solid Waste, Washington, D.C., 1990, published in *Medical Waste Management and Disposal*, Noyes Data Corporation, Park Ridge, NJ, 1991.

BIBLIOGRAPHY

AHA (1987), "Hospital Statistics," American Hospital Association, Chicago, IL.

API (1990), "Historical Overview of Solid Waste Management in the Petroleum Industry," API Discussion Paper 062, American Petroleum Institute, Washington, D.C.

_____ (1991), "The Generation and Management of Waste and Secondary Materials in the Petroleum Industry, 1987–1988," API Publication 300, American Petroleum Institute, Washington, D.C.

CMA (1990), "Hazardous Waste Surveys: 1981–1990," Chemical Manufacturers Association, Washington, D.C.

U.S. EPA (1987), "Estimating Releases and Waste Treatment Efficiencies for the Toxic Chemical Release Inventory Form," EPA 560/4-88/002, prepared by PEI and Associates, Cincinnati, OH, for Office of Pesticides and Toxic Substances, U.S. Environmental Protection Agency, Washington, D.C.

_____ (1987), "Toxic Release Inventory Reports," Office of Pesticides and Toxic Substances, U.S. Environmental Protection Agency, Washington, D.C.

_____ (1988), "Report to Congress on Solid Waste Disposal in the United States," EPA/530/SW-88/011, Office of Solid Waste and Emergency Response, U.S. Environmental Protection Agency, Washington, D.C.

_____ (1988), "Toxic Release Inventory Reports," Office of Pesticides and Toxic Substances, U.S. Environmental Protection Agency, Washington, D.C.

_____ (1989), "Toxic Release Inventory, A National Perspective," U.S. Environmental Protection Agency, Washington, D.C.

_____ (1989), "Toxic Release Inventory, Executive Summary," U.S. Environmental Protection Agency, Washington, D.C.

_____ (1989), "Toxic Release Inventory Reports," Office of Pesticides and Toxic Substances, U.S. Environmental Protection Agency, Washington, D.C.

_____ (1990), "Toxic Release Inventory Reports," Office of Pesticides and Toxic Substances, U.S. Environmental Protection Agency, Washington, D.C.

_____ (1991), "Toxic Release Inventory Reports," Office of Pesticides and Toxic Substances, U.S. Environmental Protection Agency, Washington, D.C.

SECTION II

WORKPLACE MANAGEMENT OF HAZARDOUS MATERIALS AND HAZARDOUS WASTE

4

UNDERSTANDING EXPOSURES FROM HAZARDOUS MATERIALS AND HAZARDOUS WASTES

Exposures from chemicals and waste in industrial applications can be evaluated in several ways. This chapter discusses methods typically used by industry to provide workers with information or real-time data about exposures from workplace hazardous materials and hazardous waste. These methods include the use of material safety data sheets, documented exposure limits, testing for toxicity characteristics, and medical surveillance programs. The use of chemical and waste labels, which also provides workers with information about materials in the workplace, is discussed in Chapter 8. Workplace and personal monitoring—which is integral to evaluating exposure—is addressed in Chapter 5.

MATERIAL SAFETY DATA SHEETS

All chemicals that are manufactured, imported, sold, or used in a manufacturing process must be accompanied by a material safety data sheet (MSDS), as defined in 29 CFR §1910.1200. Chemical manufacturers or importers must provide the MSDS to distributor and employers with their initial shipment of chemicals and with the first shipment after data on the MSDS has been updated. Distributors who sell to other distributors or employers, likewise, must provide the MSDS and updates to their customers. An MSDS or equivalent must be available for review by employees in the workplace and must contain, at a minimum, the following information:

- Chemical and common names—If the chemical is a single substance, the chemical and common name(s) must be identified. If the hazardous chemical is a mixture and has been tested as a whole to determine its hazards, the chemical and common names of the ingredients that contribute to the known hazards and the common name(s) of the mixture itself must be identified.

    If the hazardous chemical is a mixture that has not been tested as a whole, the following applies: chemical and common name(s) of all ingredients that have been determined to be health hazards and that comprise 1% or greater of the composition, or that are carcinogens and comprise 0.1% or greater, must be identified; the chemical or common name(s) of chemicals in the mixture that could be released in amounts greater than the permissible exposure limit must be identified; the chemical and common name(s) of all ingredients determined to present a physical hazard when present in the mixture must be identified.

- Physical and chemical characteristics—Characteristics such as vapor pressure, flash point, boiling point, melting point, specific gravity, solubility, and molecular weight must be included.

- Physical hazards—Physical hazards such as the potential for fire, explosion, and reactivity must be included. Incompatibilities and segregation practices also may be included in this part of the MSDS.

- Health hazards—The health hazards of the chemical must be included. Signs and symptoms of exposure, and any medical conditions that are generally recognized as being aggravated by exposure to the chemical must be delineated.

- Primary routes of entry—Information pertaining to primary routes of entry such as inhalation, ingestion, skin absorption, and eye or skin contact is required.

- Permissible exposure limit—The OSHA permissible exposure limit or American Conference of Governmental Industrial Hygienists threshold limit value and any other exposure limit used or recommended by the chemical manufacturer, importer, or employer preparing the MSDS must be included.

- Carcinogen or potential carcinogen—The MSDS also must include information as to whether the chemical is listed in the National Toxicology Program *Annual Report on Carcinogens* (latest edition) or has been published as a carcinogen or potential carcinogen in the International Agency for Research on Cancer *Monographs* (latest edition) or has been listed by OSHA.

- Safe handling procedures—This section of the MSDS should include items such as appropriate industrial hygiene practices, protective measures for handling the chemical during repair and maintenance of contaminated equipment, and procedures for cleanup of spills and leaks.

- Control measures—Control measures such as appropriate engineering controls, protective clothing and equipment, and work practices are to be detailed.

- Emergency and first aid procedures—These procedures should include emergency actions to be taken for exposure through all of the primary routes.
- Manufacturer's and other information—The MSDS also must include the name, address, and telephone number of the chemical manufacturer, importer, employer, or other responsible party who prepares or distributes the MSDS. Additionally, the date of MSDS preparation or the last change to it must be included.

An example of OSHA's standard MSDS form is shown in Figure 4.1.

EXPOSURE LIMITS

Exposure Limits Defined by OSHA and ACGIH

OSHA has published exposure limits for hundreds of hazardous airborne contaminants in terms of permissible exposure limits (PELs) and time weighted averages (TWAs). These limits define the maximum time weighted exposure over an 8-hour work shift of a 40-hour work week that should not be exceeded. These limits are expressed in parts per million (ppm) and/or milligrams per cubic meter (mg/m^3) and are published in 29 CFR §1910.1000.

The American Conference of Governmental Industrial Hygienists (ACGIH) defines airborne contaminant exposure similarly, and terms this exposure concentration the threshold limit value–time weighted average (TLV–TWA). TLV–TWAs are defined as a concentration to which nearly all workers can be repeatedly exposed over a normal 8-hour work day and a 40-hour work week, without adverse affects. Like OSHA, ACGIH has documented TLV–TWAs for hundreds of chemicals in the annual publication *Threshold Limit Values for Chemical Substances and Physical Agents and Biological Exposure Indices* (see references). Although not legally enforceable, ACGIH standards are considered industry standards.

In some instances, industry standards established by ACGIH differ from OSHA standards. This is typically because OSHA standards must be adjusted through regulatory amendments, which is a lengthy and time-consuming process, while ACGIH standards are updated annually based on the latest scientific data. Thus, to ensure adequate protection of workers, it is considered a good management practice for employers to abide by the ACGIH standard for a particular chemical if it is lower than the OSHA standard. If the OSHA is lower, the OSHA standard must be met.

For some chemicals, ACGIH and OSHA have documented additional exposure concentrations. These include a short-term exposure limit (STEL) and a ceiling limit (C). An STEL is a 15-minute time weighted average ex-

Material Safety Data Sheet
May be used to comply with
OSHA's Hazard Communication Standard,
29 CFR 1910.1200. Standard must be
consulted for specific requirements.

U.S. Department of Labor
Occupational Safety and Health Administration
(Non-Mandatory Form)
Form Approved
OMB No. 1218-0072

IDENTITY *(As Used on Label and List)*	Note: *Blank spaces are not permitted. If any item is not applicable, or no information is available, the space must be marked to indicate that.*

Section I

Manufacturer's Name	Emergency Telephone Number
Address *(Number, Street, City, State, and ZIP Code)*	Telephone Number for Information
	Date Prepared
	Signature of Preparer *(optional)*

Section II — Hazardous Ingredients/Identity Information

Hazardous Components (Specific Chemical Identity; Common Name(s))	OSHA PEL	ACGIH TLV	Other Limits Recommended	% *(optional)*

Section III — Physical/Chemical Characteristics

Boiling Point		Specific Gravity (H_2O = 1)	
Vapor Pressure (mm Hg)		Melting Point	
Vapor Density (AIR = 1)		Evaporation Rate (Butyl Acetate = 1)	
Solubility in Water			
Appearance and Odor			

Section IV — Fire and Explosion Hazard Data

Flash Point (Method Used)	Flammable Limits	LEL	UEL
Extinguishing Media			
Special Fire Fighting Procedures			
Unusual Fire and Explosion Hazards			

(Reproduce locally)   OSHA 174, Sept. 1985

FIGURE 4.1   Example of OSHA's Material Safety Data Sheet Form.

Section V — Reactivity Data

Stability	Unstable		Conditions to Avoid
	Stable		

Incompatibility *(Materials to Avoid)*

Hazardous Decomposition or Byproducts

Hazardous Polymerization	May Occur		Conditions to Avoid
	Will Not Occur		

Section VI — Health Hazard Data

Route(s) of Entry:	Inhalation?	Skin?	Ingestion?

Health Hazards *(Acute and Chronic)*

Carcinogenicity:	NTP?	IARC Monographs?	OSHA Regulated?

Signs and Symptoms of Exposure

Medical Conditions
Generally Aggravated by Exposure

Emergency and First Aid Procedures

Section VII — Precautions for Safe Handling and Use

Steps to Be Taken in Case Material Is Released or Spilled

Waste Disposal Method

Precautions to Be Taken in Handling and Storing

Other Precautions

Section VIII — Control Measures

Respiratory Protection *(Specify Type)*

Ventilation	Local Exhaust		Special
	Mechanical *(General)*		Other

Protective Gloves	Eye Protection

Other Protective Clothing or Equipment

Work/Hygienic Practices

* U S G P O 1986-491-529/45775

FIGURE 4.1 *Continued*

57

posure limit which should not be exceeded.[1] This limit is not a stand alone limit, and it must be included in the time weighted average. The STEL generally applies to a substance whose toxic effects are considered to be chronic, but for which there may be recognized acute effects at a particular limit. ACGIH recommends that exposure at this limit should not be repeated more than 4 times daily and that at least a 60-minute rest period between exposures should be allowed.

A ceiling limit is a maximum concentration that should not be exceeded for any period of time. Ceiling limits are established for chemicals which, at certain concentrations, could produce acute poisoning during very short exposures.

Since OSHA and ACGIH exposure limits are based on an 8-hour work day, 40-hour work week time weighted average, worker exposure may safely exceed published exposure limits for periods of time during the work day, as long as they are compensated with periods of time when exposures are below the limit, so that the 8-hour time weighted average is not exceeded. However, all factors related to a chemical exposure—including cumulative effects, frequency and duration of excursions, nature of contaminant, and others—should be considered before this type of exposure is allowed. In some cases, excursions above the permissible exposure limit or threshold limit value may not be acceptable.

The STEL and ceiling limits should not be exceeded at any time during the work day. For many chemicals, there is not enough toxicological data available for ACGIH or OSHA to publish an STEL or ceiling limit. When an STEL has not been published, ACGIH recommends that worker exposure at levels of 3 times the TLV–TWA should last for no more than 30 minutes during a work day. Additionally, if there is no ceiling limit, a concentration level of 5 times the TLV–TWA should never be exceeded. Consideration also should be given to areas that contain chemical mixtures. When calculating the exposure limit of chemical mixtures in the air, additive effects for individual components should be used when the components have similar toxicological effects.

Selected examples of chemicals that have published ACGIH and OSHA exposure limits are presented in Table 4.1. Examples of allowable exposure patterns based on the time weighted average are shown in Figure 4.2.

Odor threshold is a physical property of a chemical and has no relation to acceptable exposure concentrations. There are numerous chemicals that are toxic at levels below the odor threshold such as methylene chloride and carbon monoxide (AIHA 1989). Thus, odor—or the absence of odor—should never be used in determining acceptable workplace concentrations. In cases where workers are using air purifying (cartridge) respirators, odor (and taste) is considered a warning property indicating breakthrough of the filter material.[2] When noticed, the work area should be evacuated immediately.

---

[1] The 15-minute STEL applies to almost all chemicals; however, asbestos has a 30-minute excursion limit published by OSHA.

[2] In general, cartridge respirators are not allowed for chemicals that have permissible exposure limits below the odor threshold.

TABLE 4.1   Examples of Documented OSHA and ACGIH exposure Limits for Selected Chemicals

Chemical Compound	OSHA/ACGIH PEL/TWA (ppm)	OSHA/ACGIH STEL (ppm)	OSHA/ACGIH Ceiling (ppm)
Acetic acid	10/10	–/15	–/–
Acrolein	0.1/0.1	0.3/0.3	–/–
Benzyl chloride	1/1	–/–	–/–
Fluorine	0.1/1	–/2	–/–
Hexachloroethane	1/1	–/–	–/–
Iodine	–/–	–/–	0.1/0.1
Isopropyl alcohol	400/400	500/500	–/–
Methylamine	10/5	–/15	–/–
Naphthalene	10/10	15/15	–/–
Phenol	5/5	–/–	–/–
Sulfur dioxide	2/2	5/5	–/–
Vinyl chloride	1/5	5/–	–/–
Xylenes	100/100	150/150	–/–

Note: Values in terms of milligrams per cubic meter (mg/m^3) may exist, but are not included.

Source: Information from 29 CFR 1910 Subpart Z and ACGIH, *Threshold Limit Values for Chemical Substances and Physical Agents and Biological Exposure Indices* (1992).

Exposure Limits for Radioactive Materials

NRC has codified the practice of maintaining all radiation exposures to workers and the general public as low as reasonably achievable (known as the ALARA concept). Public and worker protection standards for specific radioactive materials and wastes have been established by OSHA, EPA, NRC, Mine Safety and Health Administration (MSHA), and other agencies (Oak Ridge Associated Universities 1988). Worker limiting requirements are categorized by body part and organ and are defined in numerous regulations, including 10 CFR 20 Subpart C, 29 CFR §1910.96, 40 CFR Part 190, 30 CFR Part 57, and other citations.

TOXICITY CHARACTERISTIC LEACHING PROCEDURE

EPA's toxicity characteristic leaching procedure (TCLP), designated as Method 1311 and presented in 40 CFR 261 Appendix II of the RCRA regulations, is a solid waste extraction procedure used to identify hazardous characteristics of solid waste contaminated with listed toxic chemicals, metals, and pesticides. Chemicals regulated as toxic by TCLP testing are listed in 40 CFR §261.24 and are presented in Table 4.2.

Regulatory limits in milligrams per liter (mg/l) are set for each chemical. If the TCLP analysis shows concentrations above the regulatory limits, the

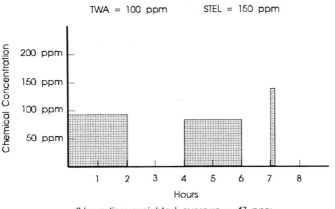

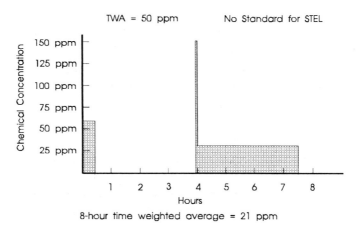

Note: A trained industrial hygienist or other chemical specialist should evaluate thoroughly the TWA, chemical additive effects, and other factors to determine the need for respiratory protection.

FIGURE 4.2 Examples of Allowable Chemical Exposure Patterns Based on Time Weighted Average.

waste must be managed as hazardous. It is expected that this list will expand in the future.

MEDICAL SURVEILLANCE

Medical surveillance programs offer information about the biological effects chemicals may be having on an individual worker. Medical surveillance is required by OSHA for workers who routinely are exposed to certain concen-

TABLE 4.2   Chemicals Regulated as Toxic by TCLP Testing

Chemical Category	Chemicals Regulated
Metals	Arsenic, barium, cadmium, chromium, lead, mercury, selenium, silver
Pesticides	Chlordane, 2,4-D, endrin, heptachlor, lindane, methoxychlor, toxaphene, 2,4,5-TP (Silvex)
Cresols	Cresol, o-cresol, m-cresol, p-cresol
Phenols	Pentachlorophenol, 2,4,5-trichlorophenol, 2,4,6-trichlorophenol
Benzenes	Benzene, chlorobenzene, 1,4-dichlorobenzene, hexachlorobenzene, nitrobenzene, 2,4-dinitrotoluene
Chlorinated compounds	Carbon tetrachloride, chloroform, 1,2-dichloroethane, 1,1-dichloroethylene, hexachlorobutadiene, hexachloroethane, tetrachloroethylene (perchloroethylene), trichloroethylene, vinyl chloride
Other toxic compounds	Methyl ethyl ketone, pyridine

trations of regulated chemicals or dusts such as lead, benzene, formaldehyde, acrylonitrile, asbestos, and coke oven emissions. Additionally, medical surveillance is required for workers who routinely wear a respirator or who take part in hazardous waste operations and emergency response activities. The employer is responsible for administering and paying for the medical examinations and other medical surveillance activities.

Tests included in a typical medical surveillance program can vary depending on specific chemical exposures. An initial questionnaire is mandatory for workers who are exposed to asbestos and generally is included in some form in most medical surveillance programs. The OSHA asbestos questionnaire includes occupational history, past medical history, chest and other illnesses, family history, smoking habits, and other pertinent information.

Other aspects of the medical surveillance program could include a physical exam, general health survey, blood count, chemistry screen, urinalysis, spirometry test, chest X-ray, audiogram, and electrocardiogram. Breath analysis may be included for workers exposed to chemicals that can be detected through this type of screening.

All exams should be performed on a regularly scheduled basis, with the time between exams dependent on types of exposures. The time of testing, such as at the end of the shift or the end of the week, often can be critical for observing accurate effects on blood or urine, so this should be taken into account when establishing the examination schedule.

Biological effects of chemical overexposure are documented by ACGIH, and can be seen in blood, urine, and respiratory functions. Effects on blood can be seen directly or indirectly through blood counts and blood screening. Indirect indicators are red and white blood cells, plasma, and other blood components. Examples of chemicals that affect the blood composition include parathion, carbon monoxide, aniline, and nitrobenzene.

Overexposure can be ascertained directly by the presence of a specific chemical in the blood stream at elevated levels. Chemicals that can be tested in this manner include cadmium, lead, toluene, and perchloroethylene.

In addition to blood screening, chemistry screening can provide information on cholesterol, triglycerides, total protein, blood urea nitrogen, calcium, phosphorus, and other chemical components. Urinalysis is also a very useful screening test. This test can provide direct detection of numerous chemicals in the urine. Examples include mercury, cadmium, lead, methyl ethyl ketone, and phenol.

Breath analysis can be used to detect the presence of numerous chemicals. Standards for acceptable concentrations of chemicals in the breath have been developed by ACGIH for chemicals such as benzene, carbon monoxide, toluene, and trichloroethylene.

Decreases in lung capacity can be determined with the spirometry test. When necessary, this test can be coupled with chest X-rays. In some cases, respiratory capacity tests may be performed more often than other parts of the medical surveillance tests, if exposure warrants.

CONCLUSION

Evaluating exposures from hazardous chemicals and hazardous waste is an important part of workplace management. Typically, exposure assessment is provided through several methods. A chemical MSDS provides useful data about the physical and hazard properties of the chemical and other important information. This or equivalent information is required to be accessible to employees at all times.

Airborne and other exposure limits published by OSHA and ACGIH provide data for the determination of respirator requirements or administrative controls for particular tasks. Testing for toxicity also can provide data needed for safe handling of wastes. Medical surveillance gives real-time data of biological effects chemicals may be having on an individual.

The next chapter discusses another important means of assessing exposures to workers—workplace and personal monitoring. Included in the chapter is information pertaining to standard test methods and instruments used to monitor indoor atmospheres.

REFERENCES

American Conference of Governmental Industrial Hygienists, *Threshold Limit Values for Chemical Substances and Physical Agents and Biological Exposure Indices*, Cincinnati, OH, 1992.

American Industrial Hygiene Association, *Odor Thresholds for Chemicals with Established Occupational Health Standards*, Akron, OH, 1989.

Code of Federal Regulations, 29 CFR 1910 Subpart Z, U.S. Department of Labor, Occupational Safety and Health Administration, Washington, D.C.

Oak Ridge Associated Universities, *A Compendium of Major U.S. Radiation Protection Standards and Guides: Legal and Technical Facts*, prepared by W.A. Mills et al., Oak Ridge, TN, 1988.

BIBLIOGRAPHY

API (1983), "Surveillance of Reproductive Health in the U.S: A Survey of Activity Within and Outside Industry," Monograph, prepared by M. Hatch et al., American Petroleum Institute, Washington, D.C.

_____ (1990), "A Case-Control Study of Kidney Cancer Among Petroleum Refinery Workers," API Publication 4504, American Petroleum Institute, Washington, D.C.

Ashford, N.A., and C.S. Miller (1991), *Chemical Exposures—Low Levels and High Stakes*, Van Nostrand Reinhold, New York.

CMA (1991), "Draft ANSI Standards for the Preparation of Material Safety Data Sheets," Chemical Manufacturers Association, Washington, D.C.

_____ (1991), "Occupational Epidemiology Resource Manual," Chemical Manufacturers Association, Washington, D.C.

Clayton, George D., and Florence E. Clayton, eds. (1991), *Patty's Industrial Hygiene and Toxicology*, Vols. 1A and 1B, 4th ed., John Wiley & Sons, New York.

Dillon, H.K., and M.H. Ho, eds. (1991), *Biological Monitoring of Exposure to Chemicals: Metals*, John Wiley & Sons, New York.

Ho, M.H., ed. (1987), *Biological Monitoring of Exposure to Chemicals: Organic Compounds*, John Wiley & Sons, New York.

Hodgson, Ernest (1988), *Dictionary of Toxicology*, Van Nostrand Reinhold, New York.

Kusnetz, S., and M.K. Hutchinson (1979), *A Guide to the Work-Relatedness of Disease*, National Institute for Occupational Safety and Health, Cincinnati, OH.

Lewis, Richard J., Sr. (1990), *Rapid Guide to Hazardous Chemicals in the Workplace*, 2nd ed., Van Nostrand Reinhold, New York.

Lipton, Sidney, and Jeremiah Lynch (1987), *Health Hazard Control in the Chemical Process Industry*, John Wiley & Sons, New York.

NCRP (1989), "Radiation Protection for Medical and Allied Health Personnel," Report No. 105, National Council for Radiation Protection and Measurements, Bethesda, MD.

NIOSH (1987), "Pocket Guide to Chemical Hazards," DHHS (NIOSH) Publication No. 85-114, U.S. Department of Health and Human Services, National Institute for Occupational Safety and Health, Washington, D.C.

NRC (1992), "Occupational Radiation Exposure at Commercial Nuclear Power Reactors and Other Facilities," 22nd Report, NUREG-0713-V11/XAB, Nuclear Regulatory Commission, Washington, D.C.

Proctor, Nick H., James P. Hughes, and Michael L. Fischman (1990), *Chemical Hazards of the Workplace*, 2nd ed., Van Nostrand Reinhold, New York.

Reichert, Richard J. (1990), "Pitfalls and Protocols for Medical Surveillance of Hazardous Waste Workers," pp. 119-129, *Proceedings of the Third Annual Hazardous Materials Management Conference/Central*, Chicago.

Sherman, Janette (1988), *Chemical Exposure and Disease: Diagnostic and Investigative Techniques*, Van Nostrand Reinhold, New York.

Technology Assessment Task Force (1990), *Reproductive Health Hazards in the Workplace*, Van Nostrand Reinhold, New York.

U.S. EPA (1984), "Biological Effects of Radiofrequency Radiation," EPA/600/8-83/026F, prepared by D.F. Cahill and J.A. Elder, eds., Health Effects Research Laboratory, Office of Research and Development, U.S. Environmental Protection Agency, Research Triangle Park, NC.

Williams, Phillip L., and James L. Burson (1989), *Industrial Toxicology—Safety and Health Applications in the Workplace*, Van Nostrand Reinhold, New York.

5

WORKPLACE AND PERSONAL MONITORING

General workplace and personal monitoring are two means of measuring exposures associated with industrial activity. Workplace monitoring provides qualitative and/or quantitative data about chemical atmospheres of a work location or area. This type of monitoring is used at hazardous waste sites for characterization of the atmosphere before entry. Additionally, general workplace monitoring is used throughout industry as a means of safe entry verification after evacuation, and for other chemical investigations resulting from unusual odors or other problems. Some continuous monitoring systems are used in the workplace to detect leaks and improper operations that cause higher than normal chemical exposures.

Personal monitoring provides quantitative data about employee exposure associated with a specific set of work tasks. The samples are taken in the employee's breathing zone to represent actual inhalation exposures. This type of sampling is used widely throughout industry to ensure permissible exposure limits are being met. This chapter discusses both general workplace and personal monitoring and provides information about standard test methods, types of instruments available for sampling, and other information pertinent to the topic.

STANDARD TEST METHODS AND PRACTICES

Standard test methods, practices, and guides for workplace and personal monitoring are defined by the American Society of Testing and Materials

(ASTM) in the *Annual Book of ASTM Standards* (see references) and include:

- D 1356-73(1991)—Standard Definitions of Terms Relating to Atmospheric Sampling and Analysis
- D 1357-82(1989)—Standard Practice for Planning the Sampling of the Ambient Atmosphere
- D 1605-60(1990)—Standard Practices for Sampling Atmospheres for Analysis of Gases or Vapors
- D 3686-89—Standard Practice for Sampling Atmospheres to Collect Organic Compound Vapors (Activated Charcoal Tube Adsorption Method)
- D 3687-89—Standard Practice for Analysis of Organic Compound Vapors Collected by Activated Charcoal Tube Adsorption Method (using Gas/Liquid Chromatography)
- D 3824-88—Standard Test Method for Continuous Measurement of Oxides of Nitrogen in the Ambient or Workplace Atmosphere by the Chemiluminescent Method
- D 4240-83(1989)—Standard Test Method for Airborne Asbestos Concentration in Workplace Atmosphere
- D 4490-90—Standard Practice for Measuring the Concentration of Toxic Gases or Vapors using Detector Tubes
- D 4532-85(1990)—Standard Test Method for Respirable Dust in Workplace Atmospheres
- D 4597-87—Standard Practice for Sampling Workplace Atmospheres to Collect Organic Gases or Vapors with Activated Charcoal Diffusional Samplers
- D 4598-87—Standard Practice for Sampling Workplace Atmospheres to Collect Organic Gases or Vapors with Liquid Sorbent Diffusional Samplers
- D 4599-86—Standard Practice for Measuring the Concentration of Toxic Gases or Vapors using Length-of-Stain Dosimeter
- D 4844-88—Guide for Air Monitoring at Waste Management Facilities for Worker Protection
- D 4861-91—Standard Practice for Sampling and Analysis of Pesticides and Polychlorinated Biphenyls in Indoor Atmospheres
- D 4947-89—Standard Test Method for Chlordane and Heptachlor Residues in Indoor Air
- E 1370-90—Guide to Air Sampling Strategies for Worker and Workplace Protection

As can be seen from the above listing, standards exist for workplace monitoring of all types of chemicals including gases or vapors, particulates, pesticides,

PCBs, asbestos, and dusts. Personal sampling methods, including passive or diffusional sampling (D 4597 and D 4598) and sampling with a personal sampling pump (D 3686), also are addressed in the standards.

In addition to ASTM standards, useful air sampling information can be found in ACGIH's handbook entitled *Air Sampling Instruments for Evaluation of Atmospheric Contaminants* (see references). This handbook addresses the following topics:

- The basics of air sampling—Included is an overview of air sampling and analysis, strategies for occupational and community air sampling, information about the measurement process, methods for calibrating air sampling instruments, and information about gas stream sampling and sampling in calibration and exposure chambers.
- Sampling for specific health hazards—Included is information pertaining to size-selective health hazard sampling, sampling airborne microorganisms and aeroallergens, and sampling airborne radioactivity.
- Sampling systems and components—Information pertaining to air movers and samplers, systems for the sampling of ducts and stacks, and unattended sampling systems is provided.
- Sample collectors—Information includes summaries of types of samplers typically used in industry and their applications.
- Direct-reading instruments—Included is information about direct-reading instruments and their applications.

The handbook has several useful tables which summarize instrument capabilities, as well as numerous photographs depicting marketed instruments.

Other useful information is published by NIOSH, OSHA, and EPA. Acceptable analytical methods for workplace applications are documented in *NIOSH Manual of Analytical Methods, OSHA Analytical Methods Manual*, and EPA's *Compendium of Methods for Determination of Toxic Organic Compounds in Indoor Air* (see references). NIOSH, OSHA, the U.S. Coast Guard, and EPA have also published information about air monitoring equipment for use at hazardous waste sites in the manual *Occupational Safety and Health Guidance Manual for Hazardous Waste Site Activities* (see references).

WORKPLACE MONITORING

Workplace monitoring can be performed using a wide range of analyzers. These include single chemical or total concentration analyzers, which are relatively easy to use and typically provide quantitative information about a single chemical or chemical species. For evaluating a multichemical environment, a more sophisticated instrument is needed.

Single Chemical or Total Concentration Analyzers

There are numerous single chemical or total concentration analyzers available for workplace monitoring. In general, these analyzers are uncomplicated and easy to operate. The next several pages present a synopsis of several analyzers of this type, with information about the operation of the instrument, chemicals that the instrument can detect, and some brief comments on the limitations of the equipment.

Chemiluminescence Analyzer for Monitoring Oxides of Nitrogen

- Instrument operation—Oxides of nitrogen (NO_x) are converted to nitric oxide (NO) and reacted with ozone to generate light emissions that are monitored by a photomultiplier tube, as shown in Figure 5.1. Nitrogen dioxide (NO_2) concentrations are determined by intermittent direct sampling of the stream (without NO_x conversion) and by subtracting the NO concentration from the NO_x concentration.
- Chemicals detected—NO_x or NO_2 can be detected in the parts per million (ppm) range.
- Limitations—Negative interferences may occur at high humidities for instruments calibrated with dry span gas. Also olefins and organic sulfur compounds, if present, will positively interfere with no detection (ASTM 1992, Vol. 11.03).

Oxygen Meter with an Electrochemical Sensor

- Instrument operation—An oxygen meter typically uses an electrochemical sensor, as shown in Figure 5.2. The electrochemical sensor has a semipermeable membrane to allow air into the cell through diffusion, and uses an electrolytic, current-conducting solution to register current changes directly proportional to the amount of oxygen in the atmosphere. The current is amplified and displayed on a meter. Alarms can be set to sound if the oxygen concentration drops below a preset percentage.

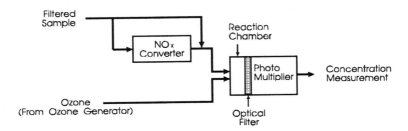

Source: Adapted from ASTM D 3824.

FIGURE 5.1 Example of a Chemiluminescence Analyzer.

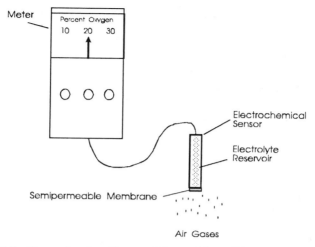

FIGURE 5.2 Example of an Oxygen Meter with an Electrochemical Sensor.

- Chemicals detected—This instrument is programmed to read the percent of oxygen in the atmosphere. Other applications for an electrochemical sensor include a sulfur dioxide sensor and a hydrogen sulfide analyzer.
- Limitations—The sensor responds to the partial pressure of oxygen, and is therefore altitude sensitive, with reduced readings at higher altitudes. The sensor in an oxygen meter may be affected by oxidants such as ozone. Carbon dioxide will poison the detector cell (NIOSH 1985).

Combustible Gas Indicator (CGI)

- Instrument operation—A CGI utilizes a sensor to measure the relative resistance changes produced by gases burning on hot filaments, one of which is coated with a catalyst. All readings on the combustible gas meter are relative to the calibrant gas, usually hexane, methane, or pentane. A CGI is used to determine whether flammable/combustible material is present in a concentration that could be dangerous. Concentrations between the lower and upper explosive limit (LEL and UEL, respectively), are considered immediately dangerous.
- Chemicals detected—Flammable gases can be measured with this instrument. The concentrations are measured in percent LEL and can be converted to ppm using vendor response curves and conversion factors.
- Limitations—Oxygen variations will affect combustion and, consequently, proper operation of the instrument. Lean and enriched mixtures will give inaccurately low and high readings, respectively. Temperature differences between calibration and use also can affect the instrument's accuracy. Additionally, silicones, halides, and lead compounds will coat the detector unit and render it inoperable (NIOSH 1985). Thus, its use in

atmospheres containing unknown vapors can be limited. Corrosive atmospheres, over time, can damage internal components of the instrument and limit its use. Under oxygen-deficient conditions, the instrument will not provide an accurate reading.

Length-of-Stain Detector Tube

- Instrument operation—The length-of-stain detector tube is a colorimetric visual indicator. It is operated by drawing a fixed volume of sample air through a tube with a squeeze bulb or small hand pump. Diffusional models (dosimeters) are also available. A length-of-stain dosimeter operates by allowing an air sample to diffuse through the tube over a 1 to 8 hour period.

  With both types of tubes, the tube contains a length of granulated resin or gel impregnated with a reactive chemical. The granules change color when specific types of air contaminants are introduced. Generally, length-of-stain detector tubes that use a pump allow the chemical concentration to be read directly on the tube, based on the length of color change. In the diffusional models, a color chart is provided by the manufacturer. An example of a hypothetical calibration graph for a length-of-stain dosimeter is shown in Figure 5.3.

- Chemicals detected—The detector tube can be used for determining the presence of most hydrocarbons, acids, bases, organic amines, and alcohols. Accuracy is typically better than ±25% of the actual concentration in the ppm range (ASTM 1992, Vol. 11.03).

- Limitations—Temperature and humidity can affect the length-of-stain color change, and calibration charts must be used properly to make these corrections. Additionally, similar chemicals can interfere positively with the detector tube's reading.

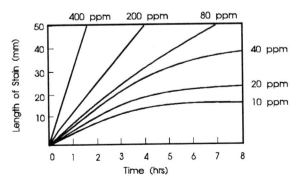

FIGURE 5.3   General Calibration Graph for Length-of-Stain Detector Tubes.

Photoionization Detector (PID)

- Instrument operation—A PID uses high-energy ultraviolet (UV) light to ionize volatile organic compounds in an air sample. Ionization of the sample produces a current that is proportional to the number of ions measured.
- Chemicals detected—The PID can measure total concentrations of many organic and some inorganic gases. Major air components such as oxygen, nitrogen, and carbon dioxide are not ionized in the process. The detector can quantify chemicals in the parts per billion (ppb) to ppm range, depending on the compound (Daisey 1987).
- Limitations—The instrument cannot detect some compounds if the probe used has a lower energy level than the compound's ionization potential. Additionally, high humidity can dampen the detector's response. If the instrument does not have a filter, charged particles can damage the internal parts. Likewise, atmospheres containing corrosive gases can cause damage unless a corrosive resistant instrument is used.

Flame Ionization Detector (FID)

- Instrument operation—The FID mixes an air sample with hydrogen and combusts the sample in a detector cell, ionizing the gases and vapors. The instrument electronically measures the current flow through the flame as the sample is burned and amplifies this measurement on an analog display.
- Chemicals detected—Total hydrocarbons are measured in the ppm range.
- Limitations—Ultra-high purity hydrogen fuel and high purity air, free from hydrocarbons, must be used to ensure accurate readings. Also, the combustion chamber is very sensitive and can be damaged by corrosive or reactive gases. The portable version of the instrument is limited in atmospheres that are oxygen deficient, since lack of oxygen can cause flame-outs (Erb, Ortiz, and Woodside 1990).

Infrared (IR) Analyzer

- Instrument operation—The IR analyzer uses a spectrometer to read chemical "fingerprints" or concentration intensities created by passing infrared frequencies through a heated air sample. The basic instrument has a fixed cell pathlength and is calibrated for one or a few predetermined chemicals.
- Chemicals detected—The instrument can detect volatile organic compounds in the spectral range of 2.5 to 15 micrometers (μm). In the most sophisticated instruments, numerous compounds can be detected in the ppb to ppm ranges (Daisey 1987).
- Limitations—This instrument cannot be used in flammable or explosive atmospheres. Additionally, excessively humid or corrosive atmospheres

can cause damage to the instrument. Water vapors and carbon dioxide can interfere with the instrument's readings (NIOSH 1985).

Portable Gas Chromatograph (GC)

- Instrument operation—A portable GC can be used with other detectors such as a PID or FID to identify and measure specific compounds. The portable model uses a packed or capillary column to concentrate and separate the compounds according to their vapor pressures. After separation, a detector quantifies the individual compounds (peaks). The identity of the compounds or peaks is qualitatively or quantitatively determined by its retention time in the GC.
- Chemicals detected—The GC can separate organic gases and vapors for quantitation with a detector. Quantitation limits are dependent on the detector used.
- Limitations—For selective quantitative results, the instrument must be calibrated with the specific analyte. The instrument may not be consistently sensitive to all organic compounds. Additionally, mixtures of polar and nonpolar compounds can cause peak superimpositions, which may require changes in column type, length of column, or operating conditions (ASTM 1992, Vol. 11.03).

Multichemical Analyzers

Multichemical analyzers have become more readily available over the last decade and are starting to be used more frequently in industry. In general, this type of analyzer is more costly than a single chemical or total concentration analyzer since it is more complex. However, many of these analyzers have the capability to monitor more than one port, which gives the user added sampling flexibility. The following paragraphs provide information about several multichemical analyzers.

On-line Mass Spectrometer (MS)

- Instrument operation—This on-line system utilizes a central MS unit and a vacuum pump system to sequentially draw samples and analyze chemicals from remote locations as far away as 1,500 feet. As many as 50 different workplace locations can be connected with chemical resistant tubing. Up to 25 chemicals can be analyzed per location. Depending on the number of chemicals analyzed and number of ports attached, the time between samples can range from a few minutes to several hours (Erb, Ortiz, and Woodside 1990).
- Chemicals detected—An on-line MS can quantify most chemicals that normally are read on a lab MS, at ppm levels. Metals cannot be quantified. The lower the detection limit required, the longer the dwell time in the analyzer. Lengthened dwell times can substantially increase the time between samples.

- Limitations—In certain cases, high chemical concentrations may not be monitored optimally by this system. Corrosive chemicals, dust, and humidity may cause damage to the system. Additionally, the system is operationally complex and may require extensive calibration and maintenance.

On-line GC/FID
- Instrument operation—The GC/FID is an automatic system designed for continuous monitoring of a wide variety of volatile organic compounds. The automated GC separates chemicals by vapor pressure, and the FID quantifies the chemicals. The system can be configured to sequentially sample up to 24 lines from 50 to 100 feet away (Coleman 1990).
- Chemicals detected—Organic vapors can be detected in the low ppm ranges.
- Limitations—The system is operationally complex and may require extensive calibration and maintenance.

On-line Fourier Transform Infrared (FTIR) Spectrometer
- Instrument operation—This instrument is commonly used for quantitative spectroscopic applications in the mid-infrared range. The instrument operates similarly to a standard IR, but uses variable pathlengths and wavelengths to allow for quantitation of numerous chemicals (EPA 1989).
- Chemicals Detected—Chemicals detected will vary, depending on set-up of instrument, but the equipment is capable of detecting the same chemicals detected by any IR spectrometer.
- Limitations—Sample conditioning, which can potentially remove chemicals of interest, is required for sample streams containing moisture or corrosive chemicals. Additionally, system complexity requires extensive calibration and may require extensive maintenance.

PERSONAL MONITORING

Personal monitoring is performed to determine actual chemical concentrations that workers are exposed to during the work day. This type of monitoring is performed periodically under typical, representative working conditions. The samples generally are taken over an 8-hour shift if chemical-use operations are continuous. In some cases, sampling may be performed over a short period of time within a shift to quantify peak exposures.

Personal sampling generally is accomplished using two methods—sampling with a personal sampling pump and passive or diffusional sampling. Sampling with a personal sampling pump generally is accepted by OSHA for documenting chemical exposures. Passive sampling typically is used for supplemental sampling and screenings. Sampling verification using OSHA methods, when specified, should be used. The following paragraphs provide information about both of these types of sampling methods.

Sampling with a Personal Sampling Pump

Sampling with a personal sampling pump is standard across industry and is documented in ASTM standards and by ACGIH. For this type of sampling, a battery powered personal sampling pump typically is used in conjunction with an air metering device and an adsorbent tube sampler such as charcoal or silica gel tube samplers. Sampling with a personal sampling pump yields accurate results since the volume of air sampled is metered and can be accurately quantified. Factors that must be considered when calculating sampled air volume include time, flow rate, pressure, and workplace temperature and humidity.

Charcoal Tube Adsorption Sampler. This sampler uses a charcoal sampling tube containing two sections of activated charcoal, as shown in Figure 5.4, and a sampling pump that draws a sample at a stable flow rate. The carbon tube is taken to a lab and desorbed into a GC using carbon disulfide (or other recommended desorber) to determine the identity and concentration of the chemical adsorbed. Based on air flow rate and time, a workplace concentration is determined.

Numerous organic chemicals can be detected using this sampling device at detection limits in the ppm range. The method is useful for determining airborne time weighted average concentrations of many of the organic chemicals listed by OSHA in 29 CFR §1910.1000 (ASTM 1992, Vol. 11.03).

There are some limitations to sampling with charcoal tube samplers. High humidity can reduce the adsorptive capacity of activated charcoal for some chemicals. Further, mixtures of polar and nonpolar compounds are difficult to recover (desorb) from activated charcoal (ACGIH 1989).

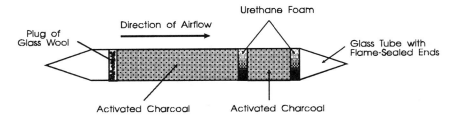

Source: Adapted from ASTM D 3686.

FIGURE 5.4 Example of an Activated Charcoal Tube.

Silica Gel Tube Sampler. This sampler is very similar to the charcoal tube sampler except that silica gel, which is an amorphous form of silica derived from sodium silicate and sulfuric acid, is used instead of charcoal. Unlike charcoal, silica gel is well suited for sampling polar contaminants since they are easily removed from the adsorbent with common solvents. Additionally, amines and some inorganic substances not suitable for charcoal sampling can be sampled with a silica gel tube sampler. The major disadvantage of this type of sampler is it will adsorb water (ACGIH 1989).

Passive or Diffusional Sampling

Passive or diffusional sampling is performed without the use of a personal sampling pump. Sampling devices consisting of badges or dosimeters are attached to the worker near the breathing zone. The dosimeters collect vapors by diffusion onto a medium such as charcoal or other adsorbent to indicate chemical concentrations in the breathing zone. The vapors diffuse through the media over time at a rate dependent on the cross-sectional area of the diffusion cavity, the diffusion coefficient, and the length of diffusion path. Diffusion factors can be determined using manufacturer's supplied calibration factors, or by applying diffusion laws. In some samplers, the adsorbent will have to be desorbed and analyzed. Other types of passive samplers are colorimetric and use length of stain or color change as the concentration indicator.

Like the charcoal tube sampler described previously, passive charcoal adsorbent tube samplers can detect numerous organic chemicals in the ppm range. Limitations are, likewise, similar.

For colorimetric badges or dosimeters, the manufacturer's color graph must be used. The main limitation of these types of samplers is that they can give only a gross assessment of chemical concentration. In addition, humidity and temperature must be taken into account when reading the calibration curves.

Other Gas and Vapor Samplers

In addition to the samplers described previously, numerous other samplers are available for collecting gases and vapors from the workplace atmosphere for analysis. Examples include gas sample tubes, glass bottles or containers, collapsed bags, and bubblers.

Gas sample tubes use air displacement as a means of collecting an air sample. Generally, an aspirator is used to sweep out the air that is in the tube and replace it with air that is to be sampled. Glass bottles or containers can be filled similarly, using air displacement, or they can be filled using water displacement. During water displacement, the glass container is filled with water and drained at the area to be sampled, allowing air into the container. This method should not be used for sampling water soluble gases. Once the sample tube or container is filled, it is capped with an impermeable cap and transported to a laboratory for analysis.

Collapsed bags provide another method for collecting air samples. The bags are rolled tightly to exclude extraneous air and are opened and filled with air at the monitoring site. For outdoor applications, if the wind velocity is sufficient to fill the bag, the bag can be held open directly into the wind. If not, a blower will be needed to blow air into the bag. For indoor applications, a small pump can be used to fill the bags. The bags are made of polyethylene or other nonreactive material.

Bubblers, including simple bubblers and bubblers with diffusers, also can be used for sampling. A bubbler with a diffuser is shown in Figure 5.5. Both types use pumping devices to pull a sample through the apparatus. Bubblers are generally easy to use and absorb gases and vapors by gas-liquid contact. Bubblers with diffusers allow for better contact than simple bubblers, but they are subject to more frequent clogging. The effectiveness of bubblers is dependent on several factors, including the size of the bubble, the length of the path through the absorbent, the rate of gas flow, transfer coefficients, and the degree of solubility of the contaminant in the absorbent.

Dust Sampler

Dust samplers are used to determine the amount of fines and particulates in the workplace atmosphere that could create a health hazard. ASTM D 4532

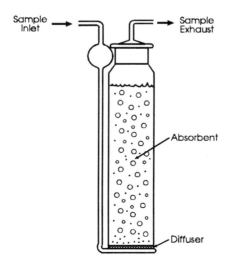

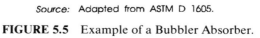
Source: Adapted from ASTM D 1605.

FIGURE 5.5   Example of a Bubbler Absorber.

specifies the use of a personal sampling pump and a sampling head consisting of an aerosol preclassifier (in the form of a cyclone) and a filter assembly for sampling respirable dust. The filter—which must be nonhygroscopic and have a collection efficiency of greater than 95%—is weighed before and after sampling to determine the collected dust weight. The sample volume is calculated and then used to calculate the mass concentration of the respirable dust of interest.

An equivalent sampling method, described by ACGIH, also uses a two-stage dust sampler. Examples of first stage collectors include cyclones, elutriators, and impactors. Second stage collectors are generally high collection efficiency filters (ACGIH 1989). Once the sample is collected and weighed, light scattering instruments can provide information on size distribution, if necessary.

Sampling for Asbestos

Sampling for airborne asbestos is required during any demolition or renovation project that involves any of the OSHA-listed asbestos minerals—chrysotile, crocidolite, amosite, anthophyllite, tremolite, and actinolite. Procedures are defined by OSHA in 29 CFR §1910.1001, Appendix A, by ASTM in D 4240, by NIOSH in Publication No. 79-127, and by ACGIH–AIHA (see references).

The sample is collected by pumping air through an open-faced filter membrane. OSHA specifies a mixed cellulose ester filter membrane designated by the manufacturer as suitable for asbestos counting. A section of the membrane is converted to an optically transparent homogenous gel, and the asbestos particles are sized and counted by phase contrast microscopy at a magnification of 400 to 500 times.

All sampling must be conducted so as to be representative of typical working conditions. Since there are OSHA worker exposure limits for times as short as 30 minutes and as long as 8 hours, selection of an appropriate sampling time is an important consideration. For the most reliable results, several samples should be taken over an 8-hour shift to allow for quantitation of peaks as well as an 8-hour time weighted average. These samples must include personal samples taken in the worker's breathing zone and also can include static samples at fixed locations, if the dust is uniformly distributed over a large area.

Sampling for Radioaction Exposure

Most sampling for worker radioaction exposure is performed using a personal dosimeter. Personal sampling must be performed on all workers who have the potential to receive in excess of 25% of the allowable radiation dose limits in any calendar quarter. Additionally, all workers who enter a high radiation area (an area with potential exposure of greater than 100 millirem in one hour) must be sampled. Some states may have other requirements.

The personal dosimeter must be processed by processors accredited through the National Voluntary Laboratory Accreditation Program (NVLAP)

(DOE 1991). Personal dosimeter performance standards are outlined in ANSI Standard N13.11. Other ANSI standards pertaining to radiation monitoring are listed in Appendix B.

Fixed and portable samplers are also available for sampling indoor and outdoor atmospheres that are potentially radioactive. These samplers include alpha particle analyzers, beta particle analyzers, gamma particle analyzers, scintillation counting systems, radon monitors, and others.

MONITORING INDOOR AIR QUALITY

Indoor air quality of nonmanufacturing areas such as office space or low chemical-use labs is a key part of environmental health. As a result of negative publicity about Legionnaire's disease, excessive radon exposure in homes and workplaces, and the phenomena known as "sick building syndrome," indoor air quality has gained national focus. Periodic monitoring of indoor air quality can help ensure that the air in a building is fresh and free from contamination. If occupants complain of specific symptoms such as headaches, dizziness, or general malaise, the complaints should be investigated immediately.

Several ASTM standards are published for monitoring indicators of adequate (or poor) indoor air quality. These include D 3824 (standard test for continuous measurement of oxides of nitrogen), D 4861 (standard practice for sampling and analysis of pesticides and PCBs), and D 4947 (standard test method for chlordane and heptachlor residues), and others.

In addition to ASTM standards, useful information can be obtained from ACGIH's guidance for the sampling of airborne microorganisms and aeroallergens, which are considered sources of indoor air contamination (ACGIH 1989). Elevated levels of carbon dioxide also contribute to poor indoor air quality and can be monitored using one of several marketed monitors.

FIELD APPLICATIONS

With numerous types of analyzers available, it is often difficult to select one that will be appropriate for a given application. ACGIH's handbook pertaining to air sampling instruments has an index of instruments, which makes researching the literature considerably easier. In addition, vendors carry literature about their products including applications of the instruments and operational information. Each type of analyzer should be investigated carefully before purchase to ensure that it meets all requirements of the application.

General Considerations

The accuracy of any sampling data is determined by numerous elements including instrument sensitivity, instrument calibration, chemical interferences, temperature and humidity factors, sample pump inconsistencies, and other factors. If approached systematically, most of the variables can be eliminated or mathematically adjusted using real-time data.

Likewise, the determination of where to collect the sample must be addressed in an organized manner. Air flow patterns, sources of emissions, vapor density of the chemical, location of workers, and other considerations should be taken into account when selecting sample locations.

When prioritizing workplace monitoring efforts, occupancy frequency versus hazard or risk must be assessed. The most frequently occupied areas will normally require more frequent monitoring unless the hazards of the chemicals in the area are low. High-risk areas such as toxic gas storage rooms are equally important, and should be included in a monitoring plan. Normally unoccupied areas such as trenches, equipment housings, and support areas also might be candidates for monitoring, since leaks and spills could go unnoticed for long periods of time in these places.

When selecting a monitoring instrument, analysis turnaround time should be taken into consideration. Charcoal adsoprtion tubes, silica gel adsorption tubes, and air and water displacement samplers require laboratory analysis and, therefore, cannot be used if real-time data is required. Additionally, concentrations are integrated over time, so peak concentrations cannot be identified with these samplers, if the sample time is an 8-hour shift.

On-line analyzers allow for immediate analytical feedback. These instruments also allow for quantitation of peak concentrations. When installed on line, the system output can be programmed to alarm at preset concentrations with the use of programmable logic controls or other means.

Leaks, Spills, and Unknown Atmospheres

For situations where a known chemical has leaked or spilled, and the concentrations in the atmosphere must be quantified, portable instruments that can monitor and quantify those chemicals are appropriate for use. If the material is flammable/combustible, a combustible gas indicator is useful for evaluating the explosiveness of the atmosphere at the point of entry and, as results allow, at the area of the spill.

For pinpointing fugitives and leaks around equipment, colorimetric detector tubes are adequate instruments. Portable FIDs and PIDs also can provide accurate, real-time data for known constituents.

Confined space atmosphere evaluation would require use of instruments to determine oxygen concentration, LEL, and hydrogen sulfide. Concentrations of known chemicals also should be evaluated prior to entry, so that proper personal protection can be specified. Guidelines under 29 CFR §1910.146 should be reviewed before initiating entry into a confined space.

For unknown atmospheres such as Superfund and other hazardous waste cleanup sites, abandoned warehouses, or rooms with unlabeled containers, more sophisticated equipment is commonly used. IR analyzers and portable GCs with a detector can be used to read multichemical atmospheres. Set-up and analytical readings are more complicated for these instruments, so skilled personnel must be on hand during the evaluation.

For all unknown atmospheres, proper protective clothing, including maximum respiratory and skin protection, must be worn until the atmosphere can be evaluated as safe for entry without such protection. The topic of personal protective clothing and respiratory protection is addressed in the next chapter.

Routine Workplace Monitoring Applications

Routine or periodic workplace monitoring of manufacturing process areas is not uncommon in industry. A portable IR or other portable on-line instrument is useful for these applications since it can be set up in an area for several days with a strip chart or other type of recorder. Once data collected is adequate to demonstrate that exposure concentrations are acceptable, the analyzer can be moved to another area. For a less expensive approach, area diffusional samplers or dosimeters can be used instead of a portable IR. For monitoring hazardous chemical and toxic gas storage rooms, continuous monitors or semicontinuous systems that speciate chemicals and toxic gases are used.

Personal Monitoring Applications

For preplanned personal monitoring, charcoal and silica gel adsorbent tubes, both utilized with a personal sampling pump, are the mainstay of the industry. Passive or diffusional samplers in the form of dosimeters or badges also are used for supplemental sampling and screening. The dosimeter is attached to the wearer at the breathing zone during an entire shift. Minimum supervision from the facility's industrial hygienist is necessary for this type of monitoring. Dosimeters also can be used in Superfund or other cleanup activities for general evaluation of chemical concentrations. In these cases, the dosimeter is taped to the outside of a worker's encapsulated suit. If cartridge respirators are approved for use while working in a cleanup area, the dosimeter can be worn in the breathing zone.

Dust samplers are used for personal monitoring of workers in operations where respirable dust is generated. This includes exposures to mining dust, cotton dust, asbestos, and other types of dust.

All asbestos removal jobs must be sampled for airborne asbestos during the demolition or renovation process. Contractors or employees who remove asbestos must be trained adequately to do this type of work. Lengthy procedures must be followed to protect the environment from contamination and the workers from exposure. Required personal protective equipment and other requirements that pertain to asbestos demolition and removal activities are discussed more fully in Chapter 6.

CONCLUSION

Numerous analyzers are available for workplace and personal sampling. These include single chemical analyzers, multichemical analyzers, diffusional sampling devices, samplers that use a personal sampling pump, gas concentration indicators, and others. Each type of analyzer has advantages and limitations, and these should be investigated and evaluated fully before purchase for a specific application. Range of chemicals detected, detection limits, correction factors, maintenance required on the instrument, and sample turnaround time all should be considered.

The last two chapters have reviewed ways to determine workplace exposure through methods that are accepted and used throughout industry. The following two chapters address methods for protecting workers. These include use of personal protective equipment, fire and life safety design features for chemical-use buildings, and effective use of ventilation for contamination control.

REFERENCES

American Conference of Governmental Industrial Hygienists, *Air Sampling Instruments for Evaluation of Atmospheric Contaminants*, 7th ed., Cincinnati, OH, 1989.

American Conference of Governmental Industrial Hygienists–American Industrial Hygiene Association, "Aerosol Hazards Evaluation Committee: Recommended Procedures for Sampling and Counting Asbestos Fibers," *American Industrial Hygiene Journal*, Vol. 36, 1975.

American National Standards Institute, "Personnel Dosimetry Performance, Criteria for Testing," ANSI N13.11, New York, 1983.

American Society of Testing and Materials, *Annual Book of ASTM Standards*, Vol. 11.03, "Atmospheric Analysis; Occupational Health and Safety," Philadelphia, 1992.

American Society of Testing and Materials, *Annual Book of ASTM Standards*, Vol. 11.04, "Pesticides; Resource Recovery; Hazardous Substances and Oil Spill Responses; Waste Disposal; Biological Effects," Philadelphia, 1992.

Coleman, Daniel R., et al., "Automatic Continuous Air Monitoring at Fixed Sites with Minicams™," pp. 114–117, *Proceedings of the Third Annual Hazardous Materials Management Conference/ Central*, Chicago, 1990.

Daisey, Joan M., "Real Time Portable Organic Vapor Sampling Systems: Status and Needs," prepared for the American Conference of Governmental Industrial Hygienists, Cincinnati, OH, Lewis Publishers, Boca Raton, FL, 1987.

Department of Energy, "The Status of ANSI N13.11—The Dosimeter Performance Test Standard," DE91 017874, prepared by C.S. Sims, Oak Ridge National Laboratory, Oak Ridge, TN, 1991.

Erb, Jeff, Evelyn Ortiz, and Gayle Woodside, "On-line Characterization of Stack Emissions," pp. 40–45, *Chemical Engineering Progress*, Vol. 86/No.5, 1990.

National Institute of Occupational Safety and Health, "Membrane Filter Method for Evaluating Airborne Asbestos Fibers," DHEW (NIOSH) Publication No. 79-127, prepared by N.A. Leidel et al., Department of Health Education and Welfare, Rockville, MD., 1979.

National Institute of Occupational Safety and Health, *NIOSH Manual of Analytical Methods*, DHHS (NIOSH) Publication No. 84-100, 3rd ed., edited by P.M. Eller for U.S. Department of Health and Human Services, Cincinnati, OH, 1984, revised 1987.

National Institute of Occupational Safety and Health, Occupational Safety and Health Administration, U.S. Coast Guard, and U.S. Environmental Protection Agency, *Occupational Safety and Health Guidance Manual for Hazardous Waste Site Activities*, DHHS (NIOSH) Publication No. 85-115, U.S. Department of Health and Human Services, U.S. Government Printing Office, Washington, D.C., 1985.

Occupational Safety and Health Administration, *OSHA Analytical Methods Manual*, Part 1, "Organics," OSHA Analytical Laboratories, Salt Lake City, UT, 1990.

Occupational Safety and Health Administration, *OSHA Analytical Methods Manual*, Part 2, "Inorganics," OSHA Analytical Laboratories, Salt Lake City, UT, 1991.

U.S. Environmental Protection Agency, *Compendium of Methods for the Determination of Toxic Organic Compounds in Indoor Air*, EPA/600/4-90/010, prepared by Engineering Science, Inc. for Atmospheric Research and Exposure Assessment Laboratory, Office of Research and Development and Quality Assurance Division, Environmental Monitoring Systems Laboratory, Research Triangle Park, NC, 1990.

U.S. Environmental Protection Agency, "Fourier Transform Infrared Spectroscopy as a Continuous Monitoring Method: A Survey of Applications and Prospects," EPA/600/D-90/003, prepared by Entropy Environmentalists, Inc., Research Triangle Park, NC, 1989.

BIBLIOGRAPHY

ACGIH (1988), *Advances in Air Sampling*, American Conference of Governmental Industrial Hygienists, Cincinnati, OH.

AIHA (1985) *Biohazards Reference Manual*, American Industrial Hygiene Association, Akron, OH.

_____ (1987) *Cotton Dust Exposures*, Vol. 2, American Industrial Hygiene Association, Akron, OH.

_____ (1990), *The Practitioner's Approach to Indoor Air Quality Investigations: Proceedings of the Indoor Air Quality International Symposium*, American Industrial Hygiene Association, Akron, OH.

ASTM (1989), *Design and Protocol for Monitoring Indoor Air Quality*, Nagda, N.L., and J.P. Harper, eds., Special Technical Publication 1002, American Society of Testing and Materials, Philadelphia.

_____ (1990), *Biological Contaminants in Indoor Environments*, Morey and Feeley, eds., Special Technical Publication 1071, American Society of Testing and Materials, Philadelphia.

DOE (1988), "The Evaluation of Four Models of Personal Air Samplers," DE88-005563, prepared by James F. Boyer and Bruce J. Held, Lawrence Livermore National Laboratory, Department of Energy, Washington, D.C.

Hanford Works (1949, declassified 1991), "Manual of Standard Procedures for 100, 200, and 300 Area Survey Work," DE92-002891, compiled by J.M. Smith, Jr., Health Instrument Operational Division, Hanford Works, Hanford, WA.

Knoll, G.F. (1989), *Radiation Detection and Measurement*, 2nd ed., John Wiley & Sons, New York.

Lodge, James P., Jr., ed. (1988), *Methods of Air Sampling and Analysis*, 3rd ed., Lewis Publishers, Boca Raton, FL.

NCRP (1988), "Measurement of Radon and Radon Daughters in Air," NCRP Report No. 97, National Council of Radiation Protection and Measurements, Bethesda, MD.

Ness, Shirley (1991), *Air Monitoring for Toxic Exposures*, Van Nostrand Reinhold, New York.

Sheldon, L.S., C.M. Sparacino, and E.D. Pellizzari (1984), "Review of Analytical Methods for Volatile Organic Compounds in the Indoor Environment," *Indoor Air and Human Health, Proceedings of the Seventh Life Sciences Symposium*, Knoxville, TN.

U.S. EPA (1985), "Measuring Airborne Asbestos Following An Abatement Action," EPA/600/4-85/049, Quality Assurance Division, Environmental Monitoring Systems Laboratory, Office of Research and Development, Research Triangle Park, NC and Exposure Evaluation Division, Office of Toxic Substances, Office of Pesticides and Toxic Substances, U.S. Environmental Protection Agency, Washington, D.C.

―――― (1986), "Standard Operating Procedures Employed in Support of an Exposure Assessment Study," Vol. 4, prepared by R.W. Handy et al., eds., for Air, Toxics, and Radiation Monitoring Research Division, Office of Monitoring, System and Quality Assurance, Office of Research and Development, U.S. Environmental Protection Agency, Washington, D.C.

U.S. EPA/NIOSH (1991), *Building Air Quality*. U.S. Environmental Protection Agency and National Institute for Occupational Safety and Health, Washington, D.C.

Yocom, John E., and Sharon M. McCarthy (1991), *Measuring Indoor Air Quality: A Practical Guide*, John Wiley & Sons, New York.

6

PERSONAL PROTECTIVE EQUIPMENT

Personal protective equipment (PPE)—including protective clothing, respiratory protection, and other protective devices such as eyewashes and emergency showers—is often needed when working in a hazardous materials or hazardous waste environment. The need for these items is based on several factors including types of chemicals and concentrations in the work area, the amount of time the worker is exposed, and activities involved in the work task.

There is much useful information pertaining to PPE published by agencies and standards organizations (see references). Examples include:

- American Conference of Governmental Industrial Hygienists (ACGIH), *Guidelines for Selection of Chemical Protective Clothing.*
- American Industrial Hygiene Association (AIHA), *Respiratory Protection: a Manual and Guideline.*
- American National Standards Institute (ANSI), "Emergency Eyewash and Shower Equipment," ANSI Standard Z358.1.
- ANSI, "Practices for Respiratory Protection," ANSI Standard Z88.2.
- American Society for Testing and Materials (ASTM), *ASTM Standards on Protective Clothing.*
- National Institute for Occupational Safety and Health (NIOSH), *Occupational Safety and Health Guidance Manual for Hazardous Waste Site Activities.*

This chapter discusses regulatory requirements and specific information about PPE selection. Special PPE requirements for asbestos removal and hazardous materials emergency response and waste cleanup operations also are addressed.

PROTECTIVE CLOTHING

The use of protective clothing is addressed by OSHA in 29 CFR §§1910.132, 133, 135, and 136. Areas covered include general requirements, eye and face protection, head protection, and foot protection, respectively. Work tasks involving the use or handling of hazardous materials must be evaluated by a knowledgeable person, such as an industrial hygienist or chemical specialist, for protective clothing requirements.

Information about the chemical to be used, such as contact hazard and the manufacturer's recommendations for protective clothing, can provide guidance when determining what protective clothing is required. Once the protective clothing needed to perform a task is identified, all workers performing these tasks must wear what is specified.

Material Selection

Typically, personal protective clothing must afford protection against varying types of chemical hazards. A single material is not resistant to all chemicals and wastes. Thus, engineering judgments must be made as to the best materials for face shields, gloves, aprons, boots, and protective suits for use during a specific work task. Common materials used in personal protective clothing include butyl, nitrile, neoprene, polyvinyl chloride, chlorinated polyethylene, viton, polycarbonates, polyvinyl alcohols, and others.

In the National Fire Protection Association (NFPA) publication *Hazardous Materials Response Handbook* (see references), there is a useful summary of chemical compatibilities for commonly used protective clothing materials. This summary combines six studies and includes compatibility tests for seven materials and over 1,000 chemicals. An overview of the tests for each of the seven materials, in terms of resistance to families of chemicals, is presented in Table 6.1.

Protective Suits

Protective suits are used for many tasks, including normal work activities, chemical emergency response, hazardous waste cleanup activities, fire emergency response, and other activities. Before donning any type of PPE, including protective suits, the user should be fully trained.

Minimum requirements for protective suits used in hazardous chemical emergencies are defined by NFPA in the following standards:

- NFPA 1991 (1990)—Vapor-Protective Suits for Hazardous Chemical Emergencies
- NFPA 1992 (1990)—Liquid Splash-Protective Suits for Hazardous Chemical Emergencies
- NFPA 1993 (1990)—Support Function Protective Garments for Hazardous Chemical Operations

TABLE 6.1 Chemical Compatibilities for Selected Protective Materials

Material	Compatibilities	Incompatibilities
Butyl	Moderate to strong acids, ammonia solutions, alcohols, inorganic salts, ketones, phenols, and aldehydes	Petroleum distillates, solvents, alkanes; limited use with esters and ethers
Polycarbonates	Weak acids, ammonium solutions, alcohols, inorganic salts, phenols, some bases and aldehydes	Aggressive petroleum distillates; limited use with ketones
Polyvinyl chloride	Moderate to strong acids, bases, ammonium solutions, inorganic salts, petroleum distillates, alcohols, alkanes, and some aldehydes	Petroleum distillates, ketones, concentrated solvents, and phenols
Neoprene	Moderate acids, bases, ammonium solutions, alcohols, inorganic salts, some solvents, some phenols, some aldehydes, and ethers	Petroleum distillates, esters; limited use with ketones
Chlorinated polyethylene	Moderate to strong acids, bases, ammonium solutions, some petroleum distillates, alcohols, inorganic salts, phenols, alkanes, and aldehydes	Limited use with ketones and solvents
Nitrile	Moderate to strong acids, bases, most ammonium solutions, some petroleum distillates, some solvents, and some alcohols	Phenols, ketones; limited use with inorganic salts and aldehydes
Viton	Acids, bases, some ammonium solutions, petroleum distillates, alcohols, inorganic salts, most solvents, and phenols	Ketones; limited use with aldehydes

Note: This listing provides general information about chemical resistant materials. The specific chemical or chemical mixture used in the workplace should be evaluated by a trained specialist for compatibility with any material used for protective clothing.

Source: Information from NFPA (1989).

Included in the requirements are:

- A product (suit) certification program that ensures that adequate inspection and testing is performed
- Documentation of materials, chemical permeation resistance, and other technical data
- Design and performance criteria for the overall suit and suit components
- Test methods for water penetration, chemical permeation and penetration, flammability resistance, abrasion resistance, cold temperature performance, tear resistance, flexural fatigue, and other tests, as appropriate

Test methods for evaluating adequacy of protective suits should follow ASTM methods for fabric evaluation. Several of these methods are presented in Appendix C. In addition to testing information for the chemical protective suit, the manufacturer also should provide user information. This information should include items such as cleaning instructions, marking and storage instructions, frequency and details of inspections, maintenance criteria, and other information.

Once a protective suit has been worn, it must be thoroughly decontaminated and inspected for signs of chemical penetration, puncture, tears, or other signs of failure. If any penetrations or signs of failure are noticed, the suit should be disposed of properly, unless the damage is limited and repairable. Repair of protective suits, if applicable, should be in accordance with the manufacturer's instructions.

An example of a fully-encapsulating protective suit is shown in Figure 6.1. This type of suit offers the highest level of skin and eye protection and can be worn in an unknown chemical environment or in a chemical environment that is known to be dangerous.

For fire emergencies, fire fighter's protective clothing—including gloves, helmet, bunker coat, bunker pants, and boots—is worn. This type of clothing does not protect against exposures to gases or vapors, chemical splashes, or chemical permeation. An example of fire fighter's bunker coat, bunker pants, and a helmet with a faceshield (before donning respiratory protection) is illustrated in Figure 6.2.

High temperature clothing that protects against exposures of short duration and close proximity to flame and radiant heat is also available. These suits are made of fire-retardant materials, with an outer layer of aluminized fabric. Like fire fighter's protective clothing, this type of clothing does not protect against exposures to chemicals (Noll, Hildebrand, and Yvorra 1988).

Radiation-contamination protective suits are also available for protection against alpha and beta particles (NFPA 1989). Use of these types of suits should be selected and worn under the guidance of a specialist.

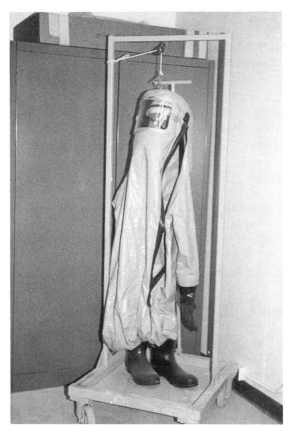

FIGURE 6.1   Fully Encapsulating Chemical Protective Suit.

RESPIRATORY PROTECTION

Respiratory protection requirements are defined by OSHA in 29 CFR §1910.134. Additional requirements are included in specific sections of 29 CFR Part 1910 that cover OSHA-listed hazardous materials such as acrylonitrile, vinyl chloride, inorganic arsenic, cotton dust, lead, and benzene. NIOSH provides a testing, approval, and certification program for respiratory protection. The Mine Safety and Health Administration (MSHA) also evaluates and approves some respirators in conjunction with NIOSH. OSHA provides standards for respirator use and fit-testing.

The need for respiratory protection is determined through sampling a worker's exposure to a specific chemical. If the permissible exposure limit as a time weighted average is expected to be exceeded during work activities, a res-

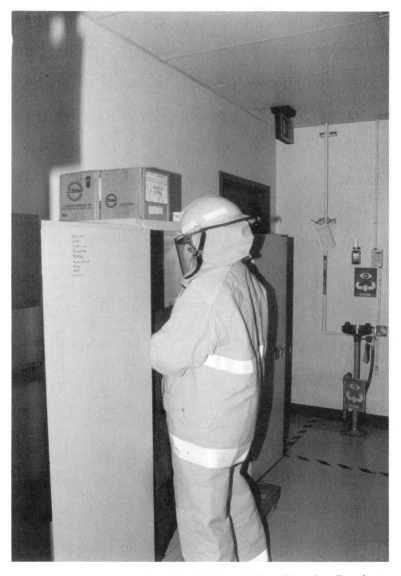

FIGURE 6.2   Protective Suit for a Fire Fighter (Before Donning Respiratory Protection).

pirator is required. Likewise, if the short-term exposure or ceiling limit is expected to be exceeded at any time, a respirator is required.

Typical respiratory protective devices include the chemical cartridge, particulate respirator, gas mask, supplied air respirator, and self-contained breathing apparatus (SCBA). Examples of a half-face chemical cartridge respirator (with goggles) and stored SCBAs (on an emergency response cart) are shown in Figures 6.3 and 6.4. Selection of a respirator is based on chemicals used in the area, concentrations, expected exposure time, and oxygen present in the atmosphere. Training is essential before use of any type of respirator. For industrial applications, OSHA requires that respirator models and applications be approved by NIOSH and, for some substances such as asbestos, by NIOSH/MSHA.

Chemical Cartridges

Chemical cartridges are air-purifying respirators and are used for respiratory protection against specific chemicals of relatively low concentrations. Since they have no air-supplying capability, the cartridges cannot be used in atmospheres containing less than 19.5% oxygen or in atmospheres immediately dangerous to life or health (IDLH). The respirators come in half-face (orinasal) or full facepiece models, with canister holders attached on the facepiece or belt. The length of time before chemical breakthrough of the canister is dependent on the concentration of the chemical in the work area. All models are designed to permit canister replacement. Table 6.2 summarizes the types of chemical cartridges and applications approved by NIOSH.

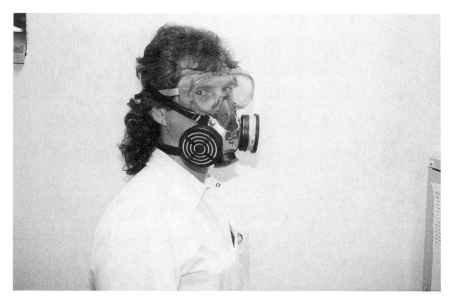

FIGURE 6.3   Half-Face Chemical Cartridge Respirator.

FIGURE 6.4   Self-Contained Breathing Apparatus (SCBAs) Stored on an Emergency Cart.

Particulate Respirators

Particulate respirators are designed for protection against dusts, mists, fumes, and other particulates. Before using this type of respirator, the atmosphere should be tested and must contain at least 19.5% oxygen and cannot be IDLH. Designs of particulate respirators include:

- Single use particulate respirators
- Particulate respirators for protection against dusts and mists
- Particulate respirators for protection against dust, fumes, and mists
- Particulate respirators with a high-efficiency filters

The first three designs listed above are approved for use in atmospheres where the permissible exposure limit as a time weighted average is equal to or greater than 0.05 mg/m^3 or 2 million particles per cubic foot. The high efficiency particulate respirator has the greatest versatility and can be used for protection against dusts, mists, asbestos-containing dusts and mists (if ap-

TABLE 6.2   Chemical Cartridge Types and Applications Approved by NIOSH

Cartridge Type	Effective Concentrations	Other Protection Included on Some Models
Ammonia	≤ 300 ppm	Methyl amine, dusts, mists
Methyl amine	≤ 100 ppm	Ammonia, dusts, mists, fumes of various metals, radionuclides, radon
Chlorine	≤ 10 ppm	Hydrogen chloride, hydrogen sulfide, organic vapor, dusts, fumes, mists, radionuclides
Hydrogen chloride	≤ 300 ppm	In general, most models provide no other protection
Chlorine, sulfur dioxide, hydrogen chloride	≤ 10 ppm ≤ 50 ppm ≤ 50 ppm	Organic vapors, hydrogen sulfide, formaldehyde, chlorine dioxide, dusts, mists, radionuclides
Organic vapor	≤ 1000 ppm by volume	Sulfur dioxide, hydrogen chloride, chlorine, pesticides, dusts, mists, mists of paints/lacquers/enamels, chlorine dioxide, formaldehyde
Mists of paints/lacquers/enamels	≤ 1000 ppm by volume	Organic vapors, pesticides
Pesticides	Specific to model	Organic vapors, mists of paints/lacquers/enamels
Vinyl chloride	≤ 10 ppm	Model provides no other protection
Other gases and vapors	Specific to model	Hydrogen sulfide, chlorine dioxide, formaldehyde, organic vapors, chlorine, dusts, mists, fumes of various metals, radon, radionuclides

Source: Information from NIOSH (1990).

proved by NIOSH/MSHA), and radionuclides. These respirators provide protection against dusts, fumes, and mists having permissible exposure limits as a time weighted average of less than 0.05 mg/m^3.

Gas Masks

Gas masks (nonpowered) are similar to chemical cartridge respirators, except they are all full-face masks and are for a one-time application in an emergency situation. Like chemical cartridge respirators, they are not effective in atmospheres of less than 19.5% oxygen or in IDLH atmospheres (NIOSH 1990).

Gas masks afford protection against a specific chemical or chemical type such as ammonia, chlorine, acids gases, organic vapors, carbon monoxide, sulfur dioxide, and pesticides. Many models have approval for protection against additional vapors, gases, and dusts or mists.

Supplied Air Respirators

Supplied air respirators are used for the same applications as those of cartridge respirators. As with the other respirators described previously, they are not approved for use in IDLH atmospheres or atmospheres containing less than 19.5% oxygen. This limitation is based on safety considerations, since the air supply or air line could fail. However, there is an exception to this standard if an auxiliary tank of air permitting escape is incorporated into the respirator system (AIHA 1991). Approved types of supplied air respirators include continuous flow, pressure demand, and continuous flow abrasive blasting.

When using a supplied air respirator, a line is attached to the user from a supply of respirable air. Full facepieces or hoods must be worn with these devices. Hose lengths and air supply pressure ranges specified by the manufacturer (and approved by NIOSH/MSHA) must be used. The maximum length of hose allowed is 300 feet and the maximum inlet pressure is 125 pounds per square inch gage (psig).

Self-Contained Breathing Apparatus (SCBA)

The SCBA is the only respirator approved by NIOSH for use in oxygen-deficient and IDLH atmospheres. SCBAs also are worn in emergency situations and when unknown atmospheres must be entered. A widely used SCBA is the open circuit pressure demand system. The open circuit pressure demand system maintains a positive pressure inside the facepiece at all times, which provides the highest level of respiratory protection. The system has a full facepiece and a facepiece- or belt-mounted regulator that regulates air flow from a bottle worn on the worker's back. Service life is generally 30 to 60 minutes.

Other types of SCBAs include open circuit demand system and closed circuit demand and pressure demand systems. The open circuit demand system maintains a positive pressure in the facepiece during exhalation, but allows for negative pressure during inhalation. Closed circuit systems allow the worker's exhalation to be filtered (chemically scrubbed) and rebreathed, while providing supplemental oxygen from a supply source. This type of system has a longer service life, but does not provide protection as adequately as the open circuit systems.

Open circuit pressure demand systems for "escape only" typically have a service life of 5 to 15 minutes. These devices are full facepieces or hoods, and do not have a large air bottle associated with them.

Closed circuit "escape only" systems usually have only a mouthpiece instead of a facepiece or hood. These systems have a service life of up to 60 minutes.

EYEWASHES AND EMERGENCY SHOWERS

ANSI Standard Z358.1 provides recommended practices for eyewashes and emergency showers. Eyewashes and emergency showers must be available in all chemical work areas and must be inspected periodically to ensure proper flow and, in the case of eyewashes, flow direction. Additionally, the area around eyewashes and emergency showers must be kept clear for easy access. Easily readable signs that mark the location of the chemical safety devices should be posted.

In most chemical-use work locations, eyewashes and emergency showers are permanently installed in the building. For field applications such as hazardous waste cleanup operations or when water service is not available due to scheduled maintenance of the process and potable water system, portable eyewash and shower units can be used. An example of these portable units is shown in Figure 6.5.

Other emergency equipment that should be available in the workplace includes a portable fire extinguisher and first aid equipment. An evacuation route should be posted clearly in the work area, and all personnel should be trained on how to respond in emergencies.

PERSONAL PROTECTION DURING ASBESTOS REMOVAL

An asbestos fiber is a mineral fiber longer than 5 micrometers (μm) with a length to diameter ratio of at least 3 to 1, which consists of chrysotile, crocidolite, amosite, anthophyllite, tremolite, or actinolite asbestos. The permissible exposure limit set by OSHA is 0.2 fibers per cubic centimeter (f/cc) for an 8-hour time weighted average and 1.0 f/cc for a 30-minute excursion limit (short-term exposure limit).

Requirements for PPE when removing asbestos are defined in OSHA 29 CFR §§1910.1001 and 1926.58 and under TSCA in §763.121. Guidance for workplace safety and health as it relates to asbestos exposure is detailed in ASTM E 849–86. Demolition and renovation workers are required to wear appropriate NIOSH/MSHA-approved respiratory protection as specified in the regulations, which includes:

- Not in excess of 2 f/cc—Half-mask air-purifying respirator, other than a disposable respirator, equipped with high-efficiency filters (filters that are at least 99.97% efficient against monodispersed particles of 0.3 μm or larger).

- Not in excess of 10 f/cc—Full facepiece air-purifying respirator equipped with high-efficiency filters.

FIGURE 6.5   Portable Eyewash and Emergency Shower Units.

- Not in excess of 20 f/cc—Any powered air-purifying respirator equipped with high-efficiency filters or any supplied-air respirator operated in continuous flow mode.

- Not in excess of 200 f/cc—Full facepiece supplied-air respirator operated in pressure demand mode.

- Greater than 200 f/cc—Full facepiece supplied-air respiratory operated in pressure demand mode equipped with an auxiliary positive pressure SCBA.

At any time, respirators assigned for higher environmental concentrations may be used at lower concentrations.

During the respirator selection process, the worker should be allowed to pick a comfortable respirator from a selection of various sizes from different manufacturers. A qualitative respirator fit test using saccharin nebulizer, irritant smoke, or odor threshold screening should be performed before initial use. Additionally, quantitative fit tests can be performed. Quantitative fit tests generally use a particle counting method. Field checks by positive and negative pressure methods should be performed before each entry into the contaminated area.

Full-body work clothing is required with hoods, boot covers, and gloves. Armcuffs of coveralls should be taped to seal them tightly against the gloves. Overboots also should be taped tightly to the legs of the coveralls. Face shields, goggles, or other appropriate protective equipment also might be warranted, depending on the respiratory protection required. The employer is responsible for cleaning, laundering, and replacement of protective clothing.

Engineering controls are needed during demolition to control the airborne particles. These include, but are not limited to:

- Complete enclosure of the area
- Wetting of the area and particularly the asbestos-containing material that is to be removed
- Exhausting the area through an air pollution control device such as a baghouse, high-efficiency particulate air (HEPA) filter, furnace exhaust filter, or wet collector.
- Controlled-access change rooms with clothes lockers and showers for decontamination after work.
- Controlled sampling which minimizes worker and public exposure to airborne asbestos fibers.

Determinations of airborne concentrations must be made following OSHA's sampling and analytical procedure specified in Appendix A to 29 CFR §1910.1001. This method is described in Chapter 5 (see page 77). Additionally, personal sampling is required at least once every six months for employees whose exposure to asbestos is expected to exceed the exposure limits (discounting wearing a respirator). Samples must be collected from within the breathing zones of employees and in areas representative of such breathing zones. Affected employees must have the opportunity to observe monitoring and must have access to the monitoring records.

Asbestos waste includes all friable asbestos, contaminated clothing discarded for disposal, and other contaminated material such as plastic bags and enclosure material. The waste must be sealed in leak-tight bags or drums prior to proper disposal.

PERSONAL PROTECTION DURING HAZMAT RESPONSE AND WASTE CLEANUP

Types of PPE acceptable for hazardous materials (HAZMAT) emergency response and hazardous waste cleanup activities have been established by EPA in guidance documents and by OSHA in 29 CFR §1910.120, Appendix B, and have been incorporated into training manuals (International Fire 1988 and Noll, Hildebrand, and Yvorra 1988). During a HAZMAT response or waste cleanup activity, there are four levels of clothing that can be worn by responders, depending on the hazard potential associated with the job performed. These include:

- Level A—This type of protection should be worn when the highest level of protection is needed for respiratory, skin, and eye protection. In general, this is used in unknown atmospheres and known high hazard atmospheres. Level A protection includes a positive pressure SCBA, a fully-encapsulating chemical resistant suit, chemical resistant inner and outer gloves, and chemical resistant safety boots. A two-way communications device such as a two-way radio or other device is also necessary. Depending on the situation, other safety clothing might be added such as a hard hat, coveralls, long cotton underwear, or a cooling unit.

- Level B—This type of protection should be selected when the highest level of respiratory protection is needed, but a lesser level of skin and eye protection is required. In general, this type of protection is used in oxygen-deficient and IDLH atmospheres, where the substances do not represent a severe skin hazard. Level B protection includes a positive pressure SCBA, chemical resistant clothing (hooded, one- or two-piece chemical splash suit, disposal chemical resistant coveralls, overalls and long-sleeved jacket, or equivalent clothing), inner and outer chemical resistant gloves, and chemical resistant safety boots, and a two-way communication device. Optional clothing could include a hard hat, long cotton underwear, or disposal boot covers.

- Level C—This type of protection should be selected when the atmosphere contaminants are known, and the concentrations have been measured and meet the criteria for use of an air-purifying respirator. Additionally, skin and eye exposure must be unlikely. Level C protection includes use of a full facepiece air-purifying respirator, chemical resistant clothing, chemical resistant inner and outer gloves, chemical resistant safety boots, and a two-way communication device. Optional clothing could include a hard hat, disposable boot covers, long cotton underwear, a faceshield, and an escape mask.

- Level D—This is type of protection should be selected when there is no respiratory hazard and when minimum skin and eye protection are required. This is primarily a work uniform, and the atmosphere must have no known hazards. Level D protection includes coveralls or other work uniform, safety glasses, and safety boots.

The type of PPE used during a HAZMAT response or waste cleanup activity should be reevaluated periodically as new information becomes available. Any change in level of PPE used should be reviewed and approved by an industrial hygienist, safety engineer, or other specialist in charge of specifying PPE. Other aspects of HAZMAT response, including incident handling techniques, are described in Chapter 18.

CONCLUSION

The need for personal protective equipment—including respirators, protective suits, gloves, eyeglasses, shoes/boots, and eyewashes and emergency showers—must be evaluated for each task performed in the workplace. In order to give guidance on the selection of personal protective clothing, NIOSH, NFPA, ACGIH, and others have documented research that includes testing of materials for permeation, stability under thermal stress, vapor and liquid penetration, and other physical properties.

Eyewashes and emergency showers should be available during activities involving chemicals. Portable units are available for use at outdoor applications or for other uses. ANSI has set standards for location, inspection, maintenance, and other aspects pertaining to eyewashes and emergency showers.

Respiratory protection is evaluated based on chemical concentrations and length of exposure. Types of respirators are approved by NIOSH or NIOSH/MSHA for protection against fumes, mists, asbestos-containing and other dusts, radionuclides, and specific chemicals. If exposures can be limited through the use of ventilation or other engineering methods, however, these methods should be evaluated and implemented where practical. The use of ventilation for control of airborne contaminants in the workplace is discussed in the next chapter. The next chapter also explores fire protection and life safety management.

REFERENCES

American Conference of Governmental Industrial Hygienists, *Guidelines for Selection of Chemical Protective Clothing*, 3rd ed., A.D. Schwope, P.P. Costas, J.O. Jackson, and D.J. Weitzman, eds., Cincinnati, OH. 1987.

American Industrial Hygiene Association, *Respiratory Protection: A Manual and Guideline*, 2nd ed., Craig E. Colton, Lawrence R. Birkner, and Lisa M. Brosseau, eds., Akron, OH, 1991.

American National Standards Institute, "Emergency Eyewash and Shower Equipment," ANSI Z358.1, New York, 1990.

American National Standards Institute, "Practices for Respiratory Protection," ANSI Z88.2, New York, 1992.

American Society for Testing and Materials, *ASTM Standards on Protective Clothing*, Philadelphia, 1990.

International Fire Service Training Association, *HAZMAT for First Responders*, Fire Protection Publications, Oklahoma State University, Stillwater, OK, 1988.

National Fire Protection Association, *Hazardous Materials Response Handbook*, Henry F. Martin, ed., Quincy, MA, 1989.

National Institute for Occupational Safety and Health, "Certified Equipment List," DHHS (NIOSH) Publication No. 90-102, U.S. Department of Health and Human Services, U.S. Government Printing Office, Washington, D.C., 1990.

National Institute for Occupational Safety and Health, "Federal Research on Chemical Protective Clothing and Equipment: A Summary of Federal Programs for Fiscal Year 1988," DHHS (NIOSH) Publication No. 89-119, U.S. Department of Health and Human Services, Washington, D.C., 1989.

National Institute for Occupational Safety and Health, Occupational Safety and Health Administration, U.S. Coast Guard, U.S. Environmental Protection Agency, *Occupational Safety and Health Guidance Manual for Hazardous Waste Site Activities*, DHHS (NIOSH) Publication No. 85-115, U.S. Department of Health and Human Services, U.S. Government Printing Office, Washington, D.C., 1985.

Noll, Gregory G., Michael S. Hildebrand, and James G. Yvorra, *Hazardous Materials: Managing the Incident*, Fire Protection Publications, Oklahoma State University, Stillwater, OK, 1988.

U.S. Environmental Protection Agency, *Standard Operating Safety Guides*, Office of Emergency and Remedial Response, Hazardous Response Support Division, Edison, NJ, 1984.

BIBLIOGRAPHY

ASTM (1989), *Chemical Protective Clothing Performance in Chemical Emergency Response*, Special Technical Publication 1037, Perkins and Stull, eds., American Society of Testing and Materials, Philadelphia.

―――― (1989), *Performance of Protective Clothing: Second Symposium*, Special Technical Publication 989, Mansdorf, Sager, and Nielsen, eds., American Society of Testing and Materials, Philadelphia.

Center for Labor Education and Research (1990), *Worker Protection During Hazardous Waste Remediation*, Hazardous Waste Training Program, University of Alabama, Birmingham, AL.

Code of Federal Register, 40 CFR Part 61 Subpart M, "National Emission Standard for Asbestos," U.S. Environmental Protection Agency, Washington, D.C.

DREO (1991), "Heat Stress Caused by Wearing Different Types of CW Protective Garment," DREO Technical Note 91-14, prepared by S.D. Livingstone and R.W. Nolan, Defence Research Establishment Ottawa, National Defence, Ottawa, Ontario.

DOE (1990), "Recent Investigation in Personal Protective Equipment (Including Respirators)," DE90 005785, Lawrence Livermore National Laboratory, Department of Energy, Livermore, CA.

―――― (1990), "Respirator Standards, Regulations, and Approvals in the U.S.A.," prepared by Bruce J. Held, Lawrence Livermore National Laboratory, Livermore, CA.

Department of the Navy (1991), "Survey of Hazardous Chemical Protective Suit Materials," NCTRF Report No. 186, Navy Clothing and Textile Research Facility, Department of the Navy, Natick, MA.

Forsberg, Krister, and Lawrence H. Keith (1989), *Chemical Protective Clothing Performance*, John Wiley & Sons, New York.

Forsberg, Krister, and S.Z. Mansdorf (1989), *Quick Selection Guide to Chemical Protective Clothing*, Van Nostrand Reinhold, New York.

Government Institutes (1987), *Asbestos in Buildings, Facilities and Industry*, Government Institutes, Rockville, MD.

Keith, Lawrence H., and Douglas B. Walters (1992), *The National Toxicology Program's Chemical Data Compendium*, Vol. 6, "Personal Protective Equipment," Lewis Publishers, Boca Raton, FL.

NIBS (1988), *Asbestos Abatement and Management in Buildings, Model Guide Specification*, 2nd ed., National Institute of Building Sciences, Washington, D.C.

U.S. EPA (1984), "NESHAPS Asbestos Demolition and Renovation Inspection Workshop Manual," EPA/340/1-85/008, prepared by GCA Corporation for U.S. Environmental Protection Agency, Washington, D.C.

_____ (1987), "Development and Assessment of Methods for Estimating Protective Clothing Performance," EPA/600/2-87/104, prepared by R. Goydan et al., U.S. Environmental Protection Agency, Washington, D.C.

_____ (1990), "Environmental Asbestos Assessment Manual, Superfund Method for Determination of Asbestos in Ambient Air," U.S. Environmental Protection Agency, Washington, D.C.

_____ (1992), "Limited-Use Chemical Protective Clothing for EPA Superfund Activities," EPA/600/R-92/014, prepared by Arthur D. Little, Inc., for Risk Reduction Engineering Laboratory, Office of Research and Development, U.S. Environmental Protection Agency, Cincinnati, OH.

7

WORKPLACE AND BUILDING SAFETY

Workplace and building safety are described in this chapter in terms of fire protection systems, life safety management, and ventilation systems. Fire protection systems are the main safety systems used at all facilities to mitigate fires, explosions, and other chemical emergencies. Life safety management, which is related to fire protection, includes means of egress, building occupancy management, and building construction. Ventilation systems are critical in maintaining adequate air exchange throughout the workplace for proper control of workplace contamination.

FIRE PROTECTION

An integral part of loss prevention and worker safety is fire protection. Fire protection equipment can range from portable extinguishers to elaborate dry chemical, foam, or sprinkler systems. Design and installation of a fire protection system is based on chemicals used in the area, room layout, and occupancy. Two useful references on the subject of fire protection include the *Fire Protection Handbook*, published by the National Fire Protection Association (NFPA) and the *Uniform Fire Code* published by the International Conference of Building Officials (ICBO)(see references). The following paragraphs detail some basic fire protection information described in these references and in other NFPA documents that can be used in industrial applications.

Portable Extinguishers

NFPA has published a standard — NFPA 10 — that outlines criteria for the selection, installation, maintenance, and testing of portable fire extinguishers.

According to the standard, fires can be classified into four categories:

- Class A—These are fires in ordinary combustible materials such as paper, wood, and plastics. These fires can be quenched with water or water solutions that absorb heat or with the use of certain dry chemicals that interrupt the combustion chain reaction. Class A fire-extinguishing agents include water, saturated steam, and multipurpose dry chemicals such as ammonium phosphate.
- Class B—These are fires in flammable or combustible liquids and flammable gases. These fires require the use of dry chemicals that interrupt the combustion chain reaction, inhibit vapor release, or displace oxygen. Class B extinguishing agents include carbon dioxide, aqueous film-forming foam, halons, and dry chemical agents such as potassium bicarbonate, potassium chloride, monoammonium phosphate, and potassium carbamate.
- Class C—These are fires in electrical equipment. These fires require nonconductive extinguishing agents such as carbon dioxide and dry chemical agents.
- Class D—These are fires in combustible metals such as magnesium, sodium, lithium, and potassium. These fires require nonreactive, heat-absorbing extinguishing agents such as graphite, dry soda ash, dry sand, and dry powder such as sodium chloride.

There are several types and sizes of portable fire extinguishers suitable for fighting the different classes of fires that can be stored in the workplace. Water-based portable extinguishers with a capacity of 2½ gallons are commonly used for Class A fires. This extinguisher weighs approximately 30 pounds. Bigger units that supply 25 to 60 gallons of water are available, but must be mounted on wheels to make them portable.

Dry chemical portable extinguishers for Class A, B, and C fires weigh from 2 to 50 pounds. As was the case for water-based systems, larger and heavier units can be mounted on wheels. Dry sand is often stored in barrels on pallets and can be moved using a fork truck or dolly.

Basic operational designs of water, dry chemical, and carbon dioxide extinguishers (NFPA, *Fire Protection Handbook*, 1991) include:

- Stored-pressure unit—This unit is used for water-based agents and dry chemicals. The unit contains the extinguishing agent and the propellant gas, which are mixed in a sealed cylinder. Once activated, the extinguishing agent can be discharged intermittently using a release lever and hose attachment.
- Pump tank—This type of design is used only for plain water and water-based agents. The unit uses a hand-operated, vertical piston pump to force water from the tank. Force, range, and duration of the discharge are operator dependent.

- Cartridge operated extinguisher—This extinguisher is used for dry chemicals. The unit utilizes a small cartridge of propellant gas attached to the shell and threaded into the nonpressurized chamber containing the dry chemical. When activated, the propellant gas is released into the chamber to pressurize the agent in the cylinder.
- Self-expelling extinguishers—These are extinguishers containing carbon dioxide. The carbon dioxide is retained as a liquid under high pressure and is self-expelling when released.

In addition to water-based, dry chemical, and carbon dioxide portable fire extinguishers, there are portable halon 1301, halon 1211, and aqueous film-forming extinguishers that are designed much like the stored-pressure water units. A portable dry powder extinguisher is available that is cartridge operated.

Fire Protection Systems

Common fire protection systems found in industrial applications include automatic sprinkler systems, carbon dioxide systems, halon systems, foam systems, and dry chemical systems. NFPA standards pertaining to these different types of systems are presented in Appendix D. Pertinent information about the various fire suppressant systems, including industrial applications and toxicity of the fire-extinguishing agent, is summarized in Table 7.1.

Fire Protection in Storage Rooms

NFPA Requirements.   NFPA has published several reference documents pertaining to fire hazards and fire protection in chemical storage areas. Numerous NFPA standards pertain to the identification of fire hazards of materials and the storage of hazardous materials. These standards are summarized in Appendix D. In addition to these standards, NFPA's *Fire Protection Guide to Hazardous Materials* has included technical data from several key standards about fire hazards of chemicals and materials (see references). The *Fire Protection Handbook* addresses storage practices for hazardous materials and wastes, including storage of flammable and combustible liquids (applications for tanks and containers), storage of gases, storage and handling of chemicals, and special protection for medical gases, pesticides, and waste handling systems.

In large chemical storage areas, segregation by chemical type or category provides the best fire protection. Separate rooms or buildings should be used to segregate oxidizing chemicals, flammable or combustible chemicals, unstable chemicals, corrosives, chemicals reactive to water or air, radioactive materials, and materials subject to self-heating. Additionally, chemicals and other hazardous materials should be stored separately from nonchemical parts and other manufacturing supplies. Examples of fire extinguishing meth-

TABLE 7.1 Industrial Fire Protection Systems

Type	Applications	Toxicity	Considerations
Automatic water sprinkler	Combustibles such as fuel oils; water soluble liquids such as acetone	Nontoxic	Can slow down rate of combustion in Class B fires; special pipe and fitting materials needed in corrosive atmospheres; severe dust explosions can damage system
Carbon dioxide	Flammable liquids such as gasoline; flammable gases such as propane; electrical fires	Asphyxiant in closed rooms	Extinguishing atmosphere short-lived; special storage requirements in extreme climates
Halons—1301 and 1211	Flammable liquids such as methyl ethyl ketone; flammable gases such as methane; flammable solids such as thermoplastics	Generally nontoxic	Clean, noncorrosive; ozone depleting potential will require replacement of these agents in the near future
Dry chemical agents	Flammable or combustible liquids in enclosed areas; electrical fires	Slightly corrosive	Can coat surfaces; ineffective if reignition occurs
Foam agents	Flammable or combustible liquids such as gasoline and alcohols	Nontoxic; biodegradable	Used on spills and tank fires; not suitable for water reactive materials; conductive, so not suitable for use on electrical fires

Source: Information from NFPA, *Fire Protection Handbook,* (1991) and NFPA, *Fire Protection Guide to Hazardous Materials,* (1991).

ods for the different chemical categories are detailed in Table 7.2.

Drums or packages of chemicals may be stored on pallets and lifted into place on storage racks by forklifts or placed in racks by an automated system. The design of storage rack systems is presented in NFPA 231 Appendix A for general storage and NFPA 231C for rack storage of materials. Although this latter standard does not address chemical storage, *per se*, it is a good guide for sprinkler design for materials of varying flammability.

In all cases, the storage areas must be clean and orderly, and fire protection equipment must be well maintained. Access to the racks and the load/unload areas must be roomy, with a clear path for egress. Aisle widths generally range from 4 to 8 feet, but should provide enough space for egress when fork trucks and other equipment is in use.

Design of rack storage systems inside buildings is based generally on storage height and commodity class. Storage height is measured from floor level to the top of the storage container resting on the top rack. In design schemes, the most flammable products and packaging require in-rack sprinklers, as well as ceiling sprinklers of adequate water supply. Horizontal barriers that cover the entire rack, including the flue spaces, also are required in many cases. A general schematic of an in-rack sprinkler system is shown in Figure 7.1. Other considerations for storage areas include total water demand (gallons per minute per square foot), rack configuration, storage to ceiling clearance, and aisle width. Specific design criteria should be researched thoroughly before sprinkler installation.

Uniform Fire Code Requirements.   Requirements for fire protection in storage rooms that contain hazardous materials also are outlined in the *Uniform Fire Code*. Requirements are defined for storage of compressed gases (Article 74), storage of cryogenic fluids (Article 75), storage of explosive materials (Article 77), and storage of flammable and combustible liquids (Article 79).

Hazardous materials are included as a discrete category in Article 80. This category includes additional requirements for the above-mentioned hazardous materials, as well as storage requirements for other hazardous materials such as flammable solids, liquid and solid oxidizers, organic peroxides, pyrophoric materials, unstable and water-reactive materials, highly toxic solids and liquids, radioactive materials, corrosives, and carcinogens, irritants, sensitizers, and other health hazard solids, liquids, and gases. For each category there is a defined exempt storage amount. Exemptions typically are based on type of storage and type of fire protection provided (i.e., unprotected by sprinkler or cabinet, stored within cabinet in unsprinklered building, stored in sprinklered building but not in cabinet, or stored in cabinet inside sprinklered building).

The next several paragraphs summarize some of the key requirements specified in the *Uniform Fire Code*. The summary is not meant to be all inclusive, but should provide a general idea of the types of requirements defined in the code. For inclusive information, the code and its companion publication, *Uniform Fire Code Standards* (see references), should be reviewed.

TABLE 7.2 Fire Extinguishing Methods by Chemical Category

Category	Examples	Fire Extinguishing Methods
Oxidizing chemicals	Potassium permanganate, potassium perchlorate, and sodium nitrate	Large quantities of water should be used as the fire extinguishing agent; fire fighters should wear self-contained breathing apparatus; area should be ventilated
Flammable/combustible chemicals	Gasoline, heavy oils, and sulfur/sulfides of sodium, potassium, and phosphorus	Large quantities of water or foam should be used as the fire extinguishing agent; sulfur and sulfide compounds could cause irritating gases
Unstable chemicals	Picric acid, TNT, and organic peroxides	Large quantities of water should be used as the fire extinguishing agent, preferably through automatic sprinklers since explosion potential exists
Corrosive chemicals	Sodium hydroxide, hydrochloric acid, sulfuric acid, and perchloric acid	Water spray is recommended in acid and alkali storage areas; toxic fumes can result in some cases; and explosion potential also may exist in some cases
Water reactive chemicals	Calcium carbide, chromyl chloride, and alkali metals	Graphite base powder or inert materials can be used for some water reactive chemicals
Air reactive chemicals	Elemental sodium and yellow phosphorus	Water followed by application of dirt or sand
Radioactive materials	Plutonium, radium, and uranium	Fire and explosion chacteristics are not affected by radioactivity of material; automatic fire protection is necessary to avoid human exposure to radioactivity; preplanning for disposal of contaminated water is necessary
Self-heating materials	Elemental sulfur and activated charcoal	Water or steam should be used only on fire area; wet material is more susceptible to self-heating than dry material

Source: Information from NFPA, *Fire Protection Handbook* (1991).

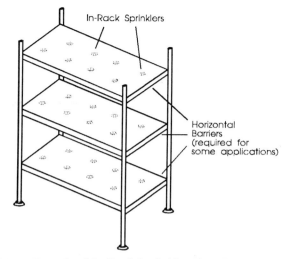

FIGURE 7.1   Eample of In-Rack Sprinklers for Flammable Products.

General Storage of Hazardous Materials.   General storage requirements address the topics of spill control, drainage, containment, ventilation, separation of incompatibles, construction of hazardous materials storage cabinets, fire-extinguishing systems, and other relevant topics. Any special requirements related to a particular material are addressed separately in discussions of that material.

Storage of Compressed Gases.   Compressed gases are required to be adequately secured to prevent falling or being knocked over. Additionally, legible operating instructions must be maintained at the operating location. "NO SMOKING" and other warning signs must be posted around enclosures that hold the compressed gas cylinders. Additional items pertain to toxic gases and highly toxic gases. These and include ventilation, use of gas cabinets and gas detection systems, acceptable rates of release, treatment systems for exhausted gas cabinets, and other topics.

Storage of Cryogenic Fluids.   Cryogenic fluids are classified in the code as flammable, nonflammable, corrosive/highly toxic, and oxidizer. These materials generally are not stored in storage rooms, but rather in aboveground, below-ground, or in-ground containers. Requirements for each type of installation is addressed in the standard. Additionally, minimum distances from the tank are specified for the different classes of cryogenics.

Storage of Explosive Materials. Explosive materials must be carefully handled and stored. Smoking, matches, open flames, or other spark-producing devices are prohibited within 50 feet of magazines. Also, dried grass, leaves, trash, and debris must be cleared within 25 feet of magazines. Explosives cannot be unpacked or repacked within 50 feet of a magazine or in close proximity to other explosives. Any deteriorated explosive must be reported to the fire chief immediately and destroyed by an expert. Numerous other requirements pertain to explosive and blasting agents, and these requirements should be investigated thoroughly before these materials are brought on site for storage.

Storage of Flammable and Combustible Liquids. Flammable and combustible liquids are addressed fully in the code. Storage room requirements include the use of in-rack sprinkler systems. Drainage systems must be designed to carry off to a safe location any anticipated spill plus a minimum calculated flow of the sprinkler system. Aisle widths must be 4 feet between racks and 8 feet for main aisles. Class I flammable liquids (which have a flash point below 100°F) cannot be stored in basements. Additional requirements for individual classes of flammable and combustible liquids and types of storage also are addressed.

Storage of Flammable Solids. Indoor storage of these materials sometimes requires explosion venting or suppression. If the material stored is over 1,000 cubic feet, the material must be separated into piles of 1,000 cubic feet. Aisle widths between piles must equal the height of the pile or 4 feet, whichever is greater. For exterior storage, flammable solids must be stored in piles no greater than 5,000 cubic feet. Aisle widths between piles must be one-half the height of the piles or 10 feet, whichever is greater. Additionally, for exterior storage applications, distance from storage to any building, property line, or public way must be at least 20 feet unless the storage area has an unpierced 2-hour fire-resistive surrounding barrier, defined as a wall extending not less than 30 inches above and to the sides of the storage area. Other requirements pertaining to ventilation, spill control, and other topics are defined.

Storage of Liquid and Solid Oxidizers. For indoor storage, some amounts of these materials require detached storage, as specified in the code. Detached storage buildings must be single story without a basement or crawl space. Requirements for distance from storage to any building, property line, or public way is dependent on the class of oxidizer. Class 4 oxidizers must be separated from other hazardous materials by not less than 1-hour fire resistive construction. Detached buildings for Class 4 oxidizers must be located a minimum of 50 feet from other hazardous materials storage. For exterior storage, maximum quantities and arrangements are specified in the code. Other requirements pertaining to safe distance to property boundary or public way, smoke detection systems, ventilating systems, and other items are addressed.

Storage of Organic Peroxides.   For indoor storage, some amounts of organic peroxides require detached storage, depending on class. Separation distance of detached storage from any building, property line, or public way is also dependent on class. Additionally, special storage requirements pertain to this class of material. For instance, 55-gallon containers cannot be stored more than one container high and a minimum 2-foot clear space must be maintained between storage and uninsulated metal walls. Containers and packages in storage areas must be closed, and other requirements apply.

Storage of Pyrophorics.   Special storage requirements apply to pyrophorics that exceed exempt amounts. These include a limit of palletized storage of 10 feet by 10 feet by 5 feet high. Individual containers cannot be stacked. Also, aisle space between storage piles must be at least 10 feet. Tanks or containers inside a building cannot exceed 500 gallons. Exterior storage and other requirements are also outlined.

Storage of Unstable (Reactive) Materials.   These materials must be stored in a detached building if the storage amounts exceed 2,000 pounds of a Class 3 material or 50,000 pounds of a Class 2 material. Most Class 4 materials also require detached storage. The floor must be liquid tight and there must be a means to vent smoke and heat in a fire or other emergency. The material should not be stored in piles greater than 500 cubic feet, and aisle width must be equal to the height of the piles or 4 feet, whichever is greater. Exterior storage allows for larger piles of 1,000 cubic feet, with aisle widths of one-half the height of the pile or 10 feet, whichever is greater. Additionally, exterior storage of unstable materials should not be within 20 feet of any building, property line, or public way unless the storage area has an unpierced 2-hour fire-resistive barrier. If the material may deflagrate, the safe storage distance is 75 feet. Other requirements apply.

Storage of Water-reactive Materials.   Like unstable materials, these materials must be stored in a detached building if the storage amounts exceed 2,000 pounds of a Class 3 material or 50,000 pounds of a Class 2 material. The floor must be liquid tight and the room must be waterproof. External storage requirements include a safe storage distance of 20 feet or an unpierced 2-hour fire-resistive barrier, except for Class 3, which requires 75 feet between the storage area and any building, property line, or public way. Other requirements include a means of venting smoke and heat, fire-extinguishing system requirements, secondary containment requirements, and others.

Storage of Highly Toxic Solids and Liquids.   In some cases, exhaust scrubbers are required in areas that store these materials to ensure that vapors from an accidental release or spill are mitigated. Additionally, these materials should be isolated from other materials by 1-hour fire-resistive construction or stored in approved hazardous materials storage cabinets. Exterior storage piles of these materials cannot exceed 2,500 cubic feet. Aisle widths between piles must be at least one-half the height of the pile or 10 feet, whichever is greater. Highly toxic liquids that liberate highly toxic vapors cannot be stored outside a building unless effective collection and treatment systems are provided. Additional requirements include specification of fire-extinguishing systems, safe distances from storage to exposures, and others.

Storage of Radioactive Materials. Areas used for the storage of radioactive materials must be provided with detection equipment suitable for determining surface-level contamination at levels that would pose a short-term hazard condition. All storage areas must be in compliance with the requirements of the Nuclear Regulatory Commission and state or local requirements. Additional requirements are defined including requirements for fire-extinguishing systems, safe distances from storage to exposures, and others. NFPA requirements for fire protection at these types of facilities are addressed in the next section.

Storage of Corrosives. General requirements for storage of corrosives include a liquid-tight floor and adequate containment. Exterior storage requirements include a distance of 20 feet from storage to any building, property line, or public way or an unpierced 2-hour fire-resistive barrier. Additionally secondary containment is required for outdoor storage.

Carcinogens, Irritants, Sensitizers, and Other Health Hazard Solids, Liquids, and Gases. A liquid-tight floor and secondary containment is required for indoor storage of these materials. For exterior storage, requirements for safe distance from storage to exposure is 20 feet or a 2-hour fire-resistive barrier. Outdoor piles of this type of material cannot be larger than 2,500 cubic feet. Aisle widths between piles must be one-half of the height of the piles or 10 feet, whichever is greater.

Fire Protection for Facilities Handling Radioactive Materials

Fire protection for facilities handling radioactive materials should be designed to minimize the spread of radioactive contamination during a fire incident. Design considerations will include items such as ventilation, fire-water drainage, and other considerations. In addition, good administrative controls, which can aid in minimizing effects of fire at any facility, are especially important for facilities that handle radioactive materials.

NFPA Standards. Fire protection practices for facilities handling radioactive materials are documented by NFPA in the following standards:

- NFPA 801 (1991), Recommended Fire Protection Practice for Facilities Handling Radioactive Materials
- NFPA 802 (1993), Nuclear Research and Production Reactors

- NFPA 803 (1993), Standard for Fire Protection for Light Water Nuclear Power Plants

These standards address key aspects of fire prevention and fire response, including fire protection systems and equipment, general facility design, administrative controls, and other topics.

Design Considerations.   Several design considerations are particularly important to facilities that handle radioactive materials. These include:

- Ventilation systems—Ventilation systems in a facility that handles radioactive materials must include capabilities for heat removal, fire isolation, and filtration of radioactive gases and particles. Fresh air inlets should be located to reduce the possibility of radioactive contaminants being introduced.
- Duct systems—Fire-resistant or fire-retardant materials should be used in ducts. Fire dampers should be provided (unless shutdown of ventilation system is not allowed) to resist spread of contaminated smoke.
- Drainage—Drainage systems should be designed to remove liquids to a safe area for testing and proper disposal. The liquid handling area should be designed to include the volume of a spill from the largest container, a 20-minute discharge from fire hoses or other suppressant system, and— for outdoor applications—a typical amount of rain or snow. The drains should be designed to minimize fire hazard and radioactive contamination of clean areas (NFPA 801, 1990).
- Fire Detection system—A fire detection system, such as a system that alarms at a fixed temperature or if there is an accelerated rate of rise in temperature, is important. This detection system will alert building occupants of fire conditions and allow for immediate evacuation. Additionally, the alarm can be used by management as a signal to rapidly implement the fire emergency plan.
- Electrical power—Electrical operations equipment including transformers, control panels, and main switches should be located well away from areas that handle radioactive and ignitible materials.
- Other design considerations—Other design considerations include emergency lighting for a means of egress, lightning protections, interior finish, and other considerations addressed as part of life safety management.

Administrative Controls.   Administrative controls are a necessary part of the fire safety program. Included in a well-managed fire prevention program are items such as a documented fire prevention program, a fire hazard analysis, a fire emergency plan, and an adequate testing, inspection, and maintenance program.

If an emergency does occur that requires response and mitigation, proper personal protective equipment must be worn at all times, including face masks, clothing that prevents the entry of radioactive materials into the body, and a self-contained breathing apparatus. This protective equipment should be specified by a trained industrial hygienist or safety specialist.

LIFE SAFETY

In addition to fire protection, there are other safety standards that generally pertain to the category of life safety. These include standards for methods of egress, structural features of the building such as handrails and stairwells, floor loads, use of building materials, fire-resistant walls, and other life safety standards. These standards are based on occupancy classification. Two useful references on this subject include the *Life Safety Code*® published by NFPA and the *Uniform Building Code* published by ICBO.

Life Safety Code®

The Life Safety Code® and accompanying handbook—*Life Safety Code® Handbook* (see references)—address requirements for life safety. Included is information pertaining to:

- Classification of occupancy and hazard of contents— Industrial, storage, and health care classifications are addressed along with other classifications that would not pertain to hazardous materials and hazardous waste management such as assembly, educational, residential and other classifications.
- Means of egress—Means of egress components, arrangement of means of egress, discharge from exits, emergency lighting, marking of means of egress, and other topics are addressed.
- Features of fire protection—Included is information on construction and compartmentation, smoke barriers, special hazard protection, and interior finish.
- Building service and fire protection equipment— Information pertaining to utilities, ventilation, heating and air conditioning, and smoke control is provided. Other topics covered include elevators, escalators, conveyors, rubbish chutes, incinerators, and laundry chutes. Additionally, fire detection, alarm and communications systems, automatic sprinklers, and other extinguishing equipment is discussed.
- New and existing occupancies (by occupancy classification)—Each occupancy classification is addressed in terms of new and existing occupancies. Included is information about general requirements, means of egress requirements, protection, special provisions, building services, and special requirements within the occupancy classification.

The Life Safety Code® has been in existence for more than 60 years and supplements NFPA's fire protection standards.

Uniform Building Code

The *Uniform Building Code* and its companion publication the *Uniform Building Code Standards* also provide standards for building and life safety in terms of occupancy. These requirements include items such as lighting, ventilation, sanitation, sprinkler and standpipe systems, fire alarms, construction type, construction height, allowable area, and special hazards.

   In addition to requirements for occupancy classifications, numerous other building standards are detailed. These include items such as fire-resistive standards for interior walls and ceilings, engineering regulations pertaining to quality and design of the materials of construction, and requirements based on types of construction.

Legal Requirements for Fire Protection and Life Safety

NFPA fire standards, the *Uniform Fire Code*, the *Uniform Building Code*, the *Life Safety Code*®, and other equivalent standards are considered model codes. Some municipalities have adopted these codes or parts of these or other codes into city ordinances or statutes. In this manner, these codes can become enforceable standards.

VENTILATION SYSTEMS

Ventilation is important in preventing buildup of toxic or flammable concentrations of chemicals stored or used in a process. Ventilation removes contaminants or flammable vapors from an area as well as providing fresh dilution air. The following paragraphs discuss the ventilation process in industrial and other applications.

Basic Models Used in Ventilation System Design

Mathematical models that predict air flow in a room or building have been developed and are used for design of ventilation systems. For applications such as small office buildings and nonchemical distribution centers, macro-level modeling is generally used. These models predict air dispersion and thermal loadings, assuming single or multiple well-mixed zones. The models use mass, momentum, energy balance, and fluid mechanics in simple straightforward equations. Typically, these models are too simplistic for chemical environments.

   More sophisticated models for predicting indoor airflows have been documented in the manual published by the American Society of Heating, Refrigerating, and Air-Conditioning Engineers (ASHRE) entitled *Building*

Systems: Room Air and Air Contaminant Distribution (see references). These models use complex turbulence equations to determine air flow patterns of a unique zone such as a cleanroom or a chemical laboratory. These complex models can be used in conjunction with industry-recognized design criteria to ensure proper removal of contaminants from the workplace.

Ventilation Methods

There are several methods used to ventilate industrial buildings. Common methods include natural ventilation, dilution ventilation, and local exhaust systems. Handbooks published by the American Conference of Governmental Industrial Hygienists (ACGIH), ASHRAE, and others provide standards for design and operation of ventilation systems that are accepted throughout industry. Two handbooks that are particularly inclusive are *Design of Industrial Ventilation Systems* by John Alden and John Kane and *Industrial Ventilation* published by ACGIH (see references). Upgrades from the standard may be required for some ventilation systems, depending on the specific application at the facility.

Natural Ventilation.   Natural ventilation is generally sufficient for warehouse buildings that house inert materials, nontoxic gases, and chemical handling articles such as sealed empty drums, spare pallets, and maintenance tools. In addition, small volume, off-the-shelf chemicals such as lubricating oils and epoxies are often stored in cabinets inside a building of this design. Design considerations for a naturally ventilated building include building structure, surrounding structures, prevailing winds, and other climatological factors.

Dilution Ventilation.   Dilution ventilation combines general exhaust with a supply air system and normally includes heat control, moisture control, and air conditioning. This type of ventilation is adequate for industrial applications that have low-level concentrations of chemical contaminants that are uniformly and widely dispersed and do not occur close to the worker's breathing zone. Design of this type of ventilation system takes into account:

- Rate of contaminant generation and concentration of vapors to be removed
- Room volume
- Effective volumetric air flow rate
- Incomplete mixing as dictated by placement of exhaust outlet and supply air inlet

Dilution ventilation requirements vary considerably from one chemical to the next, depending on permissible exposure limits as well as physical properties of a chemical. Examples of air replacement requirements for selected chemicals under uniform conditions are presented in Table 7.3.

TABLE 7.3   Examples of Air Replacement Requirements for Dilution Ventilation Systems and Selected Contaminants

Contaminant	OSHA PEL (ppm)*	Minimum Air Changes/Hour**
n-Butyl acetate	150	11
Cyclohexane	300	7
Ethyl benzene	100	18
Ethyl ether	400	5
Isoamyl alcohol	100	19
Isopropyl alcohol	400	7
Methyl chloroform	350	6
n-Propyl acetate	200	9
Tetrahydrofuran	200	9
Turpentine	100	14
Vinyl toluene	100	16

*The PEL is the permissible exposure limit based on a time weighted average (TWA), which is the employee's average airborne exposure in an 8-hour work shift of a 40-hour work week that shall not be exceeded.
**Based on standard temperature and pressure, perfect dilution, an evaporation rate of 2 gal/hr, and a room 50 ft long × 50 ft wide × 12 ft high.
Source: Information from CFR §1910.1000, Table Z-1-A.

Since requirements are varied from chemical to chemical, there is no general guideline for the number of air changes needed to keep chemical concentrations within acceptable limits. Typically, dilution ventilation systems are not adequate for exhausting some chemicals because of their low permissible exposure limits or because of high flammability. Examples include chloroform, carbon disulfide, and gasoline. Additionally, dilution ventilation systems normally are not adequate for exhausting particulates or large point source emissions. For those chemicals or emissions that require an excessive number of air changes per hour to meet regulatory or other standards, local exhaust at the process or tool will be necessary.

The number of air changes required for offices, warehouses, and other non-chemical industrial applications will vary depending on the application. Ventilation requirements for areas such as general manufacturing space, machine shops, engine rooms, boiler rooms, and laundries should be evaluated by a ventilation specialist to ensure that persons working in the area have a healthful and comfortable environment.

For boiler rooms and laundries, which are particularly hot environments, ventilation systems that use exhaust and supply air systems typically are used for heat control. The supply air is cooled and dehumidified in order to bring the inside air to an acceptable temperature and relative humidity. When possible, heat sources are enclosed and exhausted separately. Design considerations for these applications include temperature, humidity, air flow path, and potential convection and evaporative cooling effects. Other industries requiring specially designed ventilation systems for heat control include foundries, ceramic manufacturing, and coke processing.

Local Exhaust. Local exhaust is used to capture a chemical, particulate, mist, or other emission at or near its source, before the contaminant disperses into the workplace. Local exhaust systems are composed of hoods, air cleaners, and fans. Hood design and hood placement are two of the most important aspects of local exhaust design. Several commonly used hood types are presented in Figure 7.2. Applications for these selected hood type are summarized in Table 7.4.

The hood should always be located as close to the source as possible, as required capture velocity increases exponentially with distance. Also, the exhaust system should not pull air across the employee's breathing zone, since this would negate the purpose of the system. To ensure proper operation of a local exhaust system, alarm systems can be installed at locally exhausted tools or operations that have specific ventilation requirements for safe working conditions. These systems generally use audible alarms and a flashing light (for particularly noisy areas) to indicate that the exhaust has fallen below the required velocity. In some cases, the equipment can be interlocked to shut down at critical set points. In other cases, evacuation from the area is required until the problem can be resolved.

In addition to exhaust hood design and size, other design considerations (ACGIH 1992 and Alden and Kane 1982) include:

- Minimum duct velocity based on transport velocity needed for specific contaminants
- Branch duct size
- Friction losses throughout the system
- Air-flow balancing
- Fan type and pressure ratings
- Variations in temperature and humidity
- Air abatement devices or contaminant collectors
- Type and location of stacks

Functional considerations include:

- Location of workers with respect to replacement air vents
- Location of workers with respect to exhaust systems
- Heat stress associated with the operation

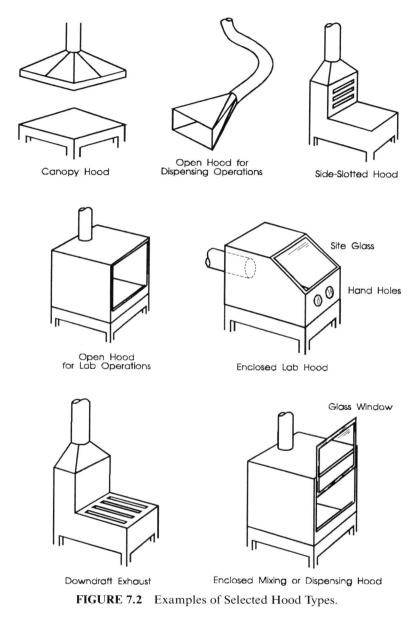

Canopy Hood

Open Hood for
Dispensing Operations

Side-Slotted Hood

Open Hood
for Lab Operations

Site Glass

Hand Holes

Enclosed Lab Hood

Glass Window

Downdraft Exhaust

Enclosed Mixing or Dispensing Hood

FIGURE 7.2   Examples of Selected Hood Types.

- Clothing required to perform the operation safely (i.e. long sleeves, arm guards, aprons) and temperature needed to maintain worker comfort
- Local lighting requirements that could add additional heat to the operation

TABLE 7.4   Examples of Selected Hood Types and Industrial Applications

Hood Type	Typical Industrial Applications
Canopy hood	Automated plating operations, degreasing operations, and hot processes
Enclosed hood or booth	Highly toxic chemical dispensing, lab operations involving infectious materials, radioactive material applications, extremely dusty operations, furnace or oven applications, spraying operations, high-agitation mixing operations
Slotted hood (side mounted)	Floor applications for operations involving fumes heavier than air, low-hazard bench operations
Open hood	Low- to medium-toxicity chemical dispensing operations, flammable chemical dispensing operations, lab operations
Slotted or open hood (downdraft)	Felting or brushing operations, grinding operations, low-level dust producing operations

Note: Some applications may require respiratory protection.

INDOOR AIR QUALITY

Since virtually all buildings utilize mechanical ventilation systems for temperature and humidity control of inside air, the maintenance of these systems for acceptable air quality in buildings is mandatory. As use of building space varies, the air supply system must meet needs and conditions specific to each application. Changes in building use from small-scale to large-scale production or from low occupancy to high occupancy may require ventilation system upgrades. If the ventilation system is inadequate for the building application, it might be evidenced in several ways including dampened inside or outside walls, high-velocity drafts through windows or doors, stagnant air in rooms or hallways, unpleasant odors, or continually unpleasant temperatures.

In addition to ventilation system inadequacies for specific building use, indoor air quality problems can be caused by other factors. These factors include environmental tobacco smoke, carpet fabrics, carpet and floor tile glues, interior paints and varnishes, and use of janitorial supplies (DOE and EPA 1991). Examples of chemical constituents potentially found in a building and the sources of contamination are shown in Table 7.5.

Odors and chemical fumes from these sources can be picked up in the ventilation system and spread to all parts of the building. In cases where fume buildup becomes extreme, sick building syndrome can occur. Occupants of such a building might experience symptoms such as headaches, dizziness or light headedness, watery eyes, wheezing, or other allergic reactions.

In addition to chemical fumes emitted from building materials, environmental tobacco smoke, and janitorial supplies, indoor air quality can be negatively affected by biological contaminants, including living organisms and the by-products of living organisms. Common biological contaminants include bacteria, fungi, pollen, molds, and algae. These contaminants can be toxic, pathogenic, or allergenic. Sources of these contaminants include stagnant drains and drip pans, uncleaned filters, and wet or humid surfaces.

Recently, radon has gained focus as a cause of indoor air quality problems. Radon is an inert radioactive gas that has been found to be present in numerous houses and buildings all over the nation. A maximum level of radon for indoor air has been established by the EPA at 4 picocuries per liter (pCi/l). Exposures over this amount are considered hazardous to the occupants of the building or house.

Radon can emanate from several different sources. These include soil and building products and materials. Radon exposure typically will vary within a house or building, with the largest exposure concentrations generally found in basements and first floors. Closed house conditions (i.e., during winter months for most states) also yield higher exposures (Belanger 1990).

TABLE 7.5   Selected Examples of Chemicals Potentially Found in an Office Building and the Source of Contamination

Contamination Source	Chemicals Potentially Found in an Office Building
Tobacco smoke	Carbon monoxide, carbon dioxide, oxides of nitrogen, ammonia, volatile N-nitrosamines, hydrogen cyanide, volatile hydrocarbons, volatile alcohols, volatile aldehydes, and ketones
Adhesives	Alcohols, ketones, and halogenated and aromatic hydrocarbons
Paints	Aromatic hydrocarbons, halogenated hydrocarbons, and aliphatic hydrocarbons
Varnishes and lacquers	Alcohols, toluene, and aromatic hydrocarbons
Fungicides, germicides and disinfectants	Formaldehyde, other aldehydes, and ammonia
Carpet	Halogenated hydrocarbons, aliphatic hydrocarbons, and aromatic hydrocarbons
Caulk	Volatile organic compounds, halogenated organics, and aliphatic and aromatic hydrocarbons
Insulation	Volatile organic compounds and aromatic hydrocarbons
Cleaning products	Ammonia and aldehydes

Source: Information from DOE and EPA (1991).

Air in buildings that are occupied by a large number of people should be monitored on a periodic basis. Complaints by the occupants should be investigated to assure that proper air quality is maintained. The fresh air intake to the circulating system should be visually inspected on a regular basis, and any change of condition should be addressed immediately. Aspects that might negatively affect the intake air include proximity of load/unload areas that can disperse excessive vehicular exhaust, proximity to cooling towers that might spray treated water into the intake system, and proximity of shrubs and trees that are sprayed periodically with pesticides or fertilizers.

CONCLUSION

Workplace safety through fire protection, life safety management, and proper ventilation is an important part of hazardous materials and hazardous waste management. Many potential problems that could result in property loss or adverse effects to human health and the environment can be averted with adequate fire protection systems, building designs, and ventilation systems.

Although fire protection, life safety management, and ventilation systems are considered to be traditional workplace management subjects, some aspects of these topics gained focus during the late 1980s and early 1990s. As an example, fire systems that use halons—which are ozone-depleting chemicals—are under evaluation for replacement with alternate chemicals. Additionally, the sick building syndrome has brought into focus ventilation system design, as well as other issues.

Thus far, this section of the book has covered a range of topics pertaining to workplace management of hazardous materials and hazardous waste including exposure assessment, workplace and personal monitoring, personal protective equipment, and workplace and building safety. In the next chapter, administrative requirements for facilities that use chemicals and generate and manage waste are discussed. These requirements include material labeling, plans and controls, employee training, material tracking, noncompliance reporting, and other administrative items.

REFERENCES

American Conference of Governmental *Industrial Hygienists, Industrial Ventilation: A Manual of Recommended Practice*, 21st ed., Cincinnati, OH, 1992.

American Society of Heating, Refrigerating, and Air-Conditioning Engineers, *Building Systems: Room Air and Air Contaminant Distribution*, Leslie L. Christianson, ed., Atlanta, GA, 1989.

Alden, J. L., and Kane, J. M., *Design of Industrial Ventilating Systems*, 5th ed., Industrial Press, New York, 1982.

Department of Energy and U.S. Environmental Protection Agency, *Sick Building Syndrome: Sources, Health Effects, Mitigation*, Baechler, M.C., et al., eds., Noyes Data Corporation, Park Ridge, NJ, 1991.

Belanger, William E., "Prediction of Long-term Average Radon Concentrations in Houses Based on Short-term Measurements," in *Proceedings of the 1990 EPA/ A&WMA International Symposium, Measurement of Toxic and Related Air Pollutants*, U.S. Environmental Protection Agency and Air and Waste Management Association, Raleigh, NC, 1990.

International Conference of Building Officials, *Uniform Building Code*, 1991 ed., Whittier, CA, 1991.

International Conference of Building Officials, *Uniform Building Code Standards*, 1991 ed., Whittier, CA, 1991.

International Conference of Building Officials, *Uniform Fire Code*, 1991 ed., Whittier, CA, 1991.

International Conference of Building Officials, *Uniform Fire Code Standards*, 1991 ed., Whittier, CA, 1991.

National Fire Protection Association, *Fire Protection Guide to Hazardous Materials*, 10th ed., Quincy, MA, 1991.

National Fire Protection Association, *Fire Protection Handbook*, 17th ed., Quincy, MA, 1991.

National Fire Protection Association, *Life Safety Code®*, NFPA 101, Quincy, MA, 1991.

National Fire Protection Association, *Life Safety Code® Handbook*, 5th ed., James K. Lathrop, ed., Quincy, MA, 1991.

National Fire Protection Association, "Recommended Fire Protection Practice for Facilities Handling Radioactive Materials," NFPA 801, Quincy, MA, 1990.

BIBLIOGRAPHY

ASHRE (1989), "Ventilation for Acceptable Indoor Air Quality," ASHRE Standard 62, American Society of Heating, Refrigerating, and Air-Conditioning Engineers, Atlanta, GA.

ASTM (1989), *Design and Protocol for Monitoring Indoor Air Quality*, N.L. Nagda and J.P. Harper, eds., American Society for Testing and Materials, Philadelphia.

_____ (1990), *Biological Contaminants in Indoor Environments*, Morley and Feeley, eds., Special Technical Publication 1071, American Society of Testing and Materials, Philadelphia.

Bond (1991), *Sources of Ignition*, Butterworth-Heinemann, Stoneham, MA.

Brooks, Bradford O., and William F. Davis (1992), *Understanding Indoor Air Quality*, Lewis Publishers, Boca Raton, FL.

Bonneville Power Administration (1984), *Ventilation in Commercial Buildings*, prepared by Seton, Johnson, & Odell, Inc., for Office of Conservation, Bonneville Power Administration, Portland, Oregon.

Burgess, W.A., M.J. Ellenbecker, and R.T. Treitman (1989), *Ventilation for Control of the Work Environment*, John Wiley & Sons, New York.

Cherry (1988), *Asbestos Engineering, Management and Control*, Lewis Publishers, Boca Raton, FL.

Cothern, C.R., and J.E. Smith, Jr., eds. (1987), *Environmental Radon*, Plenum Press, New York.

DOE (1987), "Indoor Air Quality Environmental Information Handbook: Building System Characteristics," prepared by Mueller Associates, Inc., for Department of Energy, Washington D.C.

_____ (1989), "Radon Research Program," DE89 007284, Office of Health and Environmental Research, Department of Energy, Washington, D.C.

_____ (1990), "Indoor Air Quality Issues Related to the Acquisition of Conservation in Commercial Building," prepared by M.C. Baechler, D.L. Hadley, and T.J. Marseille of Pacific Northwest Laboratory, Department of Energy, Washington, D.C.

_____ (1990), "Health Effects Associated with Energy Conservation Measures in Commercial Buildings, Vol. 2: Review of the Literature," prepared by R.D. Stenner and M.C. Baechler of Pacific Northwest Laboratory, Department of Energy, Washington, D.C.

Fisk, W.J., et al. (1987), *Indoor Air Quality Control Techniques: Radon, Formaldehyde, Combustion Products*, Noyes Data Corporation, Park Ridge, NJ.

Godish, Thad (1989), *Indoor Air Pollution Control*, Lewis Publishers, Boca Raton, FL.

Government Institutes (1991), *Fire Protection Management for Hazardous Materials: An Industrial Guide*, Government Institutes, Inc., Rockville, MD.

Heinsohn, R.J. (1991), *Industrial Ventilation: Engineering Principles*, John Wiley & Sons, New York.

Kay, Jack G., George E. Keller, and Jay F. Miller (1991), *Indoor Air Pollution: Radon, Bioaerosols, and VOCs*, Lewis Publishers, Boca Raton, FL.

Lao, Kenneth Q. (1990), *Controlling Indoor Radon*, Van Nostrand Reinhold, New York.

National Academy of Science (1985), *Building Diagnostics: A Conceptual Framework*, National Academy of Science, Washington, D.C.

NCRP (1989), "Control of Radon in Houses," NCRP Report No. 103, National Council on Radiation Protection and Measurements, Bethesda, MD.

National Research Council (1986), *Environmental Tobacco Smoke: Measuring Exposures and Assessing Health Effects*, Committee on Passive Smoking, Board on Environmental Studies and Toxicology, National Research Council, National Academy Press, Washington, D.C.

Nazaroff, W.W., and A.V. Nero, eds. (1987), *Radon and Its Decay Products in Indoor Air*, Wiley-Interscience, New York.

Stamper, Eugene, and Richard Koral (1979), *Handbook of Air Conditioning, Heating, and Ventilating*, 3rd ed., Industrial Press, New York.

U.S. EPA (1988), "Indoor Air Quality in Public Buildings," Vol. 1, prepared by L.S. Sheldon et al., for U.S. Environmental Protection Agency, Washington, D.C.

_____ (1988), "Indoor Air Quality in Public Buildings," Vol. 2, EPA-600/6-88-009b, prepared by L.S. Sheldon, et al., for Environmental Protection Agency, Washington, D.C.

_____ (1990), "Indoor Air—Assessment Methods of Analysis for Environmental Carcinogens," EPA 600/8-90/041, Environmental Criteria and Assessment Office, Office of Health and Environmental Assessment, Office of Research and Development, U.S. Environmental Protection Agency, Research Triangle Park, NC.

U.S. EPA/NIOSH (1991), "Building Air Quality," U.S. Environmental Protection Agency and National Institute for Occupational Safety and Health, American Conference of Governmental Industrial Hygienists, Cincinnati, OH.

8

ADMINISTRATIVE REQUIREMENTS FOR PROPER MANAGEMENT OF HAZARDOUS MATERIALS AND HAZARDOUS WASTE

There are numerous administrative requirements pertaining to proper management of hazardous materials and hazardous waste specified by governmental agencies. These requirements include proper container labeling, periodic training, container and containment inspections, plans and controls, material/waste tracking and scheduled reporting, and release or other noncompliance reporting. This chapter discusses administrative requirements that are defined in the regulations. In addition, management of these requirements through computer applications is addressed.

REGULATORY STANDARDS

Proper management of hazardous materials and hazardous waste has been defined through numerous administrative requirements that are applicable to manufacturing and waste management facilities. These requirements have been published in regulations promulgated by various agencies including OSHA, the Nuclear Regulatory Commission (NRC), the Department of Transportation (DOT) and EPA. EPA requirements have been promulgated under several Acts including the Clean Water Act (CWA), the Clean Air Act (CAA), Toxic Substance and Control Act (TSCA), Comprehensive Environmental Response, Compensation, and Liability Act (CERCLA), and Superfund Amendments and Reauthorization Act (SARA). Additionally, hazardous waste management requirements for treatment, storage and disposal facilities (TSD) have been promulgated under the Resource Conservation and Recovery Act (RCRA). A summary of key regulatory standards that address administrative requirements is presented in Table 8.1.

LABELING

As can be seen in Table 8.1, many agencies have labeling requirements. In some cases, when a material is regulated by more than one agency, the regulations might overlap. In general, the overlapping labeling requirements are equivalent. DOT labeling requirements are included by reference in several regulations, as appropriate. Symbols of hazards published in DOT regulations are recognized by all agencies. In addition, there is a United Nations (UN) classification system used internationally for the transport of hazardous materials. The UN classification number may be displayed at the bottom of placards or in the hazardous materials description on shipping papers.

Chemical Labeling

Chemical labeling requirements under OSHA specify that the chemical manufacturer, importer, or distributer shall ensure that each container of hazardous chemicals is tagged or marked with the following information:

- Identity of the hazardous chemical
- Appropriate hazard warnings
- Name and address of the chemical manufacturer, importer, or other responsible party

Manufacturers may put additional information on the label such as the DOT shipping label, as appropriate.

It is the responsibility of the manufacturer to ensure that all chemical containers are labeled properly before shipment. The employer has the responsibility to ensure that all chemical containers in the workplace identify the hazardous chemicals therein and display appropriate hazard warnings, as specified under OSHA. Bulk chemical tanks, process tanks, and other containers all must be labeled to meet OSHA requirements.

OSHA allows alternatives to affixing labels on stationery containers. Acceptable alternative labeling methods include signs, placards, process sheets, batch tickets, and equivalent methods. An alternative labeling method must identify the containers to which it is applicable, convey the required information, and be easily accessible to employees in their work area throughout each shift. Additionally, if hazardous chemicals are transferred from a labeled container into a portable container that is intended only for the immediate use of the employee who performs the transfer, labeling of the portable container is not required.

TABLE 8.1 Regulatory Standards Applicable to Manufacturing and Waste Management Facilities

Labeling

10 CFR §20.203	Caution signs, labeling, signals, and controls for radiation areas (NRC)
10 CFR §61.57	Labeling of low-level radioactive waste (NRC)
29 CFR §§1910.1001(j)(l), 1910.1001(j)(2), and 1926.58	Labeling of asbestos abatement and project areas (OSHA)
29 CFR §1910.1200(f)	Chemical labeling (OSHA)
40 CFR §§262.31–262.34	Waste labeling (RCRA)
40 CFR §761.40	Marking of PCBs and PCB items (TSCA)
40 CFR §763.121(k)	Labeling of asbestos abatement project areas (TSCA)
49 CFR Part 172 Subpart D	Marking of hazardous materials packages (DOT)
49 CFR Part 172 Subpart F	Placarding (DOT)
49 CFR Part 172 Subpart E	Package or container labeling (DOT)

Training

10 CFR §19.11	Posting of notices to workers (NRC)
10 CFR §19.12	Instructions to workers (NRC)
10 CFR Part 72 Subpart I	Training of personnel working in spent fuel and high-level radioactive waste areas (NRC)
29 CFR §1910.119(g)	Training for workers involved with processes storing highly hazardous chemicals (OSHA)
29 CFR §§1910.120(e) and 1910.120(p)	Training for hazardous waste operations and emergency response (OSHA)
29 CFR §1910.1200(h)	Training requirements in the hazad communication standard (OSHA)
40 CFR §61.145	Asbestos NESHAP training for handling of regulated asbestos-containing material (CAA)
40 CFR §264.16	Hazardous waste handling training for permitted TSD facilities (RCRA)

continued

TABLE 8.1 *Continued*

Training (Continued)

40 CFR §763.92	Training and periodic surveillance for asbestos-containing materials in schools (TSCA)
49 CFR §172 Subpart H	Training of individuals involved in the transportation of hazardous materials (DOT)

Inspections

10 CFR §19.14	Presence of representatives of licensees and workers during inspections (NRC)
10 CFR §61.82	Commission inspections of low-level waste land disposal facilities (NRC)
40 CFR §264.15	General facility inspections for permitted TSD facilities (RCRA)
40 CFR §264.174	Hazardous waste container inspections for permitted TSD facilities (RCRA)
40 CFR §264.191	Tank asessments/certifications for permitted TSD facilities (RCRA)
40 CFR §264.195	Hazardous waste tank inspections for permitted TSD facilities (RCRA)
40 CFR §264.226	Hazardous waste surface impoundment inspections for permitted TSD facilities (RCRA)
40 CFR §264.254	Hazardous waste pile inspections for permitted TSD facilities (RCRA)
40 CFR §264.347	Hazardous waste incinerator and associated equipment inspections for permitted TSD facilities (RCRA)
40 CFR §264.574	Hazardous waste drip pad inspections for permitted TSD facilities (RCRA)
40 CFR §264.602	Hazardous waste miscellaneous unit inspections for permitted TSD facilities (RCRA)
40 CFR §280.43–44	Release detection requirements for underground storage tank systems (RCRA)
40 CFR §720.122	Inspections conducted by EPA for TSCA information verification (TSCA)
40 CFR §761.65	PCB inspections (TSCA)

Plans and Controls

10 CFR §19.13	Notification and reports to individuals (NRC)

TABLE 8.1 *Continued*

Plans and Controls (Continued)

10 CFR Part 20 Subpart I	Storage and control of licensed material (NRC)
10 CFR Part 20 Subpart K	Waste disposal (NRC)
10 CFR §61.12(j)	Quality control program for low-level radioactive waste disposal facilities (NRC)
10 CFR §61.53	Environmental monitoring (NRC)
10 CFR Part 61 Subpart E	Financial assurances (NRC)
29 CFR §1910.120(a)–(q)	Plans and controls for emergency response and hazardous waste operations (OSHA)
29 CFR §1910.1200(e)	Hazard communication plan (OSHA)
40 CFR Part 112	Spill prevention, control, and countermeasure plan (CWA)
40 CFR §264.13	Waste analysis plan for permitted TSD facilities (RCRA)
40 CFR §264.99	Compliance monitoring program for permitted TSD facilities (RCRA)
40 CFR §264.100	Corrective action program for permitted TSD facilities (RCRA)
40 CFR §264.112	Facility closure plan for permitted TSD facilities (RCRA)
40 CFR §264.118	Facility post-closure plan for permitted TSD facilities (RCRA)
40 CFR Part 264 Subpart D	Emergency and contingency plan for permitted TSD facilities (RCRA)
40 CFR Part 264 Subpart H	Proof of financial assurance for permitted TSD facilities (RCRA)
40 CFR Part 280 Subpart H	Proof of financial assurance for underground storage tank systems (RCRA)

Material/Waste Tracking, Reporting, and Recordkeeping

10 CFR Part 20 Subpart L	Nuclear waste tracking/recordkeeping (NRC)
10 CFR §60.71	Records of receipt, handling, and disposition of radioactive waste at a geologic repository (NRC)

continued

TABLE 8.1 *Continued*

Material/Waste Tracking, Reporting, and Recordkeeping (Continued)

10 CFR §61.80	Reports of radioactive waste acceptance at land disposal facilities (NRC)
10 CFR Part 74	Material control and accounting of special nuclear material (NRC)
40 CFR §61.145	Asbestos abatement project notification (CAA)
40 CFR §259.72	Transporter notification for medical waste handling in covered states (RCRA)
40 CFR §259.78	Transporter report of medical wastes handled in covered states (RCRA)
40 CFR §§264.71 and 264.72	Manifest requirements for permitted TSD facilities (RCRA)
40 CFR §264.73 and Appendix I	Operating record for permitted TSD facilities (RCRA)
40 CFR §264.75	Biennial reporting including waste minimization reporting for permitted TSD facilities (RCRA)
40 CFR §264.76	Unmanifested waste report for permitted TSD facilities (RCRA)
40 CFR §280.34	Release detection recordkeeping requirements for underground storage tank systems (RCRA)
40 CFR §370.25	Inventory reporting—Tier I and II (SARA)
40 CFR §372.30	Toxic chemical release reporting (SARA)
40 CFR Part 716	Health and safety data reporting (TSCA)
40 CFR Part 717	Records and reports of allegations that chemical substances cause significant adverse reactions to health or the environment (TSCA)
40 CFR Part 720	Premanufacturing notification (TSCA)
40 CFR Part 761 Subpart K	PCB waste disposal records and reports (TSCA)
40 CFR §761.208	Use of manifest for PCB waste (TSCA)
40 CFR Part 763 Subpart G	Asbestos abatement project notification (TSCA)
49 CFR §172.205	Hazardous waste manifest requirements (DOT)

Release and Other Noncompliance Reporting

10 CFR §20.2201	Reports of theft or loss of licensed material (NRC)
10 CFR §20.2202	Notification of incidents (NRC)

TABLE 8.1 *Continued*

Release and Other Noncompliance Reporting (Continued)

10 CFR §20.2203	Reports of exposures, radiation levels, and concentrations of radioactive material exceeding the limits (NRC)
10 CFR §21.21	Reporting of defects and noncompliance (NRC)
40 CFR Part 61	Reporting of noncompliance with NESHAP standards (CAA)
40 CFR §117.3	Noncompliance reporting (CWA)
40 CFR §264.56	Release reporting (RCRA)
40 CFR §270.5	Noncompliance reporting by the Director (RCRA)
40 CFR §280.50	Reporting of suspected releases from underground storage tank systems (RCRA)
40 CFR §280.53	Reporting and cleanup of spills and overfills associated with underground storage tank systems (RCRA)
40 CFR §302.6	Release reporting (CERCLA)
40 CFR §355.40	Release reporting (SARA)
49 CFR §171.15	Reporting hazardous materials incidents (DOT)

Standards Applicable to Nonpermitted Waste Management Facilities or Facilities Under Interim Status (RCRA)

40 CFR §262.34	General requirements for facilities that accumulate waste for less than 90 days
40 CFR Part 265	RCRA standards for nonpermitted facilities

Employer's labels typically are used for processes and auxiliary equipment that contain chemicals such as process baths and holding tanks. Additionally, chemicals that are dispensed from large containers (i.e., 55-gallon drums) into smaller, more easily transported containers (i.e., 1 to 2 gallon containers) require an employer's label if the person dispensing the chemical is not the user. These labels are often similar to a manufacturing label, but also may include internal information such as process center and department numbers, part number, and the phone number of the chemical distribution center. The labels can be designed and printed professionally or can be developed using a PC-based software program and printed on a PC-compatible printer. The using department can be given an array of labels for use with chemicals specific to the area. An example of an employer's chemical label is shown in Figure 8.1.

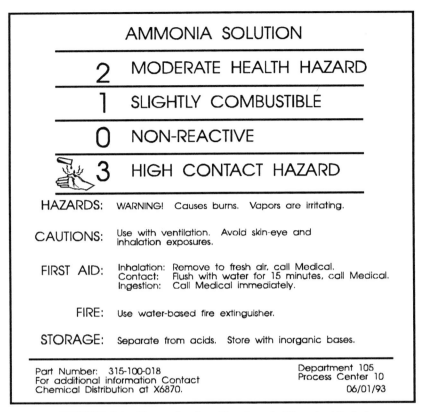

FIGURE 8.1   Example of an Employer's Chemical Label.

Hazardous Waste Labeling

Hazardous waste labeling required under RCRA is the responsibility of the hazardous waste generator. Once a container is put in use for hazardous waste storage, it must be labeled with the words "Hazardous Waste," and the hazard(s) must be specified. Another important part of hazardous waste labeling is the accumulation start date. Waste generators are not allowed to store hazardous waste for over 90 days without a permit; thus, the accumulation start date is used to track 90-day storage requirements for generators that do not have a RCRA Part B storage permit. For processes under direct operator control, the accumulation start date and 90-day storage requirements apply when the quantity stored reaches 55 gallons of hazardous waste or one quart of acutely hazardous waste. In order to ship a container, the generating facility must comply with DOT regulations specified in 49 CFR §172.304. This includes adding the facility's name, address, manifest number, and EPA waste number to the label. An example of an acceptable hazardous waste label is shown in Figure 8.2.

D.O.T. PROPER SHIPPING NAME AND I.D. NUMBER:

RQ WASTE HAZARDOUS SUBSTANCE, LIQUID, N.O.S.

(_____)

NA9188   INSERT TECHNICAL NAMES

ORM-E

HAZARDOUS WASTE

FEDERAL LAW PROHIBITS IMPROPER DISPOSAL
IF FOUND, CONTACT THE NEAREST POLICE, OR PUBLIC SAFETY
AUTHORITY, OR THE U.S. ENVIRONMENTAL PROTECTION AGENCY

GENERATOR INFORMATION:
NAME _____

ADDRESS _____ **PHONE** _____

CITY _____ **STATE** _____ **ZIP** _____

EPA / MANIFEST
ID NO. / DOCUMENT NO. _____ / _____

ACCUMULATION   **EPA**
START DATE _____ **WASTE NO.** _____

HANDLE WITH CARE!

Source: With Permission from EMED Co., Buffalo, NY.

FIGURE 8.2   Example of a Hazardous Waste Label.

As was the case with chemical labels, internal waste labels can be prepared using a PC-based software program. Waste labels can be printed as needed, with information such as accumulation start date, internal contact numbers, and other information being filled in at the time the label is printed. Or, if a printer is not available in the area, the label can be printed without the accumulation start date and contact numbers, and those items can be filled in manually when the label is put in use. Identification as to whether the waste is hazardous or nonhazardous can be preprogrammed so that labeling mistakes are minimized. Although RCRA labeling requirements do not apply to nonhazardous waste, as a good management practice this waste type should be labeled in a similar manner as hazardous waste, including the name of the waste along with the term "Nonhazardous Waste." An example of an internal waste label for a nonhazardous waste is shown in Figure 8.3.

```
┌─────────────────────────────────────────────────────────┐
│                                                         │
│              NONHAZARDOUS WASTE                         │
│  ─────────────────────────────────────────────────     │
│               PACKING MATERIAL                          │
│  ─────────────────────────────────────────────────     │
│   NO  HEALTH  HAZARD                                    │
│   COMBUSTIBLE                                           │
│   NON-REACTIVE                                          │
│   NO  CONTACT  HAZARD                                   │
│  ─────────────────────────────────────────────────     │
│   FIRE:   Use water-based extinguisher                  │
│           or Class A dry chemical extinguisher.         │
│                                                         │
│   SPECIAL INFORMATION:                                  │
│                                                         │
│   This material is  RECYCLED.                           │
│   Please segregate properly.                            │
│  ─────────────────────────────────────────────────     │
│   Accumulation Start Date:   /   /    Ship Date:  /  /  │
│   Generator (Dept./Bldg./Name):                         │
│   Name:                               Phone:_____    │
│   Internal Shipper (Dept./Bldg./Name):                  │
│   For Disposal, Contact:  Recyling Center at X6875.     │
│                                                         │
│   Comments: _____    │
└─────────────────────────────────────────────────────────┘
```

FIGURE 8.3   Example of an Employer's Waste Label for a Nonhazardous Waste.

DOT Labeling

Each person who offers a hazardous material for transportation or transports a hazardous material shall meet labeling and placarding requirements unless the material is exempt from the requirements. Special labels and placards are required for materials such as oxidizers, explosives, corrosives, poisons, combustibles, flammables, infectious substances, and others. Examples of labels acceptable for DOT shipping are presented in Figures 8.4 and 8.5. Size and color of the labels must meet DOT requirements.

Labeling of Asbestos-Containing Materials

During any demolition or renovation project involving asbestos the area must be closed to passing pedestrian traffic via signs, cones, tape, or other method, as required by OSHA and TSCA. In addition, the area must be enclosed and labeled. An example of a label used for this activity is shown in Figure 8.6. Information about asbestos removal, including required sampling and per-

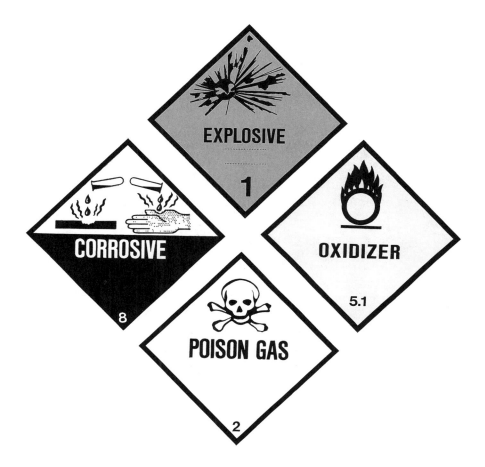

Note: Size and color must meet DOT requirements. Appropriate division number and compatibility group must be added to an explosive label. *Source:* With Permission from EMED Co., Buffalo, NY.

FIGURE 8.4   Examples of Acceptable DOT Labels.

sonal protective equipment, is presented in Chapters 5 and 6, respectively. Additional labeling requirements for products that contain asbestos and for schools containing asbestos are addressed under TSCA.

TRAINING

Training is required for employees who work with hazardous materials and hazardous waste. OSHA has three key standards that have training requirements. These are the Hazard Communication Standard, the Standard for Process Safety Management of Highly Hazardous Chemicals, and the Stand-

Note: Size and color must meet DOT requirements. *Source:* With Permission from EMED Co., Buffalo, NY.

FIGURE 8.5 Examples of Acceptable DOT Labels.

ard for Hazardous Waste Operations and Emergency Response. RCRA and DOT also define training requirements for persons working with hazardous waste and for persons involved in transporting hazardous materials, respectively.

Training Required by the Hazard Communication Standard

A widespread training requirement in the manufacturing industry is defined in OSHA's Hazard Communication Standard. This standard places requirements on chemical manufacturers, chemical distributors, and facilities that use chemicals. This standard is not applicable to workplaces that only use chemicals in the form of consumer products and the use and frequency of exposure in the workplace is the same as normal consumer use.

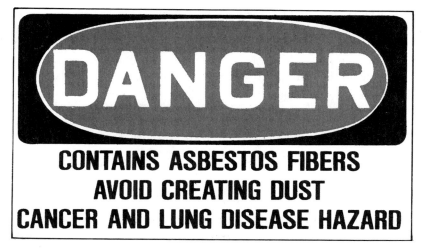

Source: With Permission from EMED Co., Buffalo, NY.

FIGURE 8.6   Example of a Warning Label for Asbestos-Containing Materials.

The Hazard Communication Standard requires employers to give workers adequate information about hazardous materials that they handle or use in the workplace. Specifically, employers are required to provide employees with information and training on hazards of chemicals used in the workplace at the time of their initial assignment and as new hazards are introduced into the workplace. This training includes information such as:

- Methods and observations to detect the presence or release of a hazardous chemical in the workplace, such as monitoring conducted by the employer, continuous monitoring devices, and visual appearance or odor of hazardous chemicals when being released
- Physical and health hazards of the chemicals in the workplace
- Protective measures required when being exposed to the chemical such as use of protective equipment, specified work practices, and emergency procedures
- Explanation of the labeling system, information and location of material safety data sheets (MSDS) or equivalent, and other information about the hazard communication program

Hazardous Waste Operations Training

Hazardous waste operations training is required under RCRA and OSHA for those employees who work with hazardous waste. RCRA regulations require that employees who work in a hazardous waste area must be trained on the hazards of the job and applicable regulations. In addition, the training program must be be designed to ensure that facility personnel are able to respond effectively to emergencies. Training should include information about emergency procedures, emergency equipment, and emergency systems. RCRA also requires that employers list all job positions that involve hazardous waste management and maintain a list of persons filling those positions. Employers must specify training requirements for each position and maintain personnel training records for each employee filling those positions until facility closure or for three years after the employee leaves that position.

The requirements under OSHA for training of workers in hazardous waste operations vary, depending on the job of the worker. Training is required for the following groups of workers:

- General site workers (such as equipment operators, general laborers, and supervisory personnel) engaged in hazardous substance removal or other activities that expose or potentially expose workers to hazardous substances and health hazardous—This group must receive a minimum 40 hours of instruction off site, and a minimum of three days actual field experience under the direct supervision of a trained, experienced supervisor.
- Workers on site only occasionally for a specific limited task (such as but not limited to groundwater monitoring, land surveying, and other tasks) and who are unlikely to be exposed over the permissible exposure limits or other published exposure limits—This group must receive a minimum of 24 hours of instruction off the site, and a minimum of one day actual field experience under the direction of a trained, experienced supervisor.
- Workers regularly on site who work in areas that have been monitored and fully characterized indicating that exposures are under permissible exposure limits or other published exposure limits, and the characterization indicates there are no health hazards or the possibility of an emergency developing—This group must receive a minimum of 24 hours of instruction off the site and a minimum of one day actual field experience under the direct supervision of a trained, experienced supervisor.
- On-site management and supervisors directly responsible for, or who supervise employees engaged in, hazardous waste operations—This group must receive a minimum of 40 hours initial training and three days of supervised field experience plus an additional eight hours of specialized training. This requirement may be reduced to 24 hours and one day for supervisors who are only responsible for employees who are not exposed to hazards over the permissible exposure limits or other published exposure limits.
- Employees who are engaged in emergency response at hazardous waste cleanup sites that may expose them to hazardous substances—This group must be trained in how to respond to such expected emergencies.

- Employees conducting operations at hazardous waste TSD facilities— This group must have 24 hours initial training to enable them to perform their assigned duties and functions in a safe and healthful manner so as to not endanger themselves or other employees.
- Employees who are engaged in emergency response at hazardous waste TSD sites—This group must be trained to a level of competence in recognition of health and safety hazards to protect themselves and other employees.

All workers and supervisors must be certified by the instructor as having successfully completed the training. Annual refresher training of at least eight hours is part of the training program for most groups. A summary of elements that might be covered during training of workers engaged in hazardous waste operations is presented in Table 8.2. Information about the incident command system used in emergency response and training required for different levels of responders (i.e., first responder, hazardous waste technician, hazardous materials specialist, and on-scene commander) is presented in Chapter 18.

Training for Process Safety Management of Highly Hazardous Chemicals

All employees who are involved in operating a process that is covered in 29 CFR §1910.119 must be trained before operating the process. The training should documented and include an overview of the process and applicable operating procedures. Emphasis must be placed on specific safety and health hazards, emergency operations including shutdown, and safe work practices applicable to the employee's job tasks. Refresher training must be provided at least once every three years to assure that the employee understands and adheres to the current operating procedures of the process.

DOT Training

DOT has outlined training for persons involved in the transportation of hazardous materials. The training requirements include general awareness training, function-specific training, safety training, and OSHA or EPA training, as applicable. Certification of training is necessary, and refersher training is required at least once every two years. If an employee changes hazardous materials job functions, that employee must be trained in the new job function within 90 days. The employee must be supervised by a properly trained employee until the training is completed. A record of training, inclusive of the last two years, must be created and maintained for each employee for as long as the employee works in a hazmat job and for 90 days thereafter.

Specialized Training

Specialized training is required for workers who handle radioactive materials and waste and for workers who remove asbestos from equipment and build-

TABLE 8.2 Summary of Elements that Might be Covered During Hazardous Waste Operations Training

General Overview

- Requirements under RCRA and OSHA
- Regulations pertaining to specific wastes such as asbestos, PCBs, benzene, and others (as applicable)
- Names of general supervisor, emergency coordinator, persons and alternates responsible for site safety and health, and other organizational information
- Lines of authority, responsibility, and communication

Hazards Review

- Comprehensive work plan
- Overview of hazards such as chemical exposures, biological exposures, radiological exposures, heat or cold, other
- MSDS information
- General safety hazards
- Medical symptoms that could indicate overexposure

Safety and Health Review

- Site-specific safety and health plan
- Site control plan
- Work practices that could exacerbate or minimize risk
- Safe use of engineering controls and equipment on site
- Air monitoring methods and equipment
- Confined space entry procedures and training
- Hazardous waste handling procedures, material handling procedures, and spill containment procedures
- Required personal protective equipment such as respirators and protective clothing and training on equipment use and inspection
- Required medical surveillance program
- Chemical data and other reference manuals available in the work area
- Hands-on review of work activities

Overview of Emergency Procedures

- Emergency and contingency plan review
- Location of and use of emergency equipment, automatic shutoffs, and emergency communication devices
- Hands-on exercises invoking plan and using emergency equipment

ings. Information about the contaminant, permissible exposure limits, personal protective equipment required, monitoring information, and other items are covered in the training.

INSPECTIONS

All TSD facilities must perform inspections as required under RCRA. These inspections must include documented inspection of all equipment on a scheduled basis, as written in the facility inspection plan and as required in various sections of the regulations.

Container Inspections

RCRA regulations require all hazardous waste management facilities to perform inspections of stored hazardous waste containers at least weekly. The containers must be visually inspected for leaks or corrosion.

Tank Inspections

Hazardous waste tank systems and associated spill containment must be inspected daily for system integrity. Inspection items should include:

- Inspection of aboveground portions of the tank system for cracks, leaks, or corrosion
- Review of data gathered from monitoring and leak detection equipment, such as pressure or temperature gauges and monitoring wells, to ensure that the tank system is being operated according to design
- Inspection of construction materials, secondary containment, and surrounding area for signs of hazardous waste releases, such as wet spots and dead vegetation

Other inspection items include the inspection of the cathodic protection system six months after installation and annually thereafter and bimonthly inspection of sources of impressed currents.

General operating requirements for hazardous waste tank systems include the use of appropriate controls and practices to prevent spills and overflows from the tank and containment system. These devices include:

- Spill prevention controls such as check valves and dry disconnect couplings
- Overfill prevention controls such as level sensing devices, high level alarms, automatic feed cutoff, or bypass to a standby tank
- Maintenance of sufficient freeboard in uncovered tanks to prevent overtopping by wave or wind action or by precipitation

Integrity testing and tank assessment by an independent, qualified, registered professional engineer is required for hazardous waste tank systems. The tank assessment should include information pertaining to structural integrity, corrosion potential, overfill and spill prevention controls, and freeboard, as applicable.

Release detection using one of several methods is required for underground storage tank systems containing petroleum products or CERCLA hazardous substances. Acceptable methods include inventory control methods, manual or automatic tank gauging, tank tightness testing, vapor monitoring, groundwater monitoring, and interstitial monitoring.

Inspection of Surface Impoundments

Surface impoundment liners must be inspected during and immediately after construction and installation. The surface impoundment itself, including dikes and vegetation surrounding the dike, must be inspected at least once a week to detect any leaks, deterioration, or failures in the impoundment. Structural integrity certification also is required.

Inspection of Incinerators

Hazardous waste incinerators have several monitoring and inspection requirements. These include:

- Continuous monitoring of combustion temperature, waste feed rate, and the indicator of combustion gas velocity specified in the permit while incinerating hazardous waste
- Monitoring of carbon monoxide on a continuous basis at a point in the incinerator downstream of the combustion zone and prior to release to the atmosphere while incinerating hazardous waste
- Upon request by the Regional Administrator, sampling and analysis of the waste and exhaust emissions to verify that the operating requirements established in the permit achieve the required performance standards
- Daily inspection of the incinerator and associated equipment such as pumps, valves, conveyors, and pipes for leaks, spills, fugitive emissions, and signs of tampering
- Weekly inspection of the emergency waste feed cutoff system and associated alarms to verify operability, unless the applicant demonstrates to the Regional Administrator that weekly inspections will unduly restrict or upset operations and that a less frequent inspection will be adequate
- Monthly operational testing

Inspection of Waste Piles, Drip Pads, and Miscellaneous Units

The amount of liquids removed from each leak detection system sump must be recorded at least weekly for waste piles and other weekly system integrity inspections are required. Drip pads must be inspected weekly and after storms to detect evidence of deterioration, malfunctions or improper operation of run-on and run-off control systems, the presence of leakage in and proper functioning of the leak detection system, and deterioration or cracking of the drip pad surface. Hazardous waste miscellaneous units must be inspected at a frequency that protects human health and the environment.

Inspection of Other Materials

Stored PCB waste containers are to be inspected at least once every 30 days for leaks or deterioration. PCB waste must be destroyed at a permitted disposal facility within one year of the waste's initial storage. All radioactive waste packaging and other materials associated with disposal of nuclear waste must be inspected according to an approved schedule specified in the facility's application for a radioactive waste disposal license.

Tanks and containers used for storage of production or process chemicals are presently not required by OSHA to be inspected on a specific schedule, although some state or local governments might require inspection of these storage units. It is a good management practice to inspect all types of tanks and containers that hold chemicals, even those not specifically regulated.

PLANS AND CONTROLS

Plans and controls for hazardous waste management facilities include a waste analysis plan, closure/post-closure plans, emergency and contingency plans, a groundwater monitoring plan, and financial assurance. Plans and controls for chemicals in the workplace include medical surveillance, emergency and contingency plans, and a hazard communication program. Facilities that store oil or petroleum distillates in aboveground and/or underground containers or tanks in specified quantities require a spill prevention, control, and countermeasure plan.

RCRA Plans and Controls

A waste analysis plan is required for all permitted facilities and includes a list of all wastes generated at a facility, the constituents of each waste, and procedures used in determining a waste's hazard. When hazardous, the RCRA classification must be included. The plan is to be updated whenever a new waste is added.

A facility closure plan is required for all facilities that are permitted as hazardous waste TSD facilities. This plan includes procedures and cost estimates for closing a facility, including waste disposal, decontamination, and an evaluation of post-closure care requirements. If post-closure care is required, a separate plan addressing these activities and costs is needed.

A contingency plan and emergency procedures also must be developed by a TSD facility. These documents should detail methods of minimizing hazards to human health or the environment from fires, explosions, or any unplanned releases. Evacuation routes should be included in the plan and the plan must be kept at the facility as well as provided to area police, fire department, hospitals, and state and local emergency response teams.

A groundwater monitoring plan for detecting any releases is required for hazardous waste surface impoundments, land treatment facilities, landfills, waste piles, and other regulated units, as required. The plan should detail the configuration of the monitoring system as well as parameters to be sampled and frequency of sampling and analysis.

Financial assurance is required of TSD facilities to provide proof that a facility can pay for or has insurance that covers any cleanup activity associated with a sudden or nonsudden release. RCRA also requires financial assurance for underground storage tanks that contain regulated substances. The financial assurance documentation must be submitted annually to EPA or the delegated state administrator.

OSHA Plans and Controls

Medical surveillance is required for employees who work in hazardous waste operations and emergency response, who are or may be exposed to hazardous substances or health hazards at or above the permissible exposure limits (disregarding the use of respirators) for 30 days or more per year, who wear a respirator for 30 days or more a year, or who are exposed to certain hazardous chemicals regulated by OSHA. Records of medical exams must include employee name and social security number, physicians written opinions, employee medical complaints, and other information. Any medical results that indicate overexposure to a chemical must be reported to the employee. Additionally, all employees have the right to review their medical records, upon request.

An emergency response plan for hazardous materials incidents must be established for facilities that have employees participating in emergency response activities. This plan should include:

- Pre-emergency planning
- Personnel roles, lines of authority, and communications
- Emergency recognition and prevention
- Safe distances and places of refuge
- Site security and control
- Evacuation routes and procedures
- Decontamination procedures
- Emergency medical and first aid procedures
- Emergency alerting and response procedures
- Critique of response and follow up
- Personal protective equipment and emergency equipment

An additional document detailing a facility's hazard communication program is required by facilities regulated under the Hazard Communication Standard. This document must describe chemical labeling practices, use of MSDS information or equivalent, the location of this information for employee access, and training. Other items required include a list of hazardous materials known to be present in the workplace, methods for informing employees of hazards of nonroutine tasks, and methods of informing contractor employers of hazards that their employees might be exposed to.

NRC Requirements

A comprehensive quality control program must be documented and approved by the NRC as part of the licensing program for low-level radioactive waste disposal facilities (NRC 1991). The program should address aspects such as:

- Organization structure
- Quality assurance activities
- Facility design control
- Procurement control
- Control of processes
- Inspections
- Control of testing
- Audits, surveillance
- Corrective actions

Records of locations of the waste at the facility must be maintained, and other administrative requirements apply.

CHEMICAL/WASTE TRACKING, REPORTING, AND RECORDKEEPING

There are several key tracking, reporting, and recordkeeping requirements associated with the management of hazardous materials. Most of these requirements are defined by EPA under RCRA, TSCA, SARA, CAA, and CWA. NRC also has requirements for radioactive waste tracking.

RCRA Requirements

Under RCRA regulations, all waste is tracked through a manifest. The manifest is a uniform, numbered form that includes information about a hazardous waste shipment, including quantities, waste codes, and generator, transporter and disposal facility information. An EPA manifest form is presented in Figure 8.7.

The receiving facility must send the generator a copy of the manifest to show receipt of the waste. Any discrepancies in weight or content must be noted by the receiving facility. Errors or discrepancies must be reconciled between the generator and receiving facility and documented to file. If discrepancies cannot be reconciled, a formal report must be sent to EPA or the delegated state authority. Additionally, if the receiving facility does not return

Please print or type *(Form designed for use on elite (12-pitch) typewriter.)* *Form Approved OMB No 2050-0039 Expires 9-30-91*

UNIFORM HAZARDOUS WASTE MANIFEST

1 Generator's US EPA ID No | Manifest Document No | 2 Page 1 of | Information in the shaded areas is not required by Federal law

3 Generator's Name and Mailing Address | A State Manifest Document Number | B State Generator's ID

4 Generator's Phone ( )

5 Transporter 1 Company Name | 6 US EPA ID Number | C State Transporter's ID | D Transporter's Phone

7 Transporter 2 Company Name | 8 US EPA ID Number | E State Transporter's ID | F Transporter's Phone

9 Designated Facility Name and Site Address | 10 US EPA ID Number | G State Facility's ID | H Facility's Phone

GENERATOR

11 US DOT Description *(Including Proper Shipping Name, Hazard Class, and ID Number)*	12 Containers No	12 Containers Type	13 Total Quantity	14 Unit Wt/Vol	I Waste No
a					
b					
c					
d					

J Additional Descriptions for Materials Listed Above | K Handling Codes for Wastes Listed Above

15 Special Handling Instructions and Additional Information

16 **GENERATOR'S CERTIFICATION:** I hereby declare that the contents of this consignment are fully and accurately described above by proper shipping name and are classified, packed, marked, and labeled, and are in all respects in proper condition for transport by highway according to applicable international and national government regulations.

If I am a large quantity generator, I certify that I have a program in place to reduce the volume and toxicity of waste generated to the degree I have determined to be economically practicable and that I have selected the practicable method of treatment, storage, or disposal currently available to me which minimizes the present and future threat to human health and the environment; **OR**, if I am a small quantity generator, I have made a good faith effort to minimize my waste generation and select the best waste management method that is available to me and that I can afford.

Printed/Typed Name | Signature | *Month Day Year*

TRANSPORTER

17 Transporter 1 Acknowledgement of Receipt of Materials

Printed/Typed Name | Signature | *Month Day Year*

18 Transporter 2 Acknowledgement of Receipt of Materials

Printed/Typed Name | Signature | *Month Day Year*

FACILITY

19 Discrepancy Indication Space

20 Facility Owner or Operator Certification of receipt of hazardous materials covered by this manifest except as noted in Item 19

Printed/Typed Name | Signature | *Month Day Year*

EPA Form 8700-22 (Rev. 9-88) Previous editions are obsolete

FIGURE 8.7 Example of an EPA Manifest.

a copy of the manifest to the generator within 45 days, the shipper must investigate the problem and file an exception notice with the agency. Under the regulations, the original manifest, returned copy, and any discrepancy reports must be kept at the facility for three years.

Hazardous waste regulations also require a large-quantity waste generating facility to submit a biennial (or, in some states, annual) report covering all hazardous waste activities. Additionally, waste minimization reports are required for most large-quantity generators.

Medical wastes were tracked for a two-year period in states participating in the Medical Waste Demonstration Program, which was set up in 1989 by EPA. The manifest was utilized for tracking these wastes, including infectious and other medical wastes. Transporters of medical wastes in the participating states were required to submit to EPA an annual report of amounts of medical waste shipped. During this timeframe, many states promulgated separate regulations that included medical waste tracking requirements. It is likely that federal regulations pertaining to medical waste tracking (and management) will be promulgated in the near future.

TSCA Requirements

Under TSCA, the manufacturer of any new chemical, as defined under the Act, is required to notify EPA of the intent to manufacture the chemical. Along with this premanufacture notice (PMN), scientific studies and other health and safety information pertaining to the hazards of the chemical must be submitted. Once submitted, the chemical receives a premanufacturing number and, when approved for manufacture, a Chemical Abstract System (CAS) Number. These numbers serve as tracking mechanisms. Additionally, once a chemical is manufactured and in use, any allegation from an employee or other person that the chemical is causing unexpected adverse reactions to health or the environment must be recorded by the facility and reported to EPA.

Asbestos abatement projects, including removal and enclosure or encapsulation of friable asbestos, are regulated under TSCA. Under the regulations, notification must be made to the Regional Asbestos Coordinator for the EPA region in which the asbestos abatement project is located. Asbestos abatement projects also are regulated under the CAA and by OSHA.

PCB waste tracking and recordkeeping requirements are specified under TSCA. Many of the requirements are similar to waste tracking requirements under RCRA. The generator, storer, transporter, and disposer of PCB waste must have an EPA identification number. The waste must be manifested, and any manifest discrepancy must be resolved within 15 days, or reported to EPA if resolution cannot be made. Additionally, if the generator does not receive the return copy of the manifest from the storer or disposer within the specified period of 35 days, then this must be reported to EPA.

In addition to manifesting requirements, TSCA regulations require the owner or operator of the disposal facility to prepare a certificate of disposal for each PCB shipment. This certificate is sent to the generator, and a copy is kept at the disposal facility. Additionally, storage and disposal facilities must maintain a written log on the disposition of all PCBs maintained at the facility.

SARA Requirements

Under Emergency Planning and Community Right-to-Know Act—also known as SARA Title III—facilities have several reporting obligations, which were reviewed briefly in Chapter 1. First, under Section 312 of SARA Title III, the facility must notify local and state governments if it stores on site any hazardous chemical in a quantity that exceeds the threshold quantity listed in the regulations. Additionally, the facility must submit either a Tier I or Tier II report to these same government entities. These reports detail chemical names (Tier II only), quantities, storage locations, associated hazards, and physical storage descriptions (i.e., type of container, pressure, temperature). This information is used for emergency planning purposes and citizen information. The submission must be updated annually.

Under Section 313 of SARA Title III, EPA has listed over 300 chemicals that the agency considers toxic. Any manufacturing facility that manufactures or process 25,000 pounds or more or otherwise uses 10,000 pounds or more of a listed chemical must file the toxic release inventory (TRI) report specifying certain release information.

Included in the report are quantities of emissions to air, wastewater, stormwater, on-site disposal facilities, transfers to off-site facilities, and other information. This report is part of the Community-Right-to-Know portion of the law and provides citizens with information about chemical releases in their immediate area. In addition, the report also provides regulators with a summary of toxic releases, which can be used to determine future regulatory focus. TRI information is available in an on-line data base accessible to the media and the public, as provided by EPA.

CAA Requirements

Notification to EPA or a delegated state authority is required 10 working days before an asbestos abatement project begins for jobs that contain friable asbestos and meet job size criteria. Information pertaining to project location, area to be abated, and date(s) of work, and other information is required. In addition, waste disposal sites accepting asbestos waste must maintain waste shipment records and meet other requirements.

Recordkeeping of sampling events, calculations of emissions, and annual reporting of these emissions is required for NESHAP chemicals. Specific sampling/reporting requirements vary from chemical to chemical.

CWA Requirements

Requirements for effluent discharge monitoring and reporting are defined in the NPDES permit. When the effluent is discharged to a publicly owned treatment works (POTW), pretreatment regulations and local requirements dictate monitoring and reporting frequency.

NRC Requirements

Transportation and tracking of radioactive material must conform to requirements in 10 CFR Part 71. Shipping papers, labeling, and packaging must meet NRC and DOT standards. Low-level radioactive waste must be accompanied by a manifest when sent to an off-site disposal facility. Transport of certain types of high-level and other dangerous radioactive waste and licensed materials outside the confines of the licensee's plant requires notification to the governors (or designees) of states through which the material will travel.

RELEASE AND OTHER NONCOMPLIANCE REPORTING

There are specific requirements for reporting accidental releases of chemicals or wastes under the various regulations. These include:

- CERCLA—Notification is required to National Response Center upon the nonpermitted release of a listed chemical in an amount that exceeds reporting threshold.
- RCRA—Notification is required to the state or local agency when their help is needed. Notification is required to the government official designated as on-scene coordinator for the geographical region (as applicable) or the National Response Center when the facility has had a release, fire, or explosion which could threaten human health or the environment. Notification to the delegated state authority is required, as written into state laws or policies. For underground storage tanks, any release must be reported to the EPA or designated state agency.
- SARA—Notification is required to the Local Emergency Planning Committee (LEPC) or their designee and the State Emergency Response Commission (SERC) upon the occurrence of a nonpermitted release of a listed chemical that exceeds the reporting threshold quantity and that leaves the facility's boundaries.
- DOT—Notification is required to the Department of Transportation if an incident occurs during the course of transportation (including loading, unloading, and temporary storage) that results in a death or injury, carrier damage exceeding $50,000, evacuation of the general public, spillage of radioactive material, the shutdown of a transportation artery, or other situations requiring notification. Notification is required to the Director, Centers for Disease Control, if the incident involves etiologic agents.

- CWA—Notification is required to the EPA Regional Administrator if a vessel or facility discharges in a single spill event reportable quantities of hazardous substances into or upon navigable waters of the United States or adjoining shorelines. Facilities that discharge wastewater must notify EPA or the delegated state authority of any activity that could result in a nonroutine or infrequent release of toxic pollutants not covered by an NPDES permit.
- CAA—Monthly reporting to the EPA Regional Administrator is required when a facility reports a noncompliance occurrence with NESHAP standards on the required annual report. Also, notification to the EPA Regional Administrator or delegated state authority is required for major upsets or releases that could threaten human health or the environment.
- NRC—Notification to the Nuclear Regulatory Commission of defects in packaging and other noncompliance issues is required by licensed radioactive waste disposal facilities.

Information required in the report includes such items as name and telephone number of reporter, name and address of facility, chemical/waste released, quantity release, possible hazards to human health or the environment, and the extent of injuries. For radioactive disposal facilities, the report will include information pertinent to the packaging defect or other noncompliance issue. In addition to these federal accidental release reporting requirements, individual permits may specify additional requirements for noncompliance reporting.

MANAGEMENT OF CHEMICAL AND WASTE ADMINISTRATIVE REQUIREMENTS THROUGH COMPUTER APPLICATIONS

Numerous computer programs or on-line data bases are currently marketed that can help in assuring compliance with chemical and waste administrative requirements. A comprehensive listing of computer applications is documented in *Environmental Software Directory* (see references).

Software applications that are marketed commercially and may be of value for meeting chemical and waste administrative requirements include:

- Regulatory requirements/updates for hazardous chemicals/hazardous waste management and compliance
- Hazardous substance release requirements, including notification, reporting quantities, and release calculations
- Report generation, including SARA Title III reports, manifests, and waste reports
- Technical data and exposure guidelines for hazardous substances
- Risk assessment and emergency planning

- Chemical/waste tracking and labeling
- MSDS management
- Waste exchange information
- Environmental data management
- Financial analysis
- Environmental auditing

Chemical information is available in diskette and/or on-line from several government sources including EPA, National Library of Medicine, National Oceanic and Atmospheric Administration, and the U.S. Army Corps of Engineer's Construction and Engineering Research Lab. These programs include TSCA management, toxicology and other chemical data, emergency response data, and established exposure limits.

In addition to purchasing software, a facility can design or have a vendor design a customized program to track chemicals from a control point such as the chemical distribution center to a specific location at a facility. Likewise, waste can be tracked from the generating department or location back to the control point. This can be achieved with a bar code scanning system or other container identification and numbering scheme.

Once designed and installed, the computer application can provide a retrievable data archive that can be maintained and updated easily. If hazard ratings and physical properties information is included on the same data base, the system can be expanded to provide labels and usage controls.

Examples of data that might be included for chemical and waste tracking and usage control are presented in Tables 8.3 and 8.4.

TABLE 8.3   Chemical/Waste Table

Part #	Description	Composition Component	Wt (%)	CAS #
315-100-001	Pumice	Pumice	98	1332-09-8
		Silica	2	Not Avail
315-100-017	Ammonia hydroxide solution	Ammonia hydroxide	12	1336-21-6
		Water	88	7732-18-5
		Inhibitor	Trace	Not Avail
614-003-005	TBDA-Tetramethyl butane diamine	TBDA	98	0097-84-7
		Amines	2	Not Avail
W315-100-001	Waste pumice (nonhazardous)	Pumice	98	1332-09-8
		Silica	2	Not Avail
WSA-095B-001	Rags/filters w/copper chloride (nonhazardous)	Copper chloride	2	1344-67-8
		Cloth media	98	Not Avail

Source: Adapted from Woodside, Stuckey, and Dalke (1991).

TABLE 8.4   Chemical Authorization Table

Dept #	Part #	Storage Quantity	Annual Usage	Unit of Meas	Storage Location
103	315-100-017	60	120	lb	C-13/16
103	315-100-316	25	100	gal	C-13/16
103	315-200-123	1	12	qt	C-13/16
103	W315-100-017	10	120	gal	D-07/06
155	315-100-001	4	50	gal	F-23/09
155	315-600-036	10	40	lb	F-23/09
155	WSA-095B-001	10	120	gal	D-07/08
168	614-003-005	10	120	gal	K-18/16
168	614-016-156	5	60	gal	K-18/16
168	315-400-007	10	40	lb	K-18/16
168	W614-003-005	5	35	lb	D-07/06

Source: Adapted from Woodside, Stuckey, and Dalke (1991).

As can be seen from the tables, chemicals and wastes can be given part numbers, with wastes being designated separately with a "W" in front of the number. Each department can then be authorized to store (order) only a specified amount of a chemical part number at any one time. An annual usage amount is also quantified. Similarly, for waste, the department may be authorized (based on engineering estimates) to generate certain wastes in specified amounts.

The system can be set up to tabulate quantities of chemicals and wastes dispensed to and received from each department. This type of chemical usage and waste generation history can be displayed by department or by part number, as shown in Tables 8.5 and 8.6.

If a department exceeds its annual chemical usage or waste generation for any part number, the application can be programmed to flag the discrepancy. Reasons for overages such as process problems or procedures not being followed can be investigated and appropriate solutions implemented.

Accurate tracking of hazardous materials and hazardous waste is important to programs such as chemical source reduction and waste minimization, which are discussed in Chapter 15. Although not required by the regulations, poster programs and other communication efforts can aid in employee awareness of waste reduction goals and safe chemical and waste handling procedures. These types of awareness campaigns can play an important role in the overall chemical and waste management program.

TABLE 8.5   Usage History by Department

Dept #	Part #	Actual Usage 1993	1992	1991	Unit of Measure
103	315-100-017	100	125	115	lb
103	315-100-316	125	90	105	gal
103	315-200-123	0	6	10	qt
103	W315-100-017	90	110	135	gal
155	315-100-001	35	45	60	gal
155	315-600-036	100	35	45	lb
155	WSA-095B-001	130	110	90	gal
168	614-003-005	65	95	150	gal
168	614-016-156	60	80	45	gal
168	315-400-007	35	40	40	lb
168	W614-003-005	35	45	20	lb

Source: Adapted from Woodside, Stuckey, and Dalke (1991).

TABLE 8.6   Usage History by Part Number

Part #	Dept #	Actual Usage 1993	1992	1991	Unit of Measure
315-100-017	103	100	125	115	lb
315-100-017	198	300	335	295	lb
315-100-017	246	25	30	35	lb
W315-100-017	377	10	10	10	lb
Total		435	500	455	lb
614-003-005	168	65	95	150	gal
614-003-005	178	120	125	120	gal
614-003-005	210	2245	2465	2590	gal
614-003-005	362	25	30	30	gal
Total		2455	2715	2890	gal
WSA-095B-001	155	130	110	90	gal
WSA-095B-001	182	110	125	115	gal
WSA-095B-001	250	480	520	590	gal
Total		720	755	795	gal

Source: Adapted from Woodside, Stuckey, and Dalke (1991).

CONCLUSION

As can be seen from the discussions in this chapter, administrative require-ments for managing hazardous materials and hazardous waste are extensive. Keeping track of all necessary documentation, including inspections, closure documentation, notification of activities to various agencies, training records, and other required documents can be a complex task. One way to manage the many administrative requirements is to have one person (or group of persons if the facility is large) at the facility be the focal point for all chemical and waste management documentation. That person should collect, inspect for ac-curacy, and maintain all required documents in a central file. This allows for easy document retrieval during an agency inspection, as well as assuring that all documents exist and are in proper form.

In addition to administrative requirements, agencies also have specified technology requirements for management of hazardous materials. The next section of the book discusses acceptable technology designs and operational guidelines for storage of hazardous materials and storage, treatment, and dis-posal of hazardous waste.

REFERENCES

Donley, Elizabeth, *Environmental Software Directory*, Donley Technology, Garrison-ville, VA, 1991.

EMED, "EMED Co, Inc., Catalog 24," EMED Graphic Communications, Buffalo, NY, 1993.

Nuclear Regulatory Commission, "Quality Assurance Guidance for a Low-Level Radioactive Waste Disposal Facility," Washington, D.C., 1991.

Woodside, Gayle, Mark Stuckey, and David Dalke, "Computerized Environmental Controls: Techniques for Streamlining Chemical and Waste Requirements," pre-sented at AIChE Spring National Meeting, Houston, TX, 1991.

BIBLIOGRAPHY

BNA (1989), *Spill Reporting Procedures Guide*, The Bureau of National Affairs, Wash-ington, D.C.

Berger, Donavee A. and Christopher Harris (1990), *The SARA Title III Compliance Handbook*, Executive Enterprises Publications, New York.

Blattner, J. Wray (1992), *The Clean Air Act Compliance Handbook*, 2nd ed., Executive Enterprises Publications, New York.

Chrismon, Randolph L., (1989), *The TSCA Compliance Handbook*, Executive Enter-prises Publications, New York.

CMA (1987), "NPDES Discharge Permitting and Compliance Issues Manual," Chem-ical Manufacturers Association, Washington, D.C.

_____ (1987), "Understanding Title III: Emergency Planning and Community Right-to-Know Videotape," Chemical Manufacturers Association, Washington, D.C.

_____ (1989), "Overview of Resource Conservation and Recovery Act Videotape," Chemical Manufacturers Association, Washington, D.C.

Frye, Russell S. (1988) *The Clean Water Act Compliance Handbook*, Executive Enterprises Publications, New York.

Government Institutes (1986), *Clean Water Act Compliance/Enforcement Guidance Manual*, Government Institutes, Rockville, MD.

_____ (1989), *Hazardous Waste Manifests Videotape*, Government Institutes, Rockville, MD.

_____ (1989), *RCRA Inspection Manual*, 2nd ed., Government Institutes, Rockville, MD.

_____ (1991), *OSHA Field Operations Manual*, 4th ed., Government Institutes, Rockville, MD.

_____ (1992), *Environmental Reporting and Recordkeeping Requirements*, 2nd ed., Government Institutes, Rockville, MD.

Jones, et al. (1989), *Occupational Hygiene Management Guide*, Lewis Publishers, Boca Raton, FL.

Kaufman (1990), *Waste Disposal in Academic Institutions*, Lewis Publishers, Boca Raton, FL.

Keith, Lawrence H., and Douglas B. Walters (1992), *The National Toxicology Program's Chemical Data Compendium*, Vol. 8, "Shipping Classifications and Regulations," Lewis Publishers, Boca Raton, FL.

Lowry (1988), *Lowry's Handbook of Right-to-Know and Emergency Planning/SARA Title III*, Lewis Publishers, Boca Raton, FL.

Neizel, Charlotte L. (1991), *The RCRA Compliance Handbook*, Executive Enterprises Publications, New York.

Phifer (1988), *Handbook of Hazardous Waste Management for Small Quantity Generators*, Lewis Publishers, Boca Raton, FL.

Stensvaag, John-Mark (1991), *Clean Air Act 1990 Amendments: Law and Practice*, John Wiley & Sons, New York.

Stensvaag, John-Mark (1989), *Hazardous Waste Law and Practice* Vol. 2, John Wiley & Sons, New York.

U.S. EPA (1986), "EPA Guide for Infectious Waste Management," EPA/530-SW-86-014, Office of Solid Waste, U.S. Environmental Protection Agency, Washington, D.C.

_____ (1989), "Hospital Incinerator Operator Training Course," EPA-450/3-89/003," Vols. 1–3, U.S. Environmental Protection Agency, Research Triangle Park, NC.

Waldo, Andrew B., and Richard deC. Hinds (1991), *Chemical Hazardous Communication Guidebook—OSHA, EPA, and DOT Requirements*, 2nd ed., Executive Enterprises Publications, New York.

Woodyard, John P., and James J. King (1992), *PCB Management Handbook*, 2nd ed., Executive Enterprises Publications, New York.

SECTION III

THE TECHNOLOGY OF MANAGING HAZARDOUS MATERIALS AND HAZARDOUS WASTE

9

TANK SYSTEMS

Any permitted hazardous waste tank system must meet design requirements defined under RCRA in 40 CFR Part 264 Subpart J. Design criteria were extensively revised in 1986 and included the requirement for secondary containment and other technology requirements. Regulation of underground storage tanks containing petroleum products and hazardous substances listed under CERCLA was initiated by EPA as a result of the Hazardous and Solid Waste Amendments (HSWA) of 1984. Regulations pertaining to these units are defined under 40 CFR Part 280. Aboveground process material and product tank systems and design criteria are not regulated under RCRA, but may be regulated by other federal, state, or local regulations.

Required or recommended design standards for aboveground and underground hazardous waste and hazardous materials tank systems include design standards not only for the tank, but also for piping and ancillary equipment such as spill and overfill prevention equipment. This chapter discusses design elements for aboveground and underground tank systems. Additionally, testing methodologies for corrosion resistance and tank integrity testing are provided, and information about tank materials and chemical compatibilities is presented.

BASIC DESIGN CONSIDERATIONS

Basic considerations in designing a tank system include material selection, structural support, operational pressure, corrosion protection, and overfill and spill prevention. Material selection is based on compatibility with wastes or process or product materials. A general overview of materials and chemical compatibilities is provided later in the chapter.

Structural support considerations include soil load bearing data, piling requirements, foundation properties, tank wall thickness requirements, integrity of seams (particularly on raised systems), and support of ancillary piping and connections. Other considerations affecting structural design include climatic conditions such as earthquake, hurricane, or flood risks and extremes of temperature. For underground tank systems, considerations include the weight of vehicular traffic, weight of structures above tanks, subsurface saturation characteristics, depth to groundwater, and depth of freeze zone.

The tank system should include pressure controls supplemented with devices such as rupture disks, pressure relief valves, and automatic shutoff devices at preset pressure changes to protect the tank against overpressurizing. Liquid level controls for load/unload operations should include a preset high level shutoff to prevent overfill and vacuum relief valves to protect the tank in negative pressure situations. Alarms for these and other abnormal operating conditions are also appropriate.

Factors affecting the potential for corrosion of an underground tank system are varied. These include soil properties such as pH, moisture content, sulfides level, and the presence of metal salts. Additionally, corrosion rates can be affected by other metal structures and interference (stray) electric current. Corrosion can occur inside the tank if the tank metallurgy is not adequate for the material stored.

Types of Aboveground Storage Tanks

There are several broad categories of aboveground storage tanks including atmospheric tanks, low pressure tanks, pressure tanks, and cryogenic tanks. Atmospheric tanks operate at atmospheric pressure, although some field applications operating at pressures less than 0.5 pounds per square inch gauge (psig) are referred to as atmospheric applications. Low pressure tanks can operate at pressures ranging from 0.5 to 15 psig. Pressure tanks operate at pressures greater than 15 psig.

Tank designs for atmospheric tanks include the open top tank, fixed roof tank, and the floating roof tank. The open top tank is often used in wastewater treatment applications for equalization or neutralization. The fixed roof tank is used for storing materials with low vapor pressure, such as methanol, ethanol, and kerosene. Floating roof tanks are used mainly for petroleum products such as crude oil and gasoline. The floating roof design reduces evaporative losses while providing increased fire protection.

Low pressure tank designs include hemispheroid and noded hemispheroid tanks, and spheroid tanks. Low pressure tanks are used for storage of volatile materials, which are gases at ambient temperature and pressure, such as pentane. Additionally, combustible chemicals such as benzene and large volume volatile liquids, such as butane, can be stored under low pressure.

Pressure tanks are typically cylindrical and are supported vertically or horizontally. These tanks are used to store high volatility materials such as liquified petroleum gas.

Cryogenic tanks are made of specialty steels with rigid low temperature specifications and are heavily insulated. Examples of materials stored in this type of tank include ammonia, liquid nitrogen, liquid propane, and other compounds that are gaseous at atmospheric temperature and pressure. Examples of several types of tanks are presented in Figures 9.1 through 9.4.

Basic Design Considerations for Underground Storage Tanks

Underground storage tanks typically are used for storing high volume flammable materials such as gasoline, diesel fuel, and heating oils. These tanks normally range from 5,000 gallons to 20,000 gallons in storage capacity and are horizontal cylindrical tanks. Underground storage tanks are especially suitable for flammable materials since diurnal temperature changes and tank breathing losses are minimized in underground applications. Other applications for underground tanks include storage of flammable product or process chemicals.

An integral part of underground storage tank design is the use of cathodic protection for prevention of tank surface corrosion. Guidelines for cathodic protection for underground tank systems are published by several standards organizations and include:

- American Petroleum Institute (API) Publication 1632 (1987)—Cathodic Protection of Underground Petroleum Storage Tanks and Piping Systems, 2nd ed.
- National Association of Corrosion Engineers (NACE) Standard RP0285 (1985)—Control of External Corrosion on Metallic Buried, Partially Buried, or Submerged Liquid Storage Systems
- NACE Publication 2M363 (1963)—Recommended Practice for Cathodic Protection
- Underwriters Laboratories (UL) Standard 1746 (1989)—Corrosion Protection Systems for Underground Storage Tanks

These guidelines address design considerations, coatings, cathodic protection voltage criteria, and installation and testing of cathodic protection systems.

Corrosion is an electrochemical phenomenon. In an electrochemical cell, the positive terminal—the anode—undergoes an oxidation reaction, while the negative terminal—the cathode—undergoes a reduction reaction. Corrosion is an oxidation reaction. Cathodic protection is a standard engineering method used to prevent external corrosion on the surface of a buried, partially buried, or submerged metal tank by allowing the metal surface to become the cathode of the electrochemical cell. A protective array of anodes is set in the soil at calculated distances from the tank to protect the tank against corrosion.

Two recognized types of cathodic protection systems are galvanic anode systems and impressed current anode systems. Galvanic anodes usually are made of magnesium or zinc and, depending on soil properties, may require

FIGURE 9.1 Example of a Double-Contained Open Atmospheric Tank.

FIGURE 9.2 Example of Pressurized Tanks.

FIGURE 9.3 Example of a Floating Roof Fuel Oil Tank with a Dome Cover.

FIGURE 9.4 Example of a Cryogenic Tank.

use of special backfill material such as mixtures of gypsum, bentonite, and sodium sulfate (NACE, RP0285 1985). The anodes are made of a less noble metal than the tank so that they will be more susceptible to undergoing an oxidation reaction. Once installed, the system provides sacrificial protection to the tank by inducing a continuous current source to inhibit electrical corrosion of the tank. Galvanic anode cathodic protection systems are limited in electrical current output. Thus, applications for these systems are limited to tanks well-insulated by a nonconductive coating that minimizes the exposed surface area of the tank.

Impressed current anode systems are made of higher grade materials such as graphite, platinum, and steel. As with galvanic anode systems, soil properties may require the use of special backfill material such as coke breeze and calcined petroleum coke (NACE, RP0285 1985). A direct current source is used in this system. The anodes are connected to a positive terminal of the power source, while the structure is connected to the negative terminal.

A survey of a cathodic protection system to verify proper operation includes measurements such as structure to soil potential, anode current, structure to structure potential, piping to tank isolation (if protected separately), or effect on adjacent structures. A regulated system must be tested within six months after it is installed and at least every three years thereafter. Some state agencies and standards organizations require or recommend more frequent testing.

Underground tanks for petroleum products are often constructed of carbon steel that has been coated at the factory. Desirable characteristics of a coating include resistance to deterioration when exposed to products stored in the tank, high dielectric resistance, resistance to moisture transfer and penetration, and others. Testing using ASTM Standard methods for determining cathodic disbonding of coatings can be performed on coated tanks and piping. These methods are presented in Appendix C. In addition to tests for disbonding of the coating, the coating should be tested for pinholes or other defects. Further, precaution must be taken by the construction personnel to ensure that the protective coat is not damaged during tank installation.

REGULATORY STANDARDS AND OTHER DESIGN CONCEPTS

Regulatory standards apply to hazardous waste tank systems and underground tank systems that store petroleum products and CERCLA hazardous substances. These standards and other design concepts that maximize environmental protection are described in this section.

Hazardous Waste Tank Systems

New tank systems for hazardous waste storage, including aboveground and underground tank systems, must be designed and installed to meet current regulatory requirements. These requirements include structural integrity assessment, inspection of installation, backfill material specifications, tightness testing, proper support of ancillary piping, corrosion protection, certification statements by tank system experts, and containment and detection systems that meet design criteria listed in the regulations.

Some of the requirements for containment and detection systems are:

- The system must be designed, installed, and operated to prevent any migration of wastes or accumulated liquid out of the system and into the environment.
- The system must be capable of detecting and collecting releases and accumulated liquids until the collected material is removed.
- The secondary containment for tanks must include either a liner external to the tank, a vault, a double-walled tank, or an equivalent device.
- A liner or vault must have enough capacity to contain the contents of the largest tank when full plus precipitation from a 25-year, 24-hour rainfall event.
- Liners, vault systems, and double-walled tanks must meet specified design criteria.
- Ancillary equipment must be provided with full secondary containment such as a trench, jacketing, double-walled piping, or other containment method that prevents migration of wastes into the environment and provides leak detection. Some exceptions apply to this requirement (i.e., if the aboveground piping is inspected daily and piping systems are equipped with automatic shutoff devices).

Any existing tank system that fails must be retrofitted to meet the established standards before the tank can be put back into service. Other existing tanks must be retrofitted to the standards by the time the tank reaches 15 years of age.

Design Concepts for Aboveground Tank Systems

In addition to RCRA requirements, there are several other design concepts that can be incorporated into any new aboveground double containment project to maximize environmental protection (Langlois, Bauer, and Woodside 1990). These include:

- Maximum inspectability/testability
- Greater than 100% capacity of secondary containment for indoor applications as well as outdoor applications
- Specialized leak detection
- Elimination of penetrations through containment

These concepts are not detailed specifically in current regulations, but are good management practices where economically achievable.

Inspectability/Testability. Installing tanks and ancillary piping aboveground is the first step toward making a tank system inspectable and testable. Raising the primary tank on beams or saddles inside the secondary containment allows for daily inspection under the tank. If the secondary containment is also raised, the system becomes 100% inspectable.

Likewise, if ancillary piping is run aboveground whenever possible, leaking joints or material failures can be easily detected. For cases where pipe in a pipe is installed, ports with leak detectors and drain valves can be installed every 50 to 200 feet for assessment of primary pipe integrity.

Tanks and piping can be tested for leaks using the methods described in the next section of this chapter. A good engineering practice is to set up a schedule for periodic testing of all tanks at the facility. Length of time between tests will vary depending upon application and materials of construction.

Greater Than 100% Capacity of the Secondary Containment. There are required RCRA standards for greater than 100% capacity of the secondary containment for outdoor liners and vaults used as secondary containment. In addition to these requirements, environmental protection can be maximized if indoor secondary containment is sized for catastrophic failure of the primary unit, plus spillage from ancillary piping. A capacity of 125% containment typically will be adequate. Use of this design concept will negate the possibility of spilled materials overflowing secondary containment and penetrating through the building floor.

Specialized Leak Detection. Standard leak detection systems, such as electronic moisture sensors or sensor floats, are used widely in double containment projects. Although usually adequate for indoor applications, outdoor leak detection can be maximized with the use of specialized systems that differentiate chemical leakage from rainwater. Examples of these devices are detectors that alarm based on pH, oxidation/reduction potential, or conductivity.

Elimination of Penetrations through Secondary Containment. Joints and penetrations are usually the weakest points in secondary containment, and are the most likely to leak should a catastrophic failure occur. Locating pumps inside the secondary containment is one way to minimize penetrations. Additionally, overhead trestles can be used to route piping over, instead of through, secondary tank walls.

Optimum Containment For Tanks, Pipes, and Pumps. As discussed earlier, optimum systems include inspectable/testable primary and secondary containment. An example of this type of tank system is a tank in a tank, with both tanks raised on beams. A tank in a pan, with tank and pan raised on beams is another option. These two types of tank systems are shown in Figures 9.5 and 9.6. Maximum environmental protection is provided by double-contained piping options including pipe-in-a-pipe on an overhead trestle, as shown in Figure 9.7. Another piping option is pipe on a rack inside a pan, building, or trench.

FIGURE 9.5   Example of Tank in a Tank Double Containment.

FIGURE 9.6   Example of Tank in a Pan Double Containment.

FIGURE 9.7   Example of Double-Contained Pipe on a Trestle.

As mentioned previously, pumps placed inside containment is the optimum design. When this design is not practical, pump drip pans can be used and are particularly useful for smaller applications. Drip pans are not adequate, however, for failures of pressurized pipes or other major failures such as diaphragm tears or broken seals. They are also less useful for outdoor applications, since rainwater must be drained or absorbed after each rain event.

Double Containment for Portable Chemical and Waste Containers.   Containment of portable chemical and waste containers during storage and when in use is becoming a widespread practice in industry. Containment options vary widely, but 100% inspectable/testable secondary containment typically is not achievable. For most drum storage areas, such as container storage rooms and chemical distribution centers, curbed and coated concrete can be used. For these applications, a coating should be selected not only for chemical compatibility, but also for its ability to withstand fork truck and other equipment loadings. Physical inspections of the coating should be made regularly to ensure that cracking or delaminating has not occurred.

Smaller staging areas for chemical and waste containers can utilize bermed and coated concrete or a containment pan. For these applications, fork truck loadings are not usually a factor. The containment area should be large enough to contain, at a minimum, spillage from the largest container stored.

Routine barrel pumping operations should be contained using, as a minimum, a drip pan under the barrel and, preferably, a pan sized to contain at least the contents of one barrel. Relatively inexpensive, open-grated portable containment pallets are marketed that hold up to four drums and have enough spill capacity to easily contain a drum failure. These pallets can be moved with a fork truck. Another portable containment unit that holds one or two drums and is moved easily from one job site to another is also available.

Finally, specially designed, very durable, portable chemical containers that hold approximately 220 to 440 gallons (4 to 8 drums) are available. This type of container is as shown in Figure 9.8. Although spill containment *per se* is not

FIGURE 9.8   Example of Portable Containers that Hold 220 to 440 Gallons.

part of the container design, some containers are enclosed with a metal casing to ensure that dropping or excessive bumping does not crack the container. This design can reduce the risk of spills during transport.

Underground Storage Tanks

RCRA Requirements for Underground Storage Tanks Containing Regulated Materials.   The problem of leakage from underground storage tanks (USTs) was addressed in the 1980s, and regulations were promulgated to identify leaking tanks, initiate cleanup activities, and set forth design standards for new USTs. RCRA performance standards for new USTs include the following:

- Tanks must be properly designed and constructed, and any portion underground that routinely contains product must be protected from corrosion.
- Piping that routinely contains regulated substances and is in contact with the ground must be properly designed, constructed, and protected from corrosion.
- Spill and overfill prevention equipment associated with product transfer to the UST must be designed to prevent a release of the product to the environment during transfers (including use of catch basins for hoses, automatic shutoffs, flow restriction, alarms, and other equipment).
- All tanks and piping must be properly installed in accordance with a nationally recognized code of practice.
- The installation must be performed by a certified installer and completed properly.

Standards for proper design and installation of underground tanks and piping are documented in several standards including:

- API RP 1615 (1987)—Installation of Underground Petroleum Storage Systems, 4th ed.
- American Society of Testing and Materials (ASTM) D 4021-86—Standard Specification for Glass-Fiber-Reinforced Polyester Underground Petroleum Storage Tanks
- UL Standard 1316—Standard for Glass-Fiber-Reinforced Plastic Underground Storage Tanks for Petroleum Products

In addition to new tank performance standards, there are requirements for existing tanks. These include monthly release monitoring of tanks using an approved release detection method and monitoring of piping on a specified basis, depending on the pipe type and existing leak detection equipment in the pipe. Any leaks found must be reported immediately and corrective action must be taken to prevent further leakage and to remediate the site. Existing, nonleaking USTs that do not meet the performance standards must be upgraded to meet these or alternative standards no later than December 22, 1998.

Additional requirements are defined for new UST systems that contain CERCLA hazardous substances. These include secondary containment performance requirements that will prevent the release of these substances to the environment at any time during the operational life of the UST system.

Maximizing Environmental Protection.   EPA's guidance manual entitled *Detecting Leaks: Successful Methods Step by Step* (see references) provides information on design concepts that maximize environmental protection for USTs. Information is provided on acceptable types of secondary containment for underground tanks, interstitial monitoring, and underground piping. Additionally, release detection methods are addressed.

Secondary Containment.   Secondary containment systems should be designed to provide an outer barrier between the tank and the soil and backfill material that is capable of holding the material long enough so that a release can be detected. For tank systems containing hazardous substances, preventing migration of substances is also a requirement. A double-walled tank, a jacketed tank, and fully-enclosed external liners can provide this type of protection. In addition, the secondary containment must have enough capacity to hold the contents of the tank (or the largest tank if more than one tank is included in the system), and should prevent precipitation or groundwater intrusion. Examples of these types of secondary containment systems are presented in Figure 9.9.

To maximize environmental protection for double-walled tanks, the second tank should be made of the same material as the primary tank. For jacketed tanks or tanks with liners, the material used should be compatible with the product stored. The material also should be sufficiently thick and impermeable to direct a release to the monitoring point and permit its detection. The permeation rate of the stored substance through the jacket or liner should be 10^{-6} centimeters per second (cm/sec) or less.

Interstitial Monitoring.   An interstitial monitoring system should be designed to detect any leak from the underground tank under normal operating conditions. In most cases, the leak detection system does not measure the leak rate, but only the presence of a leak. Interstitial monitoring systems operate to detect leaks based on one of several mechanisms such as electrical conductivity, pressure sensing, fluid sensing, hydrostatic monitoring, manual inspection, and vapor monitoring. Conductivity, fluid sensing, manual inspection, and vapor monitoring are suitable for all applications. Pressure sensing and hydrostatic monitoring are applicable only to double-walled installations.

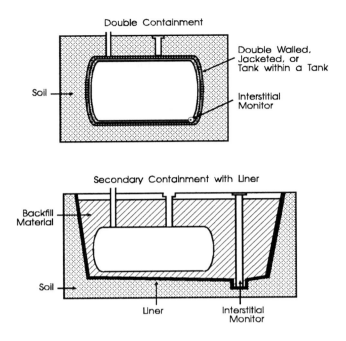

Source: Adapted from EPA (1989).

FIGURE 9.9 Examples of Secondary Containment Systems for Underground Tanks.

Because of the inability to physically inspect tanks that are underground, the interstitial monitoring system becomes the primary means of verifying tank integrity. For this reason, potential problems with the monitoring system should be identified before installation and reflected in the system's design. These potential problems might include groundwater or aboveground run-off penetration into the secondary containment from pinhole leaks in the secondary containment system or from inadequate lining. Further, faulty electrical installation can render some monitoring devices inoperable.

Other Release Detection Methods.   For existing UST systems that do not have interstitial monitoring, other release detection methods are required. These methods include inventory control, vapor monitoring, groundwater monitoring. manual or automatic tank gauging, and tank tightness testing using volumetric tests.

Product inventory control is a technique effective in finding leaks over one gallon per hour. Because of its low sensitivity, it must be combined with tank tightness testing, with the tank tightness testing occurring at least every five years until the system is retrofitted to performance standards. In general, this method requires careful tracking of inputs, withdrawals, and amounts still remaining in the tank. API RP 1621, "Recommended Practice for Bulk Liquid Stock Control at Retail Outlets," (see references) provides guidance for meeting inventory control requirements.

Vapor monitoring is used predominantly with USTs that store petroleum products. Success of this type of release detection system is dependent on site geology and soil types. Vapor monitoring systems consist of a vapor monitoring well and a vapor sensor. If a product leaks from an UST, the liquid and vapors spread throughout the surrounding soil. If the vapor sample reaches a sensor in a concentration above a predetermined set point, the system responds with an alarm.

Groundwater monitoring is a release detection method effective for petroleum and other products that have a specific gravity of less than 1.0. The method can be used in areas where the groundwater table is very near to the excavation zone of the tank. Permanent observation wells are placed close to the tank and the wells are checked periodically for evidence of free phase product on top of the water in the well.

Manual and automatic tank gauging and tank tightness testing using volumetric tests are also acceptable methods of leak detection. These methods are described in the next section of this chapter.

Design of Underground Piping.   Double-contained piping with interstitial monitoring maximizes environmental protection in underground systems and is required for uses with hazardous substances. For pressurized lines, the use of automatic flow restrictors to restrict flow in the event of a leak and automatic flow shutoff devices to stop flow when a pressure drop occurs also provide environmental protection.

Underground double-contained piping options include the use of trench liners as well as double-walled pipe. For the trench liner application, the pipe trench is lined with a flexible membrane that is impervious to the stored product. Often, the liners are thermoplastic or polymeric sheets and are at least 50 mils thick. The trench is designed sloping away from the tank so that pipe leaks can be differentiated from tank leaks.

For applications of double-walled piping, the inner and outer pipes can be made of the same material or the outer pipe can be made of a cheaper material, such as fiberglass-reinforced plastic. Fiberglass-reinforced plastic is not used for highly pressurized applications, but if compatible with the product is functional in a nonpressurized or less pressurized state. Double-walled piping is usually sloped to a containment structure, sump, or observation well that can be monitored for the presence of liquids or vapors.

TANK TESTING METHODS

Testing of tanks for corrosion resistance on a periodic basis can indicate the need for repairs and can aid in averting catastrophic failure of the system through early warning of material thinning or fatigue. Likewise, tank integrity testing, if performed according to a preset periodic schedule, can allow for early detection of leaks or pending failure. This section discusses test methodologies used for corrosion resistance and tank integrity testing.

Testing for Corrosion Resistance

There are numerous testing procedures for determining corrosion resistance of common materials used in tank systems. Some of these tests may be suitable for field use to determine the extent of corrosion in existing equipment. Others are useful for material screening during the system design phase. Tank manufacturers usually perform a variety of corrosion resistance and other tests on a tank material and make the test data available to potential customers.

Testing for Pitting and Crevice Corrosion Resistance of Stainless Steels and Related Alloys.   This test is defined in ASTM G 48–76(1980). The ASTM standard describes two test methods—Method A, which is a total immersion ferric chloride test and Method B, which is a ferric chloride crevice test. Method A is used to determine the relative pitting resistance of stainless steels and other alloys, particularly those that have been heat treated or have had surface finishes applied. Method B is useful for determining both pitting and crevice corrosion resistance of these same metals.

Standard specimens are immersed in a 10% ferric chloride solution in both tests for approximately 72 hours. For crevice testing, the specimens are attached to TFE-fluorocarbon blocks with o-rings or rubber bands to form crevices at the points of contact. The test specimens are evaluated by visual inspection, measurement of pit and crevice depth, and weight loss.

Guide for Conducting Corrosion Coupon Tests in Plant Equipment.   This guide, ASTM G 4–84, defines a method for evaluating corrosion of engineering materials under the varying conditions present in actual service. The method typically is used to evaluate materials of construction for use in similar service or as replacement or modification materials.

Size and shape of test specimens is dependent on the specific test application. The duration of exposure is based on known rates of corrosion of the materials in use, or by the convenience by which plant operations can be interrupted to introduce and remove test specimens. Evaluation of the specimen includes microscopic inspection for etching, pitting, tarnishing, scaling, and other defects. The depth of the pits can be measured and number, size, and distribution noted. A metallographic examination for intergranular corrosion or stress-corrosion cracking also may be performed.

Standard Test Method for Chemical Resistance of Protective Linings.   This method is defined in ASTM C 868–85(1990). The standard defines a procedure for testing and evaluating the chemical resistance of a protective lining applied to steel or other metals. The liner is applied and cured, and then immersed in the service solution for six months. As necessary to simulate actual conditions, heat may be applied. Color, surface gloss, surface texture, and blisters are all visually evaluated before, at interim periods, and after the completion of the test. Additionally, lining thickness is measured before and after the test to quantify the chemical attack on or the dissolution of the lining material.

Standard Methods of Testing Vulcanizable Rubber Tank and Pipe Lining. This test method is documented in ASTM D 3491–85, which describes testing and evaluation of chemical resistance of vulcanizable rubber tank lining. A test specimen of a rubber component applied to a steel plate is immersed in the service solution and heated, if required. The duration of the test should be a minimum of six months, with inspections performed every month. At the end of the test period, a visual inspection is made for changes in surface texture, evidence of cracking, blistering, swelling, delaminating, or permeation. Additionally, substrate attack or corrosion such as rusting or metal darkening is noted. Because of the length of time required for this test, prescreening of lining materials prior to this test can be performed using ASTM D 471-79(1991), which is a more convenient method.

X-Ray Fluorescence. The X-ray fluorescence (XRF) test method for determining coating thickness is defined in ASTM A 754–79(1990). A radiation detector that can discriminate between the energy levels of all radiations is used to measure the thickness of a coating and substrate exposed to an intense beam of radiation generated by a radioisotope source or an X-ray tube. The combined interaction of the coating and substrate with the beam of radiation generates X-rays of well-defined energy, which are singularly characteristic of that element.

If XRF thickness testing is performed in the field, environmental factors such as temperature, humidity, and surface cleanliness must be taken into account. Other factors including specimen size, specimen uniformity, radiation source, and radiation detector also can affect the test results.

Other ASTM Standards. Other ASTM standards, guides, and practices that address corrosion testing are documented in Volume 3.02, "Wear and Erosion; Metal Corrosion," *Annual Book of ASTM Standards* (see references). Examples of these standards include:

- G 1-90—Recommended Practice for Preparing, Cleaning, and Evaluating Corrosion Test Specimens
- G 15-90—Terminology Relating to Corrosion and Corrosion Testing
- G 46-76(1986)—Recommended Practice for Examination and Evaluation of Pitting Corrosion
- G 50-76(1984)—Recommended Practice for Conducting Atmospheric Corrosion Tests on Metals

- G 78-89—Guide for Crevice Corrosion Testing of Iron Base and Nickel Base Stainless Alloys in Seawater and Other Chloride-Containing Aqueous Environments
- G 82-83(1989)—Guide for Development and Use of a Galvanic Series for Predicting Galvanic Corrosion Performance
- G 96-90—Practice for On-Line Monitoring of Corrosion in Plant Equipment (Electrical and Electrochemical Methods)
- G 104-89—Test Method for Assessing Galvanic Corrosion Caused by the Atmosphere

Additional ASTM standards pertaining to corrosion resistance and other tank testing are presented in Appendix C.

Tank Integrity Testing

Once a tank system is installed, periodic tank integrity testing is a necessary part of the system's operation and maintenance program. A guide for selection of a leak testing method for tank and material testing and other applications is presented in ASTM E 432-91. Additionally, commonly used nondestructive test methods pertaining to tank testing are addressed in the following (see references):

- ASTM, *Annual Book of ASTM Standards*, "Nondestructive Testing," Vol. 3.03.
- American Society of Mechanical Engineers (ASME), *ASME Boiler and Pressure Vessel Code*, Section V, "Non-destructive Testing."
- API, *Design and Construction of Large, Welded, Low-Pressure Storage Tanks*, Standard 620, 8th ed.
- EPA, *Detecting Leaks: Successful Methods Step by Step*, EPA/530/UST-89/012.

Most methods for tank system testing are applicable to aboveground tanks. Tests applicable to underground tanks are limited because of inaccessibility of the tank.

Holiday Test.   The holiday test, defined in ASTM G 62-87, is used to detect pinholes, voids, or small faults that allow current drainage through protective coatings on steel pipe or polymeric precoated corrugated steel pipe. Although this test method defines the holiday test for pipeline coatings, it is also applicable to the walls and bottom of a coated tank. A highly sensitive electrical device is used in conjunction with water or another electrically conductive wetting agent to locate pinholes and thin spots (defined as holidays) in coatings of steel pipes and tanks. If electrical contact is made on the metal surface through a holiday, an alarm is activated to alert the operator of the coating flaw.

There are two methods defined for performing the test, and the thickness of the coating determines which method should apply. For thin-film coatings of 1 to 20 mils, a low-voltage holiday detector is used, which has an electrical energy source of less than 100 V d-c. This method detects pinholes and other voids, but will not detect thin spots in the coating. The high-voltage detector has an electrical energy source of 900 to 20 000 V d-c, and can be used to detect holidays and thin spots in the coating. The high-voltage detector normally is used on materials that have a coating thickness of greater than 20 mils.

Magnetic Particle Testing. Magnetic particle testing is used to detect cracks and other discontinuities near the surface of ferromagnetic materials. Applications include tank walls, tank bottom, and welds. The area to be tested is cleaned and then magnetized. Magnetic particles are applied to the surface. The particles form patterns where there are disturbances in the normal magnetic field and indicate cracks or flaws. The method is sensitive to very small discontinuities. ASTM standards which relate to this method include:

- A 275/A 275M-90—Method for Magnetic Particle Examination of Steel Forgings
- E 125-63(1985)—Reference Photographs for Magnetic Particle Indications on Ferrous Castings
- E 709-80(1985)—Practice for Magnetic Particle Examination

Ultrasonics Testing. Ultrasonics testing is defined in ASTM E 1002-86 and in the ASME Code, Section V, Article 5. This test can be used to locate pressurized gas leaks and estimate leak rates. In general, this test is considered a screening tool to be used prior to other more sensitive and time consuming tests.

This test method uses an acoustic leak detection system to detect impulsive signals that are much larger than background noise level. The ultrasonic test system provides for detection of acoustic energy in the ultrasonic range and translates energy into an audible signal that can be heard by use of speakers or earphones. The detected energy is indicated on a meter readout. Leak rates can be approximated from a formula that uses the maximum detection distance at calibrated sensitivity.

Acoustic Emission Testing. This test method can be used to monitor vessels and piping during operation to detect defects such as flaws and cracks. The tank or pipe is put under pressure, which results in a stress concentration, causing the defect to enlarge. This enlargement generates sound vibrations that can be detected using sensors located along the surface of the structure. Arrival times at the sensors are used to pinpoint the defect. In order to properly evaluate the acoustical vibrations, the tank should be in a quiescent state.

ASTM standards that pertain to this test include:

- E 750-88—Practice for Measuring Operating Characteristics of Acoustic Emission Instrumentation
- E 976-84(1988)—Guide for Determining the Reproducibility of Acoustic Emission Sensor Response
- E 1067-89—Practice for Acoustic Emission Testing of Fiberglass reinforces Plastic Resin (FRP) Tanks/Vessels
- E 1139-87—Practice for Continuous Monitoring of Acoustic Emission from Metal Pressure Boundaries
- E 1211-87—Practice for Leak Detection and Location Using Surface-Mounted Acoustic Emission Sensors

Hydrostatic Leak Testing.   Hydrostatic testing, documented in ASTM E 1003–84(1990) and API Standard 620, is a method for testing tanks with pressurized liquid. The method requires that a component be filled completely with a liquid, preferably water. Pressure is applied slowly to the liquid until the required pressure, usually between 75% and 150% of the designed operating pressure, is reached. The pressure is held for a designated period of time. Leakage can be determined by visual inspection or pressure drop indication.

Since liquid may clog small leaks, this method of testing is performed after pneumatic testing. Additionally, the test liquid temperature must be equal to or above ambient temperature or condensation can form on the outside of the tank, making visual leak detection difficult.

Pneumatic Pressure Testing.   Pneumatic pressure testing is described fully in the ASME Code. The method outlines a procedure for testing tanks and metal piping with air pressure. The test is not appropriate for fiberglass or other low-pressure materials. The empty tank or pipe is pressurized to the operating design pressure and held for a specified period of time, during which any drop in pressure is noted. Since air at high pressure can be an explosive hazard if a contaminant such as oil vapor is present, inspections should be made at a reasonable distance from the tank, using field glasses, as required.

Liquid Penetrant Testing.   Liquid penetrant test methods and practices are described in several ASTM standards and in the ASME Code, Section V, Article 6. ASTM standards include:

- E 165–91—Practice for Liquid Penetrant Inspection Method
- E 433–71(1985)—Reference Photographs for Liquid Penetrant Inspection
- E 1208–91—Test Method for Fluorescent Liquid Penetrant Examination Using the Lipophilic Post-Emulsification Process

- E 1209–91—Test Method for Fluorescent Penetrant Examination Using the Water-Washable Process
- E 1210–91—Test Method for Fluorescent Penetrant Examination Using the Hydrophilic Post-Emulsification Process
- E 1219–91—Test Method for Fluorescent Penetrant Examination Using the Solvent-Removable Process
- E 1220–91—Test Method for Visible Penetrant Examination Using the Solvent-Removable Process

The test allows for visible penetrant examination for detecting discontinuities such as cracks, openings in seams, and isolated porosity. Testing involves spreading a liquid penetrant, typically a light oil with visible or fluorescent dyes in it, over the surface to be tested. The penetrant is given time to enter open discontinuities. Excess penetrant is removed, and the liquid in any discontinuity is drawn out. The discontinuity and near surfaces are stained by the penetrant in the process. The surface can then be visually inspected for indications of surface discontinuities.

Partial-Vacuum Tests.   Partial vacuum tests on closed tanks are defined in API Standard 620. These tests are performed to ensure that the tank walls and roof meet design specifications. During the test, water is withdrawn from the tank with all vents closed until the design partial vacuum is developed at the top of the tank. Observations are made to as to when the vacuum relief valves start to open. These valves should open before the design pressure is reached. These tests are performed with the tank full, half full, and empty.

Partial vacuum testing using a vacuum box can be performed on welds of a tank bottom that rests directly on the ground. This test is accomplished by applying a solution film at the joints and pulling a partial vacuum of at least 3 psig. For this application, the vacuum box must have a transparent top.

Tank Gauging.   Manual tank gauging, commonly called static testing, is defined in EPA's guidance manual for release detection (EPA 1989). The test is effective for small-volume tanks of less than 550 gallons. The liquid level is measured in a quiescent tank at the beginning and end of a 36-hour period or other specified length of time. Any change in liquid level can be used to calculate the change in volume. Unless dramatic temperature changes occur, the liquid level change can be compared against established guidelines to determine whether any differences in the measurements are significant enough to indicate a leak.

Automatic gauging systems can be permanently installed in tanks to provide both tank integrity testing and inventory information. The system can measure the change in product level within the tank over time and can detect drops in level not associated with tank withdrawals.

Volumetric Tank Testing.   This test, defined in EPA's guidance manual (EPA 1989) is applicable mainly to underground storage tanks. The tank is tested for tightness by placing a known volume of liquid—usually water—into the tank for a period of time. The tank level is monitored for any changes during the test which might indicate leakage. For maximum sensing of level change, the tank should be overfilled so that the liquid reaches the fill tube or standpipe located above grade. Since the level changes occur in a small area, small changes in volume can be detected readily with gauging equipment.

Temperature variations must be taken into account during testing since they can affect the volume. Additionally, structural deformation resulting from filling the tank can occur, so a waiting period must be observed to ensure this effect has stabilized.

MATERIAL SELECTION FOR TANK SYSTEMS

When selecting material for a tank system, chemical compatibility and material cost are the two key considerations. Material selection must be made carefully for each application and must include expected variations in chemical or waste solutions and storage temperature and pressure ranges. Other factors to consider include normal atmospheric conditions, as well as hazardous climatic conditions such as the possibility of earthquakes and hurricanes (Woodside and Prusak 1991).

Materials selected for the primary tank should be as optimum as the budget will allow. In order to keep costs down, the secondary containment which is used only in an emergency often can be made of less expensive materials.

Material selections for tank systems must be made based on the parameters of the individual application. Test data for compatibility of specific chemicals with various metal and nonmetallic materials have been compiled by NACE into two volumes entitled *Corrosion Data Survey* (see references). Included in metal tests are:

- Iron based metals such as carbon steel and stainless steel
- Copper based metals such as copper and copper alloys
- Nickel alloys such as nickel-chrome-iron and nickel- chrome-molybdenum
- Other metals and alloys such as aluminum, silver, and titanium

Included in the nonmetals tests are materials such as carbon, glass, synthetic and natural rubber, epoxy fiberglass and other fiberglass materials, and some plastics such as polyethylene, polypropylene, and polyvinyl chloride. The temperature range for use of these materials is generally 70°F to 140°F.

Presented next is a general overview of types of chemical families that typically are used in industry and a general discussion of compatibility requirements for these chemical types, based on NACE data. Both metals and non-metallic materials are addressed. All material selections should be investigated thoroughly and/or tested before being put into service.

Acids

The iron-based metals—cast iron, carbon steel and stainless steel—are not recommended for use with weak acids. Austenitic stainless steels, such as AISI 316 and 317, and several nickel base alloys with molybdenum are suitable for some acids at varying concentrations and at low temperatures ($<200°F$). Other alloys and metals such as gold and platinum alloys, silver, tantalum, titanium, and hastelloy are also resistant to specific acids at somewhat higher temperatures. These alloys and metals typically are not used in large applications because of expense. Gaskets, valves, and other small but critical parts, however, often are made of these stronger materials since fittings and joints are particularly subject to chemical attack.

Acid-compatible nonmetals include polyester-fiberglass, glass-lined steel, and synthetic (butyl and fluorine) and natural rubber, depending on the application. Epoxy fiberglass and carbon are also compatible for some applications. Plastics such as polyethylene and polypropylene are adequate for use with some weak acids. These materials are not recommended for use with high concentrations of strong acids such as sulfuric acid, nitric acid, and hydrofluoric acid.

Alcohols

Carbon and stainless steel, cast iron, aluminum, and other metal alloys are compatible with most alcohols at low temperatures. Synthetic or natural rubber can sometimes be applied with success, but the type of rubber varies according to specific alcohol type. If the rubber is incompatible, the alcohol will dissolve it. Isopropyl alcohol is most compatible with fluorine rubber, but butyl and natural rubber can also be used at low temperatures. Other acceptable nonmetals data varied significantly according to type of alcohol.

Aldehydes

Generally, stainless steels of AISI 304 and higher are compatible with aldehydes, depending on the specific chemical. Alloys such as nickel-based alloys and copper-based alloys can also be used successfully in most applications.

Nonmetallic materials are not as suitable. Furfuryl alcohol-glass, glass-lined steel, and epoxy-asbestos-glass are compatible for various applications, but material costs are generally prohibitive. Compatibility with other nonmetals varies. Plastics such as polypropylene soften at higher concentrations and temperatures. Applications for synthetic and natural rubber are very limited.

Ammonium Solutions

Ammonium solutions are not compatible with copper based metals. Stainless steel and nickel-based alloys are compatible with most ammonium solutions. Many solutions are compatible with aluminum.

Synthetic and natural rubber are resistant to almost all ammonium solutions at low temperatures. Glass-lined steel is resistant to the more aggressive solutions such as ammonium fluoride. Polychloroprene, polyethylene, and polypropylene are compatible for many less aggressive solutions.

Ammonia is compatible with most metals, and is commonly handled in carbon steel. Urea is compatible with AISI 304 stainless steel and above.

Caustics

Carbon steel and AISI 304 and 316 stainless steel are acceptable for most caustic solutions, particularly at low temperatures. Nickel and nickel based alloys can also be used.

Natural and butyl rubbers are generally compatible with caustics. Other materials such as polychloroprene, polypropylene, epoxy fiberglass, and polyvinyl chloride also are generally acceptable.

Petroleum Distillates and Off-Shore Applications

Most petroleum distillates are compatible with a wide variety of metals, with carbon steel being the preferred material for tank systems throughout a petroleum facility. Synthetic and natural rubbers generally are inadequate. Polyethylene and polypropylene also are inadequate because the aromatic hydrocarbons tend to cause varying degrees of material swelling, softening, and stress cracking. Some glass, epoxy, or fiberglass materials have proven compatible, but petroleum distillate tank applications normally are too large for these materials to be selected from a structural standpoint.

Off-shore piping and other applications must be resistant to seawater. Cupro-nickel alloys are compatible for these uses.

CONCLUSION

There are numerous design considerations that must be addressed when designing a tank system. These include proper foundation, material selection, cathodic protection, and secondary containment. Tank systems became more tightly controlled in the 1980s when regulations were promulgated for hazardous waste tank systems and underground tank systems that store petroleum products and hazardous substances. Requirements were established for testing and certification of existing systems for both types of tank systems. Testing for corrosive resistance is important before selecting a tank material, and testing for tank integrity is important once the tank system is installed. Numerous test methods for both types of testing have been defined by national standards organizations.

Regulations also define technology standards for new tank systems. These standards include secondary containment for hazardous waste tanks and underground systems storing hazardous substances. In addition, cathodic protection is required for underground tank systems made of coated metal. Retrofitting of existing tanks to meet technology standards will be required in the near future, as published in the regulations.

Other technology standards established by EPA include hazardous waste treatment technology standards for land-restricted wastes. The next chapter discusses treatment technologies currently used in the industry that are recognized by EPA as being effective for treatment of land-restricted wastes and/or other hazardous wastes. Included are physical treatment/separation techniques, solidification/stabilization, chemical treatment, biological treatment, and thermal destruction. Chapter 11 discusses disposal technology standards and regulations pertaining to the disposal of land-restricted wastes.

REFERENCES

American Petroleum Institute, "Design and Construction of Large, Welded, Low-Pressure Storage Tanks," 8th ed., API Standard 620, Washington, D.C., 1990.

American Petroleum Institute, "Recommended Practice for Bulk Liquid Stock Control at Retail Outlets," 4th ed., API RP 1621, 1987.

American Society of Mechanical Engineers, *ASME Boiler and Pressure Vessel Code*, Section V, "Non-Destructive Testing," New York, 1986.

American Society for Testing and Materials, "Nondestructive Testing," *Annual Book of ASTM Standards*, Vol. 3.03, Philadelphia, 1992.

American Society for Testing and Materials, "Wear and Erosion: Metal Corrosion," *Annual Book of ASTM Standards*, "Wear and Erosion; Metal Corrosion," Vol. 3.02, Philadelphia, 1992.

Langlois, Kelvin E., Chris Bauer, and Gayle Woodside, "Double Contained Wastewater Treatment Tanks," presented at Water Pollution Control Federation Annual Conference, Washington, D.C., October 1990.

National Association of Corrosion Engineers, "Control of External Corrosion on Metallic Buried, Partially Buried, or Submerged Liquid Storage Systems," NACE Standard RP0285-85, Houston, TX, 1985.

National Association of Corrosion Engineers, *Corrosion Data Survey, Metals Section*, 6th ed., Houston, TX, 1985.

National Association of *Corrosion Engineers, Corrosion Data Survey, Nonmetals Section*, 5th ed., Houston, TX, 1975.

U.S. Environmental Protection Agency, *Detecting Leaks: Successful Methods Step by Step*, EPA/530/UST-89/012, Office of Underground Storage Tanks, Office of Solid Waste and Emergency Response, Washington, D.C., 1989.

Woodside, Gayle, and John J. Prusak, "Above Ground Storage: Double Containment Strategies for Today and Tomorrow," *Proceedings from Annual Air and Waste Management Association Annual Conference*, Vancouver, June, 1991.

BIBLIOGRAPHY

ASME (1986), *ASME Boiler and Pressure Vessel Code*, Section VIII, "Rules for Construction of Pressure Vessels," American Society of Mechanical Engineers, New York.

ASME/ANSI, (1987) "Chemical Plant and Petroleum Refinery Piping," ASME/ANSI Standard B31.1, American Society of Mechanical Engineers, New York, and American National Standards Institute, New York.

ASTM (1988), *Galvanic Corrosion*, Special Technical Publication 978, H.P. Hack, ed. American Society of Testing and Materials, Philadelphia.

———— (1989), *Effects of Soil Characteristics on Corrosion*, Special Technical Publication 1013, Chaker and Palmer, eds., American Society of Testing and Materials, Philadelphia.

———— (1990), *Corrosion Testing and Evaluation: Silver Anniversary Volume*, Special Technical Publication 1000, Baboian and Dean, eds., American Society of Testing and Materials, Philadelphia.

———— (1991), *Acoustic Emission: Current Practice and Future Directions*, Special Technical Publication 1077, Sachse, Roget, and Yamaguchi, eds., American Society of Testing and Materials, Philadelphia.

Cole, Mattney G. (1992), *Underground Storage Tank Installation and Management*, Lewis Publishers, Boca Raton, FL.

De Renzo, D.J. (1985), *Corrosion Resistant Materials Handbook*, 4th ed., Noyes Data Corporation, Park Ridge, NJ.

Ecology and Environment, Inc., and Whitman, Requardt, and Associates (1985), *Toxic Substance Storage Tank Containment*, Noyes Data Corporation, Park Ridge, NJ.

Gangadharan, et al. (1988), *Leak Prevention and Corrective Action for Underground Storage Tanks*, Noyes Data Corporation, Park Ridge, NJ.

Jawad, Maan H., and James R. Farr (1989), *Structural Analysis and Design of Process*, John Wiley & Sons, New York.

LeVine, Richard, and Arthur D. Little, Inc. (1988), *Guidelines For Safe Storage and Handling of High Toxic Hazard Materials*, American Institute of Chemical Engineers-Center for Chemical Process Safety, New York.

NACE (1975), "Control of Internal Corrosion in Steel Pipelines and Piping Systems," National Association of Corrosion Engineers, NACE Standard RP0175, Houston, TX.

Rizzo, Joyce A. (1991), *Underground Storage Tank Management: A Practical Guide*, Government Institutes, Rockville, MD.

Rizzo, Joyce A., and Albert D. Young (1990), *Aboveground Storage Tanks: A Practical Guide*, Government Institutes, Rockville, MD.

Schwendeman, Todd G., and H. Kendall Wilcox (1987), *Underground Storage Systems—Leak Detection and Monitoring*, Lewis Publishers, Boca Raton, FL.

U.S. EPA (1986), "Underground Tank Leak Detection Methods: A State-of-the-Art Review," EPA/600/2-88/001, prepared by IT Corporation for Hazardous Waste Engineering Research Laboratory, Office of Research and Development, U.S. Environmental Protection Agency, Washington, D.C.

———— (1987), "Soil-Gas Measurement for Detection of Subsurface Organic Contamination," Environmental Monitoring Systems Laboratory, U.S. Environmental Protection Agency, Washington, D.C.

_____ (1988), "Analysis of Manual Inventory Reconciliation," prepared by Midwest Research Institute for the Office of Underground Storage Tanks, Office of Solid Waste and Emergency Response, U.S. Environmental Protection Agency, Washington, D.C.

_____ (1988), "Common Human Errors in Release Detection Usage," prepared by Camp Dresser & McKee, Inc., for U.S. Environmental Protection Agency, Washington, D.C.

_____ (1988), "Evaluation of Volumetric Leak Detection Methods for Underground Fuel Storage Tanks," Vol. 1, EPA/600/2-88/068a, prepared by Vista Research, Inc., for U.S. Environmental Protection Agency, Washington, D.C.

_____ (1988) "Review of Effectiveness of Static Tank Testing," prepared by Midwest Research Institute for the Office of Underground Storage Tanks, Office of Solid Waste and Emergency Response, U.S. Environmental Protection Agency, Washington, D.C.

_____ (1988), "Standard Practice for Evaluating Performance of Underground Storage Tank External Leak/Release Detection Components and Systems," prepared by Radian Corporation for Environmental Monitoring Systems Laboratory, U.S. Environmental Protection Agency, Washington, D.C.

_____ (1989), "Soil Vapor Monitoring for Fuel Tank Leak Detection—Data Compiled for Thirteen Case Studies," prepared by On-Site Technologies for Environmental Monitoring Systems Laboratory, U.S. Environmental Protection Agency, Washington, D.C.

_____ (1989), "Volumetric Tank Testing: An Overview," EPA/625/9-89/009, U.S. Environmental Protection Agency, Washington, D.C.

Young, Albert D. (1990), *Corrective Response Guide for Leaking Underground Storage Tanks*, Government Institutes, Rockville, MD.

Whitlow, R. (1990), *Basic Soil Mechanics*, 2nd ed., John Wiley & Sons, New York.

Woodside, Gayle, and John J. Prusak (1992), "Above Ground Storage: State-of-the-Art Systems," *Proceedings from Annual Air and Waste Management Association Annual Conference*, Kansas City, June, 1992.

10

HAZARDOUS WASTE
TREATMENT TECHNOLOGIES

Technologies for treatment of hazardous waste have played an important role in hazardous waste management during the recent decade. Treatment standards for hundreds of wastes have been established under RCRA, and are driving the use of treatment technologies that can achieve high-efficiency contamination removal or that can stabilize the waste to reduce the level of leachable contaminants.

EPA and others have published numerous technical reports and manuals on hazardous waste treatment (see references), including:

- EPA, "Compendium of Technologies Used in the Treatment of Hazardous Waste," EPA/625/8-87/014.
- EPA, "Overview of Metals Recovery Technologies for Hazardous Waste," EPA/600/D-91/026.
- EPA, *Solvent Waste Reduction*.
- EPA, "Treatment Technology Background Document," PB91-160556.
- Jackman and Powell, *Hazardous Waste Treatment Technologies*.
- Water Environmental Federation (formerly Water Pollution Control Federation), "Hazardous Waste Treatment Processes," Manual of Practice FD-18.

This chapter describes the major types of treatment technologies currently used in industry, including physical treatment/separation techniques, solidification/stabilization, chemical treatment, biological treatment, and thermal destruction.

PHYSICAL TREATMENT/SEPARATION TECHNIQUES

Numerous physical treatment or separation techniques exist that typically are applied to wastewaters. These treatment technologies separate the contaminant from the waste stream so that it can be reclaimed, reused, or disposed of. Physical treatment/separation techniques include sedimentation/clarification, evaporation, distillation, extraction, stripping, carbon adsorption, ion exchange, and others.

Sedimentation/Clarification

Sedimentation/clarification is a process by which hazardous and nonhazardous grits, fines, and other suspended solids are removed from the waste stream through gravity settling. The process typically is used as a pretreatment step or in conjunction with another treatment process such as chemical or biological treatment. Usually, sedimentation/clarification is performed in a gravity settling clarifier, pond, or basin. Examples of a gravity settling pond and a clarifier are shown in Figure 10.1. A gravitational settling system provides enough residence time to allow the heavier suspended solids to gravitationally settle to the bottom of the system. These solids are removed and thickened periodically prior to disposal.

Flocculation, commonly used in clarifiers to enhance sedimentation, is a physical/chemical process in which small particles agglomerate to form larger particles. The large particles, because they are heavier than the smaller particles, settle more effectively in the clarifier or sedimentation basin. Flocculation of certain types of particles can be induced by slow agitation of the wastewater, without the addition of a flocculating agent. However, a flocculating chemical usually is added to the wastewater to promote flocculation of the smaller particles. The flocculants adhere readily to suspended solids and to each other to form larger particles with greater density, which settle better. The settled solids thicken at the bottom of the clarifier and, in some clarifier designs, are used to form a sludge blanket that becomes a filtration mechanism within the clarifier.

Clarifiers typically are used in chemical precipitation and biological treatment processes to remove precipitated metal solids and suspended biological solids. To prevent the sludge blanket from becoming too thick or heavy, part of the sludge blanket is removed continuously or intermittently from the system and thickened prior to disposal.

In settling basins or impoundments, the solids removed from the liquid typically are allowed to accumulate until the effective treatment volume of the basin and the solids removal efficiency is reduced. The accumulated solids then are removed from the impoundment by dredging, with the liquid in the basin, or by dewatering the impoundment and removing the solids with some type of excavating equipment such as a dragline with a clamshell. It is usually necessary to take the basin out of service while the solids are being removed.

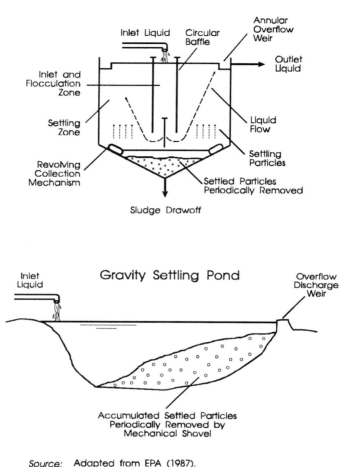

Source: Adapted from EPA (1987).
FIGURE 10.1 Examples of Sedimentation/Clarification Systems.

Evaporation

Evaporation can be used to separate volatile compounds from nonvolatile components and often is used to remove residual moisture or solvents from solids or semisolids. Thin-film evaporators and dryers are examples of evaporation equipment used for this type of application. Some evaporators are also appropriate for aqueous solutions.

Thin-Film Evaporators.   There are two types of thin-film evaporators commonly used in industrial applications. The first type introduces feed material into the center of a rotating heated conical receiver. Centrifugal force causes the feed to travel to the outer edge of the conical receiver where it is collected and drawn off as residue. During the process, the heat causes the volatile components to be driven from the feed. These volatile components are condensed on a chilled surface of the evaporator and collected as distillate.

The second type of thin-film evaporator, termed a wiped-film evaporator, introduces feed material on a heated wall of a cylinder. Rotating wiper blades continuously spread the feed along the inner wall of the cylinder to maintain uniformity of thickness and to ensure contact with the heated surface. The volatile components are driven off and collected on an internal chilled condenser surface. The condensate or distillate is removed continuously. At the end of the process, the residual becomes dry and heavy and drops to the bottom of the unit for removal. The wiped-film evaporator is best suited for treatment of viscous or high-solids content feed.

Dryers. Drying, another type of evaporation technique, is suited for waste streams of very high solids content. Several common types of dryers are vacuum rotary dryers, drum dryers, tray and compartment dryers, and pneumatic conveying dryers.

Vacuum Rotary Dryer. The vacuum rotary dryer is a batch system that uses a cylindrical rotating unit and agitator blades to mix or agitate the waste stream during the drying process. A vacuum is applied and maintained during the process to remove liquids during agitation. The process is terminated when the solids are dried to a specified moisture content.

Drum Dryer. The drum dryer is a continuously-operated unit that uses rotating heated drums for the evaporation contact area. This method of drying can be used with materials high in volatiles. In several designs, the volatiles can be withdrawn and collected for reuse, as appropriate.

Tray and Compartment Dryer. A tray and compartment dryer is a batch unit that uses a stationary tray or compartment to dry the waste, generally before transport for disposal or further treatment. Some units can be mounted on removable trucks.

Continuously-Operating Pneumatic Conveying Dryer. A continuously-operating pneumatic conveying dryer is used for applications similar to the tray and compartment dryer. In this case, however, drying is performed in conjunction with grinding, as the solids are conveyed and dried within the unit (Perry and Green 1984).

Aqueous Evaporation. Evaporation also can be used in the treatment of wastewaters containing salts and dissolved solids. Generally, evaporation used in this way is considered a concentration technique instead of a treatment method. In these cases, evaporation minimizes the volume of waste requiring disposal.

One type of aqueous evaporation occurs outdoors and utilizes a lined pond or open tank with spray nozzles designed to spray the wastewater into the air as a fine mist. Mechanical aerators also are used for this purpose, especially in ponds. The spraying effectively increases the surface area of the wastewater, which enhances evaporation. As the wastewater is recycled through the spray system, the salt or solids content is increased as the wastewater volume is reduced. The brine that forms can be withdrawn for proper disposal when a certain concentration or volume reduction is achieved.

Another type of aqueous evaporation can be accomplished in a closed process vessel that uses steam to evaporate the liquid into a water vapor, which is ultimately condensed and may be reused. The concentrated liquid is collected for further treatment or disposal. An example of this type of evaporation unit is shown in Figure 10.2.

Distillation

Distillation is a technology that is used to separate contaminants from solvent waste streams in order to produce a purified solvent stream for reuse. The technology takes advantage of differing vapor pressures or boiling points between the product and the impurity. If the product has a higher vapor pressure than the impurity, it will volatilize before the impurity volatilizes, leaving the impurity behind. When the product is cooled and condensed, it is in a purer form. There are two major types of distillation. The first is batch distillation, also termed differential distillation. The second is continuous fractional distillation, also termed continuous multistage distillation.

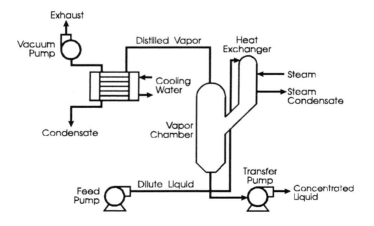

Source: EPA (1987).

FIGURE 10.2   Example of an Aqueous Evaporation Unit.

Batch Distillation. Batch distillation typically is used for small lots of solvents that have varied compositions and vapor pressures. The process uses a heated evaporation chamber or boiler to vaporize the product, which is then withdrawn and condensed for reuse. As the material's volatile, high vapor pressure component is depleted from the boiler, the boiling point temperature of the mixture rises, and more impurities with lower vapor pressures begin to evaporate. Once the impurities in the distillate become excessive, the batch process ends. The batch process also can be considered complete when the boiler reaches a predetermined temperature. A schematic of the batch distillation process is presented in Figure 10.3.

Continuous Fractional Distillation. If a product is contaminated with a material of similar vapor pressure, the separation process is more difficult and continuous fractional distillation is required. In the first stage of this process, the desired product and the impurity volatilize at or near the same temperature. However, the condensed vapor will be richer in the material of the higher vapor pressure. Continuous fractional distillation takes advantage of this phenomena by redistilling the enriched condensate during subsequent distillation stages. This further concentrates the material of higher vapor pressure. This process is continued until the desired product purity is obtained. Continuous fractional distillation is accomplished in large tray columns or packed towers. An example of a tray column system is shown in Figure 10.4.

Continuous fractional distillation cannot be used to treat liquids with high viscosity at high temperatures or liquids with a high solids concentration. Additionally, polyurethanes and inorganics are not suitable waste streams for

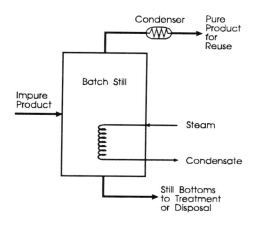

Source: Adapted from EPA (1987).

FIGURE 10.3 Example of a Batch Distillation Unit.

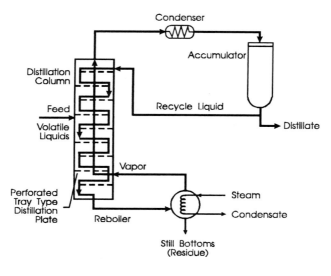

Source: EPA (1987).

FIGURE 10.4   Example of a Continuous Fractional Distillation System.

this technology. The usefulness of the technology also can be limited if the constituents in the waste stream form an azeotrope or if suspended solids in the mixture are thermally reactive (EPA 1987).

Extraction

Extraction is a technology that can be used when distillation proves ineffective. Mixtures that form azeotropes or contain substances of similar vapor pressures or boiling points may be candidates for extraction rather than distillation. Two types of extraction include solvent extraction, also termed liquid extraction, and supercritical fluid extraction.

Solvent Extraction.   Solvent extraction involves the mixing of a waste stream, which contains a hazardous constituent to be extracted, with an extraction fluid. Extraction occurs when the hazardous constituent, termed the solute, has greater solubility in the extraction fluid, termed the solvent, than in the waste stream. Mixing of the waste stream with the solvent allows mass transfer of the solute from the waste stream to the solvent. The solvent selected for the extraction procedure must be immiscible in the waste stream. As such, the waste stream and the solute-enriched solvent, or extract, will form distinct phases. A continuous gravity decanter can be used to separate and decant the extract.

The extract is sent to a recovery unit such as a distillation unit to remove the concentrated hazardous constituent from the solvent. The solvent can then be reused. The remaining waste stream, now devoid of the hazardous constituent, is either treated or recycled to the originating process, depending on the application.

Extraction is accomplished in two steps, with solvent/waste contact occurring first and phase separation of the liquids occurring second. The process can be operated as a batch or continuous process. Examples of types of equipment used for solvent/waste contact include mixing tanks, spray columns, perforated plate and baffle towers, and centrifugal contactors. Phase separation typically is accomplished using decanters, gravity settling tanks, or other equipment. Waste streams suitable for treatment using solvent extraction include hydrocarbon-bearing waste streams generated by the petroleum and petrochemical industries.

Supercritical Fluid Extraction.   Supercritical fluid extraction (SFE), an innovative extraction method, uses a pressurized or supercritical gas such as carbon dioxide as the solvent. Supercritical fluids make excellent solvents because of their comparatively low viscosities and high diffusivities. Thus, they greatly increase the efficiency of the extraction process through improved mass transfer rates. Additionally, the separation process of the hazardous constituent from the extract is simplified since supercritical fluids are gaseous at room temperature, thus the solute can be easily concentrated for reclamation, reuse, or disposal, as appropriate. Compounds that have been extracted successfully using SFE include aliphatic hydrocarbons, alkenes, simple aromatics such as benzene and toluene, polynuclear aromatics, and phenols (EPA 1991).

Stripping

Stripping is a common method for removing low concentrations of volatile organics and inorganics from wastewater. Stripping is accomplished by passing air or another stripping gas such as heated nitrogen through a liquid stream. Additionally, steam can be used as the stripping agent.

Air/Gas Stripping.   Air/gas stripping can be accomplished using one of two methods, sparging or countercurrent stripping. Factors important in the removal of organics from wastewater using this technique include temperature, pressure, air to water ratio, and surface area available for mass transfer.

Sparging.   Sparging is the simplest form of air stripping and typically is used to remove insoluble organics from water. Sparging is accomplished in a tank equipped with an air supply header in the bottom. Air is forced through small holes or nozzles to create bubbles that rise to the top of the tank. The volatile organics are stripped from the liquid phase and are transferred to the gas phase. The tank must be sized to provide adequate residence time to strip the organics from the wastewater. If necessary, finer bubbles can be generated to increase the mass transfer efficiency of the system. In general, this type of system is not as efficient as a countercurrent stripping system, which is described next. Additionally, off-gas typically is not captured during this type of stripping, and air emissions requirements may not permit the use of this type of stripping for some applications.

Countercurrent Stripping.   Stripping using a countercurrent stripping system is accomplished in a column, with countercurrent gas and liquid contact. Wastewater is introduced at the top of the column and withdrawn at the bottom. Air or another stripping gas is introduced at the bottom and exhausted at the top. The column height and diameter are sized to provide sufficient contact time for adequate contaminant removal. If necessary, the wastewater can be recirculated through the column several times to achieve a treatment standard.

When stripping insoluble organics from wastewater, the liquid phase controls the rate of mass transfer. To achieve the desired removal rates, the stripping column is flooded and operates as a bubbler tower. When stripping relatively soluble organics from wastewater, the gas phase controls the rate of mass transfer. For these cases, the column is operated like a spray chamber and the liquid is sprayed into the column. At very dilute concentrations, both phases of mass transfer rates are important. In order to increase the efficiency of a spray column, the gas/liquid surface area can be increased by packing the column. Packing media provides a surface for the liquid droplets to travel downward, which increases the gas/liquid contact time. Usually a demister is used on spray columns to filter water droplets from the exhaust. The diameter of the spray stripper is sized to minimize column flooding during operation. An example of a countercurrent packed-tower system is shown in Figure 10.5.

Stripping is best suited for aqueous organic waste with relatively high volatility and low water solubility, such as chlorinated hydrocarbons and aromatics (EPA 1987). As organic solubility increases, the temperature of the stripping operation must also increase. Hot air or another method for increasing temperature can be used to increase the efficiency of the stripping operation.

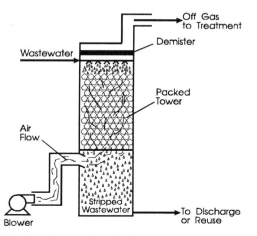

FIGURE 10.5   Example of a Countercurrent Packed Tower.

Steam Stripping. Steam stripping typically is used to strip more concentrated volatile organic compounds from aqueous wastewaters. Steam stripping is similar to continuous fractional distillation except that steam, rather than reboiled bottoms, provides the direct heat to volatilize organic vapors. The steam containing the concentrated volatile constituents is collected as the overhead from the stripper column, and then is condensed by cooling it in a condenser drum. Depending upon the characteristics and concentration of the volatile constituent in the condensate, it can be phase separated and reused, reclaimed, or disposed of. In some cases, the entire condensate stream may require additional treatment before disposal. When it is practical to separate the volatile constituent from the condensed steam, the condensate may be recycled to the steam generation process.

Waste streams suitable for steam stripping include aqueous wastes contaminated with chlorinated hydrocarbons, aromatics, ketones, alcohols, and high-boiling-point chlorinated aromatics (EPA 1987). The steam stripping process is performed in a closed system and has no continuous air emission potential until the point at which the condensed steam/volatile constituent is managed. At this point, air emission controls are required for the volatile constituent.

Decantation

Decantation is a process that can be used to separate two immiscible liquids of different densities. A gravity decanter that can be used to separate two phases in a continuous process is depicted in Figure 10.6. The denser liquid phase settles to the bottom of the tank and is drawn off through a bottom drain while the lighter phase floats on top and is drawn off via an overflow drain. For optimum performance, the liquid/liquid interface is equidistant from the bottom and the overflow drains and the influent line is located at the midpoint of the system near the interface.

For a specific application, the decantation unit must be sized to provide adequate residence time. Optimum residence time depends on the time required for the two immiscible liquids to separate into distinct phases with a

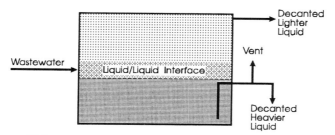

FIGURE 10.6   Example of a Decantation System.

well-defined interface. Careful consideration must be given to potential surges in flow from upstream processes. Additionally, a vacuum breaker or atmospheric vent is required on the bottom drain to prevent siphoning.

Decantation is a simple technology that is inexpensive and easy to operate. A sight glass installed on the side of the decantation tank can allow for visual inspection of the system's performance. Common uses for decantation units include oil and water separation, solvent and water separation, and separation of oils, solvents, and other immiscible liquids from groundwaters.

Carbon Adsorption

Carbon adsorption is a separation technology commonly used for removing dilute organics from a waste stream. The efficiency of this technology is measured in terms of the mass of organics adsorbed per unit mass of carbon. Carbon adsorption is most efficient for removing low concentrations of nonpolar organics with high molecular weights and boiling points and low solubilities, such as chlorinated hydrocarbons and aromatics.

Activated carbon for waste treatment applications typically is used in two physical forms—as a powder, which is often referred to as PAC (powdered activated carbon), and in a granular form. PAC generally is applied to the wastewater in a mechanically agitated tank or other type of solid/liquid contactor that maximizes contact between the powder and the liquid phase. The PAC is then separated from the liquid by gravity sedimentation or filtration. Granular carbon treatment systems contact the liquid and carbon in flow-through fixed beds. These beds can be upflow or downflow columns or can be open beds such as those used for sand filters. Activated carbon has an exceptionally large surface area per unit volume and a network of microscopic pores where adsorption takes place. Pore walls provide the surface layer of carbon molecules essential for adsorption. Since carbon best adsorbs contaminants that are not water soluble, as the water solubility of a contaminant increases, the ability for that contaminant to be adsorbed by carbon decreases. With extremely water soluble contaminants, treatment with activated carbon may not be practical or feasible.

Adsorption occurs in a predictable wave-like pattern. Organics are adsorbed first at the area of initial contact with the carbon. This is the initial mass transfer zone. Once the the carbon sites in this zone are saturated, the mass transfer zone moves through the carbon column along the direction of flow, until much of the carbon is saturated and breakthrough occurs. This concept is illustrated in Figure 10.7. At breakthrough, the concentration of organics in the wastewater stream begins to increase exponentially since available carbon sites are rapidly depleted, and organics begin to pass through the system unadsorbed.

The carbon will adsorb some chemicals preferentially to others, so this must be taken into account for wastewaters that have more than one organic

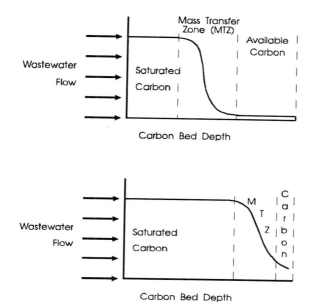

FIGURE 10.7   Adsorption Patterns in a Carbon Adsorption Unit.

contaminant. In addition, the preferentially adsorbed chemical can displace or desorb other chemicals once saturation has occurred. Thus, the chemicals adsorbed less readily likely will be the first to indicate breakthrough.

Carbon adsorption systems require either replacement of spent carbon or in situ regeneration of the carbon. In situ regenerative systems usually consist of two or three carbon beds. One or two beds treat the wastewater while the other is regenerated. In order to regenerate carbon beds used for adsorbing volatile organics, the bed must be heated to the boiling point of the adsorbed organic. This causes the organic to desorb as a vapor. The volatilized organic can be vented to an incinerator for thermal destruction, or it can be sent to a condenser for collection as a liquid. Over time, high boiling point organics that may be present in the wastewater can collect on the carbon as a "heel." For this reason, the carbon periodically should undergo an extended regeneration that can volatilize and desorb the heel. Carbon beds used for adsorbing non-volatile organics are regenerated similarly except that solvents are used in place of steam. After desorption, the contaminant is separated from the solvent through extraction or another separation method.

Carbon systems that are not regenerated in situ typically have three beds, one that is off-line for carbon replacement and two that are operated in series. The first bed is the primary treatment unit and the second bed acts as a polishing unit. The first bed is monitored continuously for breakthrough and is taken off-line for carbon replacement when breakthrough occurs. The second bed then becomes the primary treatment unit and the third bed, which has new carbon, becomes the polishing bed. When breakthrough occurs in the second bed, the third bed becomes the primary treatment unit and the first bed becomes the polishing bed. This "round robin" mode of operation is illustrated in Figure 10.8.

Carbon that is not regenerated in situ usually is regenerated using incineration. This type of regeneration is very effective in driving off all organics, but the carbon deteriorates in the process and loses some its surface area. Facilities with treatment systems that require large usage rates of activated carbon typically will have a thermal regeneration system on site. Facilities having smaller systems will send the exhausted carbon off site to a vendor for regeneration. For some very small systems, the carbon usage may be too low for any type of regeneration to be economically practical, and the carbon may be disposed of as a waste. This alternative is minimized whenever possible.

Ion Exchange

Ion exchange is a process in which hazardous or other contaminant ions are removed from an aqueous solution and are replaced or exchanged with non-

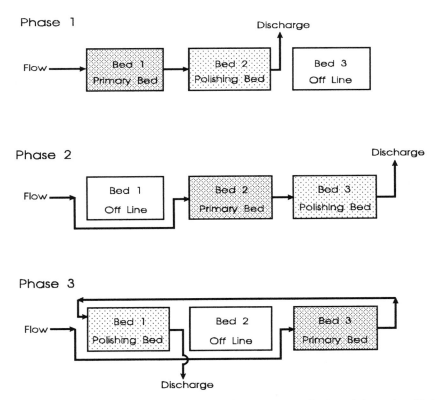

FIGURE 10.8   Round Robin Operation of a Three-Bed Carbon Adsorption Unit.

hazardous ions. This is accomplished by passing the aqueous stream through a resin bed in a packed column. Ion exchange resin is a substrate that contains mobile ions. Cation exchange resins exchange positive ions such as sodium or hydrogen for other positive ions in the solution, and anion resins exchange negative ions such as hydroxyl or chloride ions with other negative ions in the solution. These exchange ions are bonded loosely to the resin. As the hazardous ion passes over the resin, it is preferentially adsorbed, replacing the hazardous or other contaminant ion with the nonhazardous ion.

Examples of contaminants removed using cation ion exchange systems include hazardous metals and other constituents such as calcium and magnesium. Anion exchange systems are capable of removing contaminants such as sulfates, carbonates, and fluorides. In addition, some systems have demonstrated the ability to remove low-level radioactive materials such as strontium (DOE 1990). It is typically more cost effective to treat highly concentrated waste streams by another method.

Most ion exchange resins have the capability of being regenerated. The ion exchange resin is regenerated using a concentrated reagent such as sodium chloride solution for sodium- or chloride-based cation resins. Acid regenerants such as hydrochloric or sulfuric acid are used to regenerate hydrogen-based cation resins, and alkali solutions such as sodium hydroxide solutions are used to regenerate hydroxide-based anion resins. The high concentration of ions is sufficient to drive off the hazardous ions in concentrated solution. This waste stream is sent to a treatment system or to disposal. After regeneration, the resin column is flushed with clean water to remove excess regenerant. Some resin beds are not capable of being regenerated. Additionally, some adsorbed hazardous contaminants will produce toxic waste streams that are more difficult to dispose of than the spent resin. In these cases, the resin is disposed of as waste.

Ion exchange resins must be protected from suspended solids and oils and grease, which can clog the resin column and impede system efficiency. In addition, some waste streams that have ions of similar valence, which compete during the exchange process, will not be treated at the same efficiency as streams that have one predominant hazardous ion.

Novel Separation Processes.   Several novel separation technologies are available or are being explored for use in industry and other applications. These include crystallization from the melt, sublimation, adsorptive-bubble separation methods, and novel solid-liquid separation processes such as cross-flow electrofiltration and surface-based solid-liquid separation involving a second liquid. These and other novel processes are detailed in *Perry's Chemical Engineers' Handbook* (see references).

SOLIDIFICATION/STABILIZATION

Solidification/stabilization is used to immobilize a waste, make it insoluble, or detoxify it to a less hazardous state. Many solidification/stabilization technologies involve mixing liquid wastes with solids to produce a solid waste with physical characteristics such as structural integrity or decreased leachability. This process can make a waste more suitable for disposal in a landfill.

In general, the term "solidification" is used when the process involves physical restructuring of the waste to make the waste more manageable and perhaps less leachable, but does not chemically change the hazardous constituents in any way. The term "stabilization" is used when the process involves chemical reactions that convert the waste to a less hazardous state. Because many of the treatment processes involve both solidification and stabilization, the combination term is used widely, and will be used with the following technology descriptions.

Lime-Based and Cement-Based Processes

Lime-based and cement-based processes are effective treatment methods for wastes that contain multivalent metal ions. A reaction between a stabilizing medium such as lime or cement and the soluble metal ions promotes the precipitation of insoluble metal hydroxides. Additives enhance the curing reactions that solidify the mixture and bind the waste particles to the solids matrix.

This treatment method is suitable for wet wastes and sludges, which are mixed with reagents to form a gel. As the mixture is allowed to cure, fibrils are formed as silicate compounds hydrolize. These interlocking fibrils bind with various hydration products in the resultant solidified mass.

The solidification/stabilization reaction can be retarded by certain waste materials. Organics, sulfates, chlorides, borates, and silt prolong the setting and curing time and/or weaken the bond strength of the material (EPA 1987 and EPA 1991). A number of additives and special reagents have been developed to improve the solidification/stabilization process. These additives and reagents enhance the process by absorbing excessive free liquid or by increasing silicate concentrations to increase binding characteristics.

Solidification/stabilization techniques involving lime and cement-based technologies are attractive since raw materials are plentiful and relatively inexpensive. Additionally, processing equipment is readily available, and dewatering equipment is not required since the process absorbs liquid. The key disadvantage of this type of treatment is the increase in waste volume due to the addition of reagents and additives.

Other Solidification/Stabilization Methods

Other solidification/stabilization methods that are not used as widely as lime-based and cement-based processes include:

- Thermoplastic process—The thermoplastic process is a process during which wastes are blended with a molten thermoplastic material such as bitumen, asphalt, polyethylene, or polypropylene, causing the wastes to be bound in the plastic matrix. The process is compatible with heavy metals and radioactive materials. The process cannot be used with organic wastes since organics are driven off in high temperatures. Additionally, the process is unsuitable for hygroscopic wastes, which will absorb moisture and cause the thermoplastic to crack, increasing waste leachability. Strongly oxidizing contaminants, anhydrous inorganic salts, and aluminum salts, likewise, are unsuitable for this type of treatment. Process equipment and materials can be expensive and there is some potential for air pollution (Jackman and Powell 1991 and EPA 1987).

- Polymerization process—Polymerization uses catalysts to convert a monomer or a low-order polymer of a particular compound to a more stable polymer. This process treats organics including aromatics, aliphatics, and other oxygenated monomers such as vinyl chloride and acrylonitrile. This technology has been used at spill sites (EPA 1987).

- Reactive polymer process—The reactive polymer process is a process during which wastes are mixed with a polymerizing catalyst and reactive monomers such as urea-formaldehyde, phenolics, epoxides, polyesters, or vinyls. The process traps the waste in an organic matrix. The technology is compatible with radioactive materials, acid wastes, and has some use with heavy metals. Organic solvents and oils may retard setting of the polymers. The waste must be containerized prior to disposal in a landfill (Jackman and Powell 1991).

- Encapsulation—Encapsulation is a process during which wastes are solidified by drying and then encapsulated in an impermeable, protective coating. The encapsulation material is usually an organic polymer such as polyethylene, polyurethane, or fiberglass/epoxy. This technology reduces leaching and can be used for treatment of heavy metals, low-level radioactive materials, halides, and sulfates (Jackman and Powell 1991).

- Glassification/Vitrification—Glassification/vitrification is a process during which wastes are fused into glass or ceramics using very high temperatures. Thus, this process is only applicable to wastes that are stable at very high temperatures. Since glass is not leachable by water, this process provides for safe disposal without secondary containment. This technology has been carried out in situ for immobilizing low-level radioactive wastes in soils (EPA 1987).

Solidified/stabilized wastes must be tested prior to disposal in a landfill to ensure that the waste meets land disposal treatment standards. EPA's toxicity characteristic leaching (TCLP) procedure can be used to determine the stabilized waste's leachability.

CHEMICAL TREATMENT

Chemical treatment technologies are used widely throughout industry. Treatment methods are varied and include processes that are simple and inexpensive such as neutralization or chlorination, as well as processes that require large capital outlays and high energy usage such as wet air oxidation. Other commonly used chemical treatment technologies include various types of chemical oxidation/reduction and chemical precipitation.

Neutralization

Neutralization is used to treat waste acids and waste caustics (bases) in order to eliminate their reactivity and/or corrosiveness. The process is performed in a well-mixed tank system and entails the addition of an alkali (caustic) for waste acid neutralization or, vice versa, the addition of an acid for neutralization of waste caustics. If possible, waste acids and waste caustics or off-spec materials should be mixed together during the neutralization process to avoid using product-quality chemicals. In addition, trained personnel should review the treatment mixtures to ensure that the formation of new compounds that are more toxic or more hazardous than the original compounds does not occur.

Chemical Oxidation/Reduction

Chemical oxidation/reduction technologies include chlorination, ozonation, electrolytic recovery, and other processes that are used for selected applications. In general, these processes are used to convert toxic waste contaminants to a less toxic state.

Chlorination.   Chlorination is a widely used chemical oxidation technology. It is accomplished by adding chlorine to wastewaters. The chlorine attacks double carbon bond chemicals to render them to a less toxic form. The main advantage of this process is that it is inexpensive and reliable. The main disadvantage is that it produces undesirable end products. When chlorine is dissolved in water, a chloride ion and a hypochlorite ion are formed. These reactive chlorine ions have the ability to substitute for hydrogen in organics, which leads to the formation of toxic chlorocarbons such as chlorophenols and trihalomethanes. Chlorine is very effective in treating organic compounds, alcohols, and phenols. Additionally, chlorination is used to treat wastewater streams containing dilute cyanide concentrations.

Chlorination can also be accomplished with the addition of chlorine dioxide instead of chlorine. Since chlorine is a dangerous gas and a highly hazardous chemical, as defined by OSHA, chlorine dioxide oxidation provides an advantage over chlorination. Chlorine dioxide treatment is also advantageous since it does not form trihalomethanes when oxidizing organics. The oxidized end products from organic reactions with chlorine dioxide tend to be more readily biodegradable than end products formed by chlorination with chlorine. A disadvantage is that chlorine dioxide is not suitable for treatment of waste streams containing cyanide.

Chlorine dioxide can be generated at the point of use by reacting sodium chlorite with sodium hypochlorite and hydrochloric acid. In large-scale production, chlorine dioxide is produced by reacting sodium chloride with hydrochloric or sulfuric acids (Jackman and Powell 1991).

Ozonation. Ozonation is the process of injecting a wastewater stream with ozone through sparging or bubbling. Ozone provides an effective oxidation reaction for treating organic-containing wastewaters. When used to treat organics, ozonation provides a primary treatment to transform non-biodegradable constituents into biodegradable constituents. Specific organics treated by ozonation include alkanes, alcohols, ketones, aldehydes, phenols, benzene and its derivatives, and other organic contaminants. Breakdown products typically include acids and aldehydes (Jackman and Powell 1991).

The advantage of ozonation over chlorination—particularly chlorination with chlorine—is the elimination of undesirable end products. Ozone that is not consumed in the oxidation reaction decomposes to oxygen. The disadvantage of ozonation is its operating costs, particularly the electrical costs associated with ozone production.

Ozone is only slightly soluble in water. Thus, factors that affect the mass transfer between the gaseous ozone and liquid phase are important to the ozonation process. These factors include system temperature, pressure, contact time, contact surface area (bubble size), and pH.

Ozonation can be enhanced as an oxidation technique by the addition of ultraviolet (UV) radiation. With the addition of UV, ozonation can be effective in treating chlorinated compounds and pesticides. In addition, metal ions—such as iron, nickel, chromium, and titanium—can act as catalysts for ozonation, as can ultrasonic mixing. The combination of ultrasonic and metal catalysts provides for the best overall performance in many applications.

Since ozone does not completely oxidize most organic compounds, it often must be followed by a treatment process that will complete the oxidation of organic compounds, such as biological treatment. For some organic compounds at low concentrations in wastewater, ozone can be used to achieve essentially complete oxidation.

Ozonation is also an effective technology for treating cyanide wastes over a wide range of concentrations, since ozone can readily oxidize cyanide to cyanate. Cyanogen gas is a reaction intermediate in the reaction. Further oxidation of cyanate to carbon dioxide and nitrogen occurs slowly. As with treatment of other compounds, cyanide treatment with ozonation can be accelerated with the addition of UV radiation. Ozonation/UV technology has the potential to oxidize completely even the most difficult to treat cyanide complexes.

The main treatment limitation of ozonation is that the process is non-selective when competitive species are in the waste stream. Additionally, the sludges and solids generated from the process typically are not treated readily (EPA 1991).

Electrolytic Recovery. Metal ions can be reduced and removed from solution through electrolytic processing in an electrolytic cell. The cell consists of an anode and a cathode immersed in the wastewater. As the metal ions are reduced, they are collected on the anode as the base metal, as illustrated in Figure 10.9. The metal-depleted waste stream can then be discharged or can be sent to another treatment process for further treatment, if necessary. The main advantage of electrolytic recovery is that no sludge is generated from the process. The base metal collected on the anode can be removed and reclaimed directly.

The electrolytic recovery process is limited to solutions containing cadmium, chromium, copper, lead, tin, zinc, gold, and silver. These metals are high in the electromagnetic series, which allows them to be reduced readily and deposited on a cathode. Metals such as aluminum and iron are not suitable for this type of processing. In addition, dilute solutions of all metals are difficult to process because the transfer rates from cathode to anode are very low. The capital equipment necessary to process dilute solutions, along with high operating costs, make this alternative noncompetitive when compared to other treatment methods such as chemical precipitation or selective ion exchange. Metals that are accumulated in sludges and more concentrated waste streams are better candidates for electrolytic recovery.

Other Oxidation/Reduction Processes. Other innovative chemical oxidation/reduction processes can be used for treatment of selected waste streams, although these processes tend to have limitations associated with them (EPA 1987). These include:

- Oxidation by hydrogen peroxide—This treatment technology is based on the addition of hydrogen peroxide to oxidize organic compounds. Since hydrogen peroxide is unstable, it readily serves as an oxidizing agent by

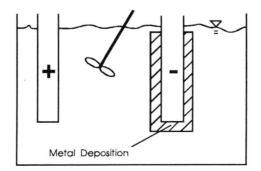

FIGURE 10.9 Example of an Electrolytic Recovery System.

giving up its extra oxygen molecule. However, the reaction from hydrogen peroxide is nonspecific and, in some cases, may be explosive depending on the constituents of the waste stream being treated.

- Oxidation by hypochlorite—This treatment process uses sodium hypochlorite or calcium hypochlorite to oxidize organic wastes. Depending on the contaminants in the waste stream, this process may produce toxic chlorinated organic by-products.

- Catalytic dehydrochlorination—This treatment technology is based on the reaction of polychlorinated hydrocarbons with high-pressure hydrogen gas in the presence of a catalyst. The waste stream should be analyzed to ensure that tars or sulfur compounds are not present since catalysts can become deactivated from these impurities.

- Alkali metal dechlorination—This process is used to displace chlorine from chlorinated organic compounds, such as PCBs and chlorinated hydrocarbons, contained in oils and liquid wastes. The waste is processed in a reactor with alkali metal reagents. Moisture can adversely affect the process.

- Alkali metal/polyethylene glycol (A/PEG) dechlorination—This innovative treatment technology was developed by research sponsored by EPA, and it uses A/PEG reagents to dechlorinate PCBs and oils. The alkali metal ion is held in solution by the large polyethylene glycol anion. The process can be accomplished under both ambient and high temperatures.

Wet Air Oxidation

Wet air oxidation is a process by which compressed air or oxygen is injected into a heated and pressurized system to oxidize organic contaminants and some inorganics, such as cyanide, in wastewater or dilute waste streams. This treatment process provides an economical method of treating wastewaters and wastes containing greater than 85% water, which are too dilute to incinerate and which contain toxics that cannot be treated biologically. In general, wet air oxidation provides a method of primary treatment for wastewaters or waste streams that are subsequently treated by conventional methods.

The process typically is operated at temperatures ranging from 175°C to 325°C and pressures ranging from 300 to 3000 pounds per square inch (psi). Typical residence time is one hour. The process generally is applied to waste streams containing dissolved or suspended organics in the range of 500 to 50,000 mg/l. Below 500 mg/l, the rates of wet air oxidation of most organic constituents are too slow for efficient application. For waste streams more concentrated than 50,000 mg/l, incineration may be more applicable (EPA 1991).

During the treatment process, waste streams are heated under pressure and then injected into a tubular reactor with simultaneous injection of compressed air or oxygen. At high temperature and pressure, oxygen has a high solubility and can oxidize numerous contaminants readily. The amount of air introduced into the system is dependent on the concentration of compounds that can be oxidized in the waste stream. Oxygen addition can be controlled by maintaining a constant dissolved oxygen concentration at the reactor outlet. The operating pressure must be great enough to maintain water in a liquid phase. A liquid reaction medium provides the advantage of buffering temperature changes, which allows for a more controlled reaction.

Wet air oxidation can be used to treat a wide range of wastes at a treatment efficiency of 99% or greater. Examples of contaminants in dilute waste streams and wastewaters that are effectively treated using this process, and the less toxic oxidation by-product(s) produced after treatment include:

- Hydrocarbons—By-products are carbon dioxide and water.
- Nitrogen compounds—By-products are elemental nitrogen, nitrate, or ammonia.
- Halogens—By-products are acid halides.
- Sulfur—By-products are inorganic sulfates.
- Phosphorus Compounds—By-products are phosphates.
- Metals—By-products are metal oxides.
- Cyanide—By-products are carbon dioxide and nitrogen.

The wet air oxidation process can be costly to operate because of a high energy demand. To reduce energy demand and resulting operating costs, a heat exchanger can be used to preheat the wastewater inlet to the reactor with the reactor effluent. In some cases ionic catalysts may be added to improve oxidation efficiency while reducing the system operating temperature. Care must be taken when selecting the catalyst in order to maximize system operation. Removal of the catalyst from the effluent typically is required to prevent contamination and to keep operating costs at a minimum.

The operating parameters adjusted in the wet air oxidation process to achieve the desired treatment efficiency include temperature, pressure, oxygen concentration, and residence time. Materials of construction for wet air oxidation systems include stainless steel, nickel, and (for extremely corrosive wastes containing heavy metals) titanium alloys. Equipment components for the system include a high-pressure pump, a heat recovery heat exchanger, an auxiliary heat exchanger, a compressed air source, a tubular reactor, an effluent cooling heat exchanger, and a blow-down tank that separates dissolved gases and suspended solids from the liquid. The vented gas from the blow-down tank may require a scrubber or other emission control device. The effluent liquid may require post treatment for residual organics or metals.

Chemical Precipitation

Metals often are removed from wastewater streams by an oxidation reaction, which converts the metal to a relatively insoluble metal oxide, hydroxide, or sulfide. The insoluble metal precipitates out of the wastewater stream as a sludge or sediment that can be dewatered in a centrifuge or filter press and collected for disposal or, more preferably, for reclamation. Examples of metals that are precipitated readily with a hydroxide reactant include copper, nickel, chrome, iron, and zinc.

A schematic of the treatment units used in a typical chemical precipitation process or treatment system is shown in Figure 10.10. As can be seen from the figure, the wastewater treatment system utilizes a large equalization tank to provide surge capacity for the treatment plant. This tank also is needed to dampen extremes in pH or high chemical loadings that might create a major upset to the system. Treatment plants operate best when changes in operational parameters and waste stream quality are made gradually.

In the next unit, the mixing tank, the chemical reaction takes place. Lime, sodium hydroxide, ferrous hydroxide, and magnesium hydroxide are common reactants used in the precipitation process. The tank should be well

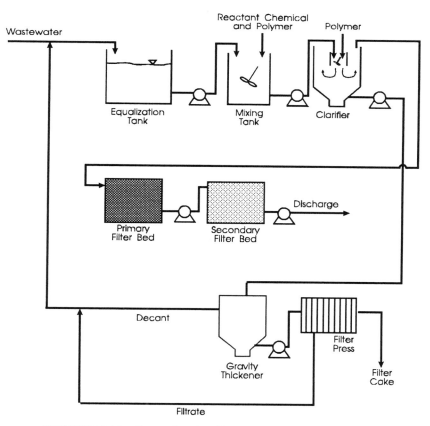

FIGURE 10.10 Example of a Chemical Precipitation System.

mixed to enhance the reaction efficiency and to keep the solids suspended. If the tank is undermixed, sediment will coat the bottom of the tank. In addition to the chemical reactant, a polyelectrolyte can be added, which has polar sites to coagulate the precipitant fines. This helps produce a larger floc that will settle more readily in the clarifier.

After initial mixing, the wastewater is pumped into the reaction zone of the clarifier where it continues to be mixed, and additional polymer is added. As the wastewater flows upward to exit the clarifier, the floc settles. Treated wastewater then exits via an overflow weir. The treated wastewater can be collected in a post-clarification tank and the pH adjusted, as necessary, to meet a discharge standard. Typically, the wastewater is filtered through a sand filter to remove any suspended solids that were not removed in the clarifier.

Sludge is removed periodically from the clarifier and transferred to a gravity thickening tank. During the thickening process the solids settle to the bottom of the tank and some of the water in the sludge is forced to the top of the tank. This water, or supernatant, is decanted and returned to the equalization tank. The thickened sludge is removed from the gravity thickening tank and further dewatered in a plate and frame filter press or centrifuge. The filtered water is returned to the equalization tank and the compacted sludge, or filter cake, is collected for disposal or reclamation.

In addition to precipitation as metal oxides or hydroxides, metals can be precipitated as metal sulfides. Sulfide salts of metals such as mercury, cadmium, and lead have much lower aqueous solubilities than their corresponding oxides or hydroxides, and thus lower effluent concentrations of these metals can be achieved with sulfide precipitation. The treatment process sequence is the same as as that described for hydroxide precipitation. Sulfide reagents (which substitute for lime in the process) include hydrogen sulfide, sodium sulfide, and sodium metabisulfite.

Chemical precipitation processes can be affected by the presence of complexing agents that prevent the metals from precipitating in the clarifier. Additionally, the presence of oxidizers can break down and reduce the effectiveness of the polyelectrolyte and allow excessive precipitant fines to escape the clarifier. Wastewaters containing complexing agents or oxidizers require segregation and pretreatment prior to treatment through a precipitation process.

BIOLOGICAL TREATMENT

Hazardous contaminants, as well as nontoxic constituents, can be removed from wastewaters using biological treatment. Microorganisms, in the form of bacteria, are integral to these types of treatment systems. In particular, heterotrophic bacteria, which utilize organic material as a source of carbon for cell growth and energy, are used to treat wastewater streams containing organic wastes. These bacteria are the most predominant and effective bacteria used in activated sludge treatment systems. Additionally, autotrophic bacteria, which utilize carbon dioxide as a source of carbon and which oxidize mineral compounds as a source of energy, are present in biological treatment systems.

Four distinct phases characterize microbial bacteria population growth in a batch biological system (Jackman and Powell 1991). In the first phase, the acclimation phase, the bacteria become acclimated to their environment and start to produce enzymes that are needed to break down complex organics present in the waste stream. In the second phase, the growth phase, the bacteria population grows at a logarithmic rate as a result of the ample supply of organic food or substrate. In the third phase, the limited growth phase, existing substrate cannot support a continuously increasing population, and organisms begin to die. These dead bacteria, in turn, become substrate for remaining bacteria. In the fourth phase, the endogeneous phase, a decreasing substrate supply causes the bacteria to metabolize the polysaccharide slime layer that coats their cell walls, thus creating a denser cell. When the slime layer decreases to the point where there is no adhesion, individual particles do not agglomerate.

Suspended growth treatment systems, such as activated sludge systems, can be maintained optimally when the cell density reaches maximum thickness while maintaining a thin slime layer. During this process, if cells collide, they agglomerate and settle. Supported growth treatment, such as trickling filters and rotating biological contactors, are operated in the logarithmic growth phase where the system supports a dense, sticky slime.

Biological treatment is divided into two major classes—aerobic biological treatment and anaerobic biological treatment. Aerobic processes, which use oxygen, convert organics into carbon dioxide and water. Anaerobic processes, which are devoid of oxygen, convert organics into methane and carbon dioxide.

Conventional systems normally used for treatment of municipal wastewater—including activated sludge, trickling filters, rotating biological contactors, aerated lagoons, facultative lagoons, and anaerobic treatment systems—also can be used to treat industrial organics. Refinements and innovations to conventional systems have been developed to allow for optimal treatment of both municipal and industrial waste streams. Factors that influence the biological treatment process include:

- Concentration of organics—This is measured as 5-day biochemical oxygen demand (BOD_5), chemical oxygen demand (COD), or total organic carbon (TOC). The waste stream should be at least 90% water.
- Amount of nutrients in the system—Primary nutrients include nitrogen, phosphorus, sulfur, potassium, calcium, and magnesium. A deficit of these nutrients is detrimental to system performance. Additionally, secondary nutrients such as iron, copper, chromium, and zinc are needed in trace amounts for optimum performance.

- Temperature—Optimal temperature for aerobic biodegradation of a waste stream with both municipal and industrial wastewaters is from 20°C to 40°C. Anaerobic systems typically are operated in the range of 40°C to 50°C.
- pH—The acceptable pH range for microorganisms is from 5 to 10, with a pH range of 6 to 8 being optimum.
- Treatability of toxic inorganics—Some toxic inorganics are treatable in a biological treatment system. Examples include arsenic, cyanide, and heavy metals such as lead and zinc. Dissolved metal ions will inhibit enzyme action in the treatment system. Toxicity is concentration-dependent, and typically low concentrations of many toxicants can be tolerated.
- Biodegradability of organics—Organics have varying degrees of biodegradability. Alkenes, alkynes, alcohols, and aldehydes typically have high biodegradability; alkanes, aromatics, amines, and nitro-groups have medium biodegradability; halogenated compounds have low biodegradability.
- Oxygen concentration—For aerobic processes to operate efficiently, dissolved oxygen must be greater than 1 mg/l; for anaerobic processes, no oxygen can be present.
- Hydraulic retention time—Optimal retention time is specific to the system dimensions and the treatment process.
- Solids reaction time—Optimal reaction time is dependent on treatment efficiency requirements.
- Shock loadings of BOD, suspended solids, and toxics—Microorganisms usually cannot tolerate shock loadings. Systems generally can accommodate increased loadings in a controlled manner.

Aerobic Suspended Growth—Activated Sludge Process

The activated sludge process is a widely used biological process for treatment of industrial wastewaters containing organic wastes. This treatment process is sensitive to total suspended solids concentrations greater than 1% and shock loadings of concentrated organics. To protect the activated sludge system, an equalization tank is used to dampen fluctuations in flow and concentration, and a primary clarifier is used to remove gross solids, grit, and oil and grease.

After primary treatment, the wastewater is routed to an aeration basin where enzymes and microbial metabolism break down the organic content. Dissolved oxygen concentrations are maintained above 1 mg/l through surface aeration devices or sparges. After aeration, the wastewater passes into a clarifier. Using gravity settling, the microbial mass, or sludge, is separated from the the treated wastewater. Most of the sludge is recycled to the aeration basin to maintain a source of acclimated bacteria. Excess sludge is drawn off and sent through a post-treatment process such as thickening, dewatering, or digestion. An example of the activated sludge process is shown in Figure 10.11.

Aerobic Supported Growth Systems

Aerobic supported growth systems include trickling filters and rotating biological contactors (RBCs). Unlike suspended aerobic growth systems, the biological culture in these systems is developed as a film culture attached to a solid surface. Treatment occurs as organic-contaminated wastewater is passed over the surface, as with trickling filters, or the surface is rotated through the wastewater, as with RBCs. These systems are considered to be aerobic, although a portion of the microbial film may become anaerobic at the film-solid interface. Periodically, a portion of the film culture sloughs off. This happens because the film becomes too heavy to remain attached to the media or because the microorganism population exceeds the feed supply and the organisms metabolize their polysaccharide slime coating and lose their adhesive properties. This sloughing process generates excess sludge, which is subsequently wasted.

Trickling Filters.   The trickling filter process includes primary clarification, equalization, and treatment through the trickling filters. Primary clarification is a critical step since gross solids and suspended particulates can clog the trickling filter system. Nutrients are added to the process at the equalization tank. The equalization tank is used to dampen fluctuations in flow and concentration, just as it is for suspended growth systems, although trickling filters generally are less susceptible to shock from fluctuations. Should extremes in organic loadings occur, sloughing of the biomass may become excessive.

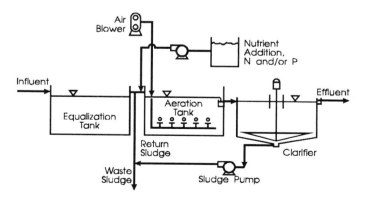

Source: EPA (1987).

FIGURE 10.11   Example of an Activated Sludge Process.

The packing media used in trickling filters is usually plastic or a material such as metallurgical coke or slag. The packing material allows for a large surface area in a small volume while providing interstitial space to prevent blockage by the microbial film. Plastic media provides the advantage of higher void space and less weight. The wastewater trickles through the packing media, and then goes through a secondary clarifier before discharge.

The advantage of the trickling filter over other wastewater treatment systems is that aeration occurs by induced draft and no power is needed for agitation and aeration. However, because of the relatively short residence time of the wastewater in the filter, the treatment efficiency is less than that of an activated sludge process. Filtered effluent can be recycled through the system after secondary clarification to achieve higher removal efficiencies. Additionally, trickling filters can be operated in series, or followed by another secondary treatment.

RBCs.   The RBC treatment process is similar to that of the trickling filter except that the trickling filter is replaced by a round support medium such as a disc, lattice network, or a screen container of plastic blocks. The microbial film grows on this medium, which is partially submerged and which rotates slowly on a horizontal axis in a tank of wastewater. This motion maintains the bacteria in an aerobic condition. The main advantage to an RBC is that it can be operated in a small space. In general, RBCs require only 10% of the space needed for a trickling filter, and have relatively high treatment efficiencies (Jackman and Powell 1991).

Anaerobic Process

The anaerobic process flow is similar to the activated sludge process. A well-mixed tank or digester treats waste with microbial growth, and a clarifier is used to separate the biosolids from the treated wastewater. A major portion of the biological sludge is recycled, and the excess sludge is wasted. Although aerobic processes can accommodate only small amounts of suspended solids, this process is able to accommodate relatively high suspended solids loadings, as well as relatively high BOD loading rates.

The anaerobic process breaks down organics by a symbiotic relationship between two classes of bacteria (Jackman and Powell 1991). Acid-forming bacteria first convert complex organics into simpler, lower molecular weight carboxylic acids, alcohols, carbon dioxide, and water. Then, methane-forming bacteria convert the carboxylic acids to methane gas and carbon dioxide. The acid-forming bacteria are less sensitive to changing conditions and grow rapidly while the methane-forming bacteria are very sensitive to changing conditions and grow very slowly.

This phenomena dictates that methane-forming bacteria are rate controlling and that the performance of the system is based on the generation of methane. In an imbalanced system, the methane-forming bacteria will not keep up with the acid-forming bacteria and the system pH will begin to drop.

Lime or bicarbonate must be added for the system to regain balance, and waste feed must be diverted until balance is restored.

The ability of the anaerobic process to treat higher concentrations of organics and suspended solids makes this type of treatment a good option for sludges and other more concentrated waste streams. Anaerobic biological treatment is not suitable for wastewaters such as aromatics and long-chain hydrocarbons because the anaerobic bacteria do not produce the enzymes needed to break down these complex organics.

THERMAL DESTRUCTION

The most common form of thermal destruction is incineration, which is used as a treatment technology for hazardous waste, medical waste, and nonhazardous solid waste. Other types of thermal destruction technologies include pyrolysis, waste destruction in industrial boilers or blast furnaces, and molten salt thermal destruction. Wet air oxidation, which was discussed earlier in this chapter as a chemical oxidation process, also can be considered a thermal destruction technology.

Incineration

Incineration, also termed combustion, is regulated under several statutes. Acceptable efficiencies, emission guidelines, and/or other standards are delineated in 40 CFR 264 Subpart O and 40 CFR Part 270 (RCRA) for hazardous waste incinerators, 40 CFR 60 Subpart C for municipal waste combustors (CAA), and 40 CFR §761.47 for PCB treatment (TSCA). Some states also have separate regulations pertaining to incinerators for hazardous and nonhazardous waste.

Incineration is a process of burning, resulting from the rapid oxidation of substances. Factors that affect the combustion process include:

- Waste feed composition and products formed by the reactants
- Energy transferred within the combustion system and the amounts of auxiliary fuel added
- Temperature of the process
- Thermodynamic and kinetic factors
- Heat transferred during the process and heat distribution within the system
- Turbulent mixing
- Residence time of the waste feed stream in the system

Proper operation of an incineration system must take into account all of these factors in order to achieve complete combustion and maximum material destruction (EPA 1989).

When complete combustion occurs, the waste element is converted into a simpler, nonhazardous product. Examples of conversion products are presented in Table 10.1.

Products of incomplete combustion can form if the incineration system is not operated optimally. These include carbon monoxide, soot, and organic breakdown products which vary based on constituents in the waste stream.

Key components to the combustion process include the fuel source, oxidization component, and the waste. The fuel is the energy source for the process and generally is provided by materials with carbon-carbon or carbon-hydrogen bonds such as coal, fuel oils, or natural gas. The oxidization component, which is normally air, reacts with the fuel or waste (if the waste has a fuel component) to transform the potential energy in the fuel into thermal energy. Additionally, a substance that does not participate chemically in the combustion process may be present. These substances include excess oxygen (i.e., the amount of oxygen above 100% of theoretical air), nitrogen, moisture, or inorganic constituents in the waste. When present, these substances alter the process physically in that they can serve as a thermal sink to limit overall temperature rise during the process.

The waste stream can alter the combustion process significantly by virtue of its fuel value or lack thereof. Combustibility of a waste stream is mainly dependent on the individual chemical constituents, but reactions between chemicals within a waste stream also can affect the combustion process. EPA has performed experiments on over 100 hazardous waste constituents in order to develop an incinerability ranking system (EPA/600/J-90/496 1990). Originally, EPA used heat of combustion to rank incinerability of organic hazardous constituents. Recently, ranking based on thermal stability in the post-flame zone has been established.

There are several types of incinerators, including rotary kilns, fluidized bed incinerators, multiple-chamber incinerators or combustors, liquid injection incinerators, and others. Although the internals of the systems are somewhat different, they all utilize the same combustion principles and usually provide:

- Two or more incineration chambers with burners, or the equivalent
- Waste feed charging systems, including solid and liquid input systems
- Ash removal systems
- Heat recovery systems
- Pollution control devices such as baghouses or scrubbers

An example of a dual chamber incinerator is presented in Figure 10.12.

Mobile/transportable incineration treatment has been used at Superfund and other cleanup sites (EPA/540/2-90/014 1990). The process has demonstrated effectiveness for use with contaminated soils, sludges, and liquids that contain halogenated and nonhalogenated volatiles, halogenated and nonhalogenated semivolatiles, PCBs, and dioxins and furans.

TABLE 10.1   Examples of Products of Complete Incineration

Waste Constituent	Product of Complete Incineration
Copper	Copper oxide
Sodium	Sodium hydroxide
Potassium	Potassium hydroxide
Fluoride	Hydrogen fluoride or fluorine
Chloride	Hydrogen chloride or chlorine
Carbon	Carbon dioxide

Source: Information from EPA (1989).

Pyrolysis

Pyrolysis is the chemical decomposition of waste brought about by heating the material in the absence of oxygen. This thermal destruction process is performed in a two-chamber system. In the primary chamber, the wastes are heated, separating the volatile components, such as combustible gases and water vapor, from the nonvolatile ash, such as metals and salts. In the secondary chamber, volatile components are burned under the proper air, temperature, time, and turbulence to destroy any remaining hazardous components. Direct or indirect heating can be used.

Systems utilizing pyrolysis can be designed for batch burning of drummed or containerized material or for continuous processing of flowable solids and

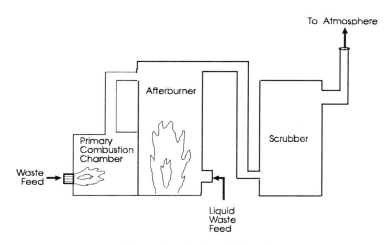

FIGURE 10.12   Example of a Dual Chamber Incinerator.

liquids. The hot combustion gases from the secondary chamber can be passed through a boiler to recover energy. This technology is suitable for treatment of viscous liquids, sludges, solids, high-ash material, salts. Metals or halogenated wastes that are not conducive to conventional incineration also may be treated using pyrolysis (EPA 1987).

Other Thermal Destruction Methods

Some industrial boilers can use limited amounts of wastes as supplemental fuels. Likewise, blast furnaces, which may reach up to 3400°F, may use high heat content hazardous waste to supplement the fuel requirements for the unit. Permitting of systems that burn hazardous waste is regulated under RCRA in 40 CFR Part 270.

Molten salt destruction uses combustion as a method for burning organic material while, at the same time, sorbing the objectionable by-products of that combustion from the effluent gas stream (EPA 1987). This process of simultaneous combustion and sorption is accomplished by mixing the air and waste into a pool of molten sodium carbonate. The melt is maintained at temperatures between 1500°F to 2000°F, which causes the hydrocarbons of the organic matter to be oxidized to carbon dioxide and water, while elements such as phosphorus, sulfur, arsenic, and halogens react with the sodium carbonate. These by-products are retained in the melt as inorganic salts and eventually build up and must be removed in order that the molten bed remain fluid and retain its ability to absorb acidic gases. An ash concentration in the melt of up to approximately 20% by weight is acceptable.

This method can be used to treat low ash or high chlorine content wastes. Low water content is required and the molten salt produced can be corrosive. The neutralization of acid gases results in the formation of other salts that can change the fluidity of the bed, which would require frequent replacement of the material. Used salts must be landfilled.

CONCLUSION

There are many types of hazardous waste treatment technologies. Use for specific waste streams is determined by many factors, including waste constituents and concentrations, waste amounts, removal efficiency requirements, energy demands, and others. During the 1970s and 1980s, there was much focus in the United States to develop waste treatment technologies that provided maximum removal efficiencies in order to meet increasingly stringent wastewater effluent limits, air emission guidelines, land-based disposal restrictions, and other requirements.

Another focus during the 1980s was design and operation of land disposal facilities, including land treatment facilities, landfills, and underground injection systems. These disposal technologies are discussed in the next chapter. It is likely that treatment and disposal technologies will remain important during the 1990s and beyond. In addition, the topic of pollution prevention, which incorporates concepts of source reduction and other waste minimization methods, is considered integral to waste management. This topic is discussed in Chapter 15.

REFERENCES

Department of Energy, "Decontamination of Low-Level Wastewaters by Continuous Countercurrent Ion Exchange," DE90 011077/XAB, Oak Ridge National Laboratory, Oak Ridge, TN, 1990.

Jackman, Alan P., and Robert L. Powell, *Hazardous Waste Treatment Technologies*, Noyes Data Corporation, Park Ridge, NJ, 1991.

Perry, Robert H., and Don Green, *Perry's Chemical Engineers' Handbook*, 6th ed., McGraw-Hill, New York, 1984.

U.S. Environmental Protection Agency, "Compendium of Technologies Used in the Treatment of Hazardous Waste," EPA/625/8-87/014, Center for Environmental Research Information, Cincinnati OH, 1987.

U.S. Environmental Protection Agency, "Incinerability Ranking Systems for RCRA Hazardous Constituents," EPA/600/J-90/496, Risk Reduction Engineering Laboratory, Cincinnati, OH, 1990.

U.S. Environmental Protection Agency, "Incineration of Solid Waste," EPA/600/J-89/531, Risk Reduction Engineering Laboratory, Cincinnati, OH, 1989.

U.S. Environmental Protection Agency, "Mobile/Transportable Incineration Treatment," EPA/540/2-90/014, Risk Reduction Engineering Laboratory, Cincinnati, OH, 1990.

U.S. Environmental Protection Agency, "Overview of Metals Recovery Technologies for Hazardous Waste," EPA/600/D-91/026, prepared by PEI Associates, Inc., for Risk Reduction Engineering Laboratory, Cincinnati, OH, 1990.

U.S. Environmental Protection Agency, *Solvent Waste Reduction*, prepared by ICF Consulting Associates for U.S. Environmental Protection Agency, published by Noyes Data Corporation, Park Ridge, NJ, 1990.

U.S. Environmental Protection Agency, "Treatment Technology Background Document," PB91-160556, Office of Solid Waste, Washington, D.C., 1991.

Water Environment Federation (formerly Water Pollution Control Federation), "Hazardous Waste Treatment Processes," Manual of Practice FD–18, prepared by Task Force on Hazardous Waste Treatment, Alexandria, VA, 1990.

BIBLIOGRAPHY

Arozarena, M.M., et al. (1990), *Stabilization and Solidification of Hazardous Wastes*, Noyes Data Corporation, Park Ridge, NJ.

API (1988), "Evaluation of the Treatment Technologies for Listed Petroleum Refinery Wastes," API Publication 4465, American Petroleum Institute, Washington, D.C.

ASTM (1989), *Environmental Aspects of Stabilization and Solidification of Hazardous and Radioactive Wastes*, Special Technical Publication 1033, Cote and Gilliam, eds., American Society of Testing and Materials, Philadelphia.

Bonner, T., et al. (1981), *Hazardous Waste Incineration Engineering*, Noyes Data Corporation, Park Ridge, NJ.

Breton, M., et al. (1988), Treatment Technologies for Solvent Containing Wastes, Noyes Data Corporation, Park Ridge, NJ.

Burton, Dudley J., and K. Ravishankar (1989), *Treatment of Hazardous Petrochemical and Petroleum Wastes*, Noyes Data Corporation, Park Ridge, NJ.

Conner, Jesse R. (1989), *Chemical Fixation and Solidification of Hazardous Wastes*, Van Nostrand Reinhold, New York.

DOE (1990), "Evaluation of Prospective Hazardous Waste Treatment Technologies for Use in Processing Low-Level Mixed Wastes at Rocky Flats," DE91 002535/XAB, prepared by S.C. McGlochlin et al., EG and G Rocky Flats, Inc., Golden, CO for Department of Energy, Washington, D.C.

_____ (1990), "Technologies to Remediate Hazardous Waste Sites," DE90 011946/ XAB, prepared by J.W. Falco, Battelle Pacific Northwest Labs, Richland, WA for Department of Energy, Washington, D.C.

_____ (1991), "Application of High Level Waste-Glass Technology to the Volume Reduction and Immobilization of TRU, Low Level, and Mixed Wastes," DE91 009319/XAD, prepared by D.F. Bickford et al., Westinghouse Savannah River Co., Aiken, SC, for Department of Energy and American Nuclear Society, Washington, D.C.

_____ (1991), "Long-Term Durability of Polyethylene for Encapsulation of Low-Level Radioactive, Hazardous, and Mixed Wastes," DE92 000486, Brookhaven National Laboratory, Department of Energy, Upton, NY.

Freeman, H. (1989), Standard Handbook of Hazardous Waste Treatment and Disposal, McGraw-Hill, New York.

Holden, Tim et al. (1989), *How to Select Hazardous Waste Treatment Technologies for Soils and Sludges*, Noyes Data Corporation, Park Ridge, NJ.

McArdie et al. (1988), *Treatment of Hazardous Waste Leachate*, Noyes Data Corporation, Park Ridge, NJ.

NSF (1986), "In Situ Electrochemical Method for Removing Environmental Pollutants," NSF/ECE-86001, prepared by F. Walsh, Tracer Technologies, Inc., Sommerville, MA, for National Science Foundation, Washington, D.C.

NTIS (1992), "Best Demonstrated Available Technology for Pollution Control and Waste Treatment: January 1989–January 1992," PB92-802164, National Technical Information Service, Department of Commerce, Springfield, VA.

_____ (1992), "Biological Industrial Waste Treatment: January 1980–February 1992," PB92-802347, National Technical Information Service, Department of Commerce, Springfield, VA.

_____ (1992), "Medical Wastes: January 1980–December 1991," PB92-801760, National Technical Information Service, Department of Commerce, Springfield, VA.

_____ (1992), "Mixed Wastes: January 1985–December 1991," PB92-801703, National Technical Information Service, Department of Commerce, Springfield, VA.

_____ (1992), "Waste Treatment: Chemical Industry: January 1980–March 1992," PB92-802784, National Technical Information Service, Department of Commerce, Springfield, VA.

———— (1992), "Waste Treatment by Reverse Osmosis and Membrane Processes: January 1980–January 1992," PB92-802172, National Technical Information Service, Department of Commerce, Springfield, VA.

———— (1992), "Waste Water Treatment, January 1980–November 1991," PB92-801562, National Technical Information Service, Department of Commerce, Springfield, VA.

Patterson, J.W. (1985), *Industrial Wastewater Treatment Technology*, 2nd ed., Butterworth Publishers, Stoneham, MA.

Reynolds, J.P., R.R. Dupont, and Louis Theodore (1991), *Hazardous Waste Incineration Calculations: Problems and Software*, John Wiley & Sons, New York.

Theodore, Louis and J.P. Reynolds (1987), *Introduction to Hazardous Waste Incinerators*, John Wiley & Sons, New York.

U.S. Congress (1989), "Partnerships under Pressure: Managing Commercial Low-Level Radioactive Waste, OTA-0-426, Office of Technology Assessment, U.S. Government Printing Office, Washington, D.C.

———— (1991), "Long-Lived Legacy: Managing High-Level and Transuranic Waste at the DOE Nuclear Weapons Complex," OTA-BP-O-83, Office of Technology Assessment, U.S. Government Printing Office, Washington, D.C.

U.S. EPA (1986), "Best Demonstrated Available Technology (BDAT) Background Document for FOO1-F005 Spent Solvents," Vol. 1, EPA/530-SW-86-056, U.S. Environmental Protection Agency, Washington, D.C.

———— (1986), "Handbook for Stabilization/Solidification of Hazardous Waste," EPA/540/2-86/001, U.S. Environmental Protection Agency, Washington, D.C.

———— (1986), "Microbiological Decomposition of Chlorinated Aromatic Compounds," EPA/600/2-86/090, U.S. Environmental Protection Agency, Washington, D.C.

———— (1987), "Municipal Waste Combustion Systems Operation and Maintenance Study," EPA/340/1-87/002, prepared by Allen Consulting and Engineering for U.S. Environmental Protection Agency, Washington, D.C.

———— (1988), "High Temperature Thermal Treatment for CERCLA Waste: Evaluation and Selection of On-Site and Off-Site Systems," EPA/540/X-88/006, U.S. Environmental Protection Agency, Washington, D.C.

———— (1988), "Technologies for the Recovery of Solvents from Hazardous Wastes," EPA/600/J-88/365, prepared by R.A. Olexsey et al., Hazardous Waste Engineering Research Laboratory, U.S. Environmental Protection Agency, Cincinnati, OH.

———— (1988), "Technology Screening Guide for Treatment of CERCLA Soils and Sludges," EPA/540/2-88/004, U.S. Environmental Protection Agency, Washington, D.C.

———— (1989), "Activated Sludge Treatment of Selected Aqueous Organic Hazardous Waste Compounds," EPA/600/D-89/271, prepared by M.K. Koczwara, J.E. Park, and R.J. Lesiecki, Risk Reduction Engineering Laboratory, U.S. Environmental Protection Agency, Cincinnati, OH and Cincinnati University, Cincinnati, OH.

———— (1990), "Chemfix Technologies, Inc. Solidification/ Stabilization Process, Clackamas, Oregon: Technology Evaluation Report," EPA/540/5-89/011A, prepared by E.F. Barth, Office of Emergency and Remedial Response, U.S. Environmental Protection Agency, Washington, D.C.

―――― (1990), "Incineration Research Facility," EPA/600/J-90/274, prepared by R.C. Thurnau and C.R. Dempsey, Risk Reduction Engineering Laboratory, U.S. Environmental Protection Agency, Cincinnati, OH.

―――― (1990), "Results from the Stabilization Technologies Evaluated by the SITE Program," EPA/600/D-90/232, prepared by P.R. de Percin, Risk Reduction Engineering Laboratory, U.S. Environmental Agency, Cincinnati, OH.

―――― (1990), "Site Demonstration of the Chemfix Solidification/Stabilization Process at the Portable Equipment Salvage Company Site," EPA/600/J-90/021, prepared by E.F. Barth, Risk Reduction Engineering Laboratory, U.S. Environmental Protection Agency, Cincinnati, OH.

―――― (1990), "Sixteenth Annual Hazardous Waste Research Symposium: Remedial Action, Treatment and Disposal of Hazardous Waste," Risk Reduction Engineering Laboratory, CERI-90-04, U.S. Environmental Protection Agency, Cincinnati, OH.

―――― (1990), "Slurry Biodegradation." EPA/540/2-90/016, U.S. Environmental Protection Agency, Washington, D.C.

―――― (1990), "Solvent Extraction Treatment," EPA/540/2-90/013, prepared by Science Applications International Corporation for U.S. Environmental Protection Agency, Cincinnati, OH.

―――― (1991), *Encyclopedia of Environmental Control Technology*, Vol. 4, "Hazardous Waste Containment and Treatment," EPA/600/D-91/088, U.S. Environmental Protection Agency, Cincinnati, OH.

―――― (1991), "Review of Treatment for Hazardous Waste Streams," EPA/600/D-91/088, prepared by D.W. Grosse, Hazardous Waste Engineering Research Laboratory, U.S. Environmental Protection Agency, Cincinnati, OH.

―――― (1991), "Treatment of Hazardous Waste with Solidification/Stabilization," EPA/600/D-91/061, prepared by C.C. Wiles, Risk Reduction Engineering Laboratory, U.S. Environmental Protection Agency, Cincinnati, OH.

11

DISPOSAL TECHNOLOGIES

Disposal of hazardous materials or hazardous waste is necessary when alternate methods such as recycling or treatment are not applicable or are cost prohibitive. Some of the more common disposal technologies are land treatment, landfill, and underground injection. Land treatment facilities and landfills are regulated under RCRA. Underground injection control systems are regulated under RCRA and under the Safe Drinking Water Act (SDWA). Land disposal restrictions are defined under RCRA and apply to facilities permitted to dispose of wastes using any disposal technology. This chapter gives an overview of the major types of disposal technologies used by industry and includes a discussion of regulations pertaining to the technology, key design elements, facility management practices, and facility closure. In addition, information pertaining to land disposal restrictions is provided and disposal of radioactive waste is addressed.

LAND TREATMENT

Land treatment is regulated under RCRA, and regulations that specifically address this technology are detailed in 40 CFR 264 and 265 Subpart M, 40 CFR 267 Subpart E, and 40 CFR Part 270. Land treatment typically is used to convert waste into a nonhazardous state through degradation, transformation, or immobilization.

Design Elements

General Considerations.   There are several general considerations that should be addressed during the design of a land treatment facility. These include the characteristics of the waste, the hydrogeological and climatological characteristics of site, the waste/soil interactions, and other considerations.

Waste Characteristics.   Waste characteristics such as chemical, physical, and biological properties should be evaluated to determine appropriateness of land treatment technology for a given waste. Information from agency documents and technical literature can provide useful case studies that can serve as a prescreening method for determining expected treatment efficiencies for selected wastes.

Site Characteristics.   Site characteristics need to be evaluated carefully during the design phase of the facility in order to ensure that the waste does not migrate to underlying groundwater and in order to optimize the facility. Characteristics that are pertinent include the site's hydrogeology, groundwater table characteristics, and climate. Hydrogeology information should include information about the geological strata, aquifers, soil type and other information. Groundwater table characteristics include items such as the location of perched zones and groundwater flow patterns. Climate characteristics include freeze/thaw and wet/dry cycles, temperature variations, and other climate-related characteristics.

Waste/Soil Interaction.   Testing must be performed to determine adequacy of waste/soil interaction prior to selecting land treatment as a disposal technology. Bench and field testing can be used to determine final fate of contaminants in soil as well as assimilative capacity of soils for waste constituents, degradation factors, immobilization, and potential pathways of contaminants to air, surface water, and groundwater.

Other Considerations.   Other items to consider when designing a land treatment facility include soil erosion properties, proximity of the facility to surface water, proximity of users of groundwater and surface waters, air quality considerations, and the potential health risks caused by human exposure to waste constituents.

Design Features.   The specific design and layout of a land treatment facility will be based on terrain, soil type, expected loading rates, and the types of wastes being treated. The main feature of the treatment facility is the treatment zone. Other components include a monitoring system, runoff management system, waste staging area, road system, and security. An example of a layout

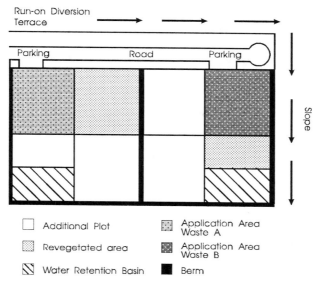

FIGURE 11.1   Example of a Lay-out for a Land Treatment Facility.

of a land treatment unit is shown in Figure 11.1. As shown in the figure, the treatment unit is divided into sections. If more than one waste type is applied at the treatment facility, the use of a separate section for each waste is common in order to simplify loading rate calculations and recordkeeping.

Treatment Zone.   The treatment zone consists of two sections—a zone of incorporation and the lower treatment zone. The maximum depth of the entire treatment zone, as specified in the regulations, is 5 feet.

The lower boundary of the lower treatment zone should be at least 3 feet above the highest seasonal level of the water table. The depth of the lower boundary also should be well above any perched water table, and it should be adjusted to accommodate water table variations in a particular section, if necessary (DOE 1987). An example of a profile of a land treatment unit incorporating these features is shown in Figure 11.2.

Other Design Components.   Monitoring of the system is an integral part of the design and must include methods for assessing leachate/contamination migration as well as treatment efficiency. Monitoring methods used to make this assessment are discussed in the next section.

Run-on and run-off management is an important part of the design. Run-on control can be achieved using terracing and other diversion mechanisms. The design also should include features that eliminate or minimize run-off

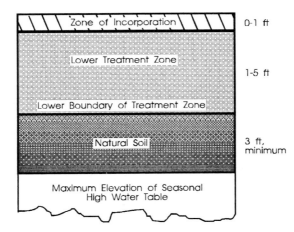

FIGURE 11.2   Example Profile of a Land Treatment Facility.

from the facility. If run-off does occur, it should be diverted to water retention basins or water filtration units. An NPDES permit for discharge of the run-off typically is required, and associated discharge monitoring and treatment requirements, as necessary, are incorporated into the permit.

Roads must be designed to withstand the vehicular loads of heavy equipment. In addition, they should be placed to allow easy access to the treatment areas and waste staging areas. Security should include, at a minimum, a fence or other barrier that protects the facility from entry by unauthorized persons. There must be a means to control entry into the facility, and "Danger—Unauthorized Personnel Keep Out" signs must be posted at each entrance to the facility.

Monitoring Program

The monitoring program for a land treatment facility is integral to its operation. Guidance on the monitoring program is given in EPA's documents SW–874, "Hazardous Waste Land Treatment" and SW–846, "Test Methods for Evaluating Solid Waste: Physical/Chemical Methods" (see references). Elements of a comprehensive monitoring program include waste monitoring, unsaturated zone and groundwater monitoring, run-off monitoring, treatment zone monitoring, vegetation monitoring, and air quality monitoring.

Waste Monitoring.   Waste streams must be checked periodically for changes in composition. A waste monitoring plan should be developed to ensure allowable cumulative applications are met and that limiting waste constituents stay at uniform levels during application.

Unsaturated Zone Monitoring.   Two aspects of the unsaturated zone must be monitored in order to determine whether hazardous constituents have migrated out of the treatment zone. The soil itself should be analyzed from soil cores and soil-pore liquid should be analyzed using monitoring devices such as lysimeters. The samples are collected at a frequency defined in the permit, and are compared to background samples to determine whether there is a statistically significant change over background values for constituents monitored.

Groundwater Monitoring.   All land treatment facilities are required by regulation to have a groundwater monitoring program. Details for groundwater monitoring system design and installation are presented in the next chapter. Groundwater monitoring supplements unsaturated zone monitoring.

Run-off Monitoring.   When NPDES permits are issued for discharge of run-off from the land treatment facility, monitoring requirements outlined in the permit must be met. Normally, a flow measuring device and an automated sampler will be required in order to acquire flow-proportional samples. For run-off management systems that contain and treat the run-off before discharge, composite (grab) sampling and analysis prior to discharge normally is required.

Treatment Zone Monitoring.   Monitoring the treatment zone is necessary in order to evaluate soil pH, moisture content, and degradation rates of organic and other waste constituents. Additionally, rates of accumulation are evaluated to assess loading rates and to estimate the active life of the facility.

Vegetation Monitoring.   Where food chain crops are grown, analysis of the vegetation at the land treatment facility will ensure that harmful quantities of metals or other toxic constituents are not accumulating in the plants. This is normally a part of the post-closure requirements.

Air Monitoring.   Air monitoring may be required to assure that waste constituents are not leaving the site through wind dispersal. Additionally, personal sampling or another means of evaluating worker exposure to volatile constituents may be part of the overall land treatment facility monitoring program.

Operating Records

As part of a comprehensive operating record, and as a good management practice, numerous records pertaining to the facility's operation should be maintained. Examples of these records are presented in Table 11.1.

A good management practice is to maintain all records pertaining to land treatment at the facility for the life of the facility. A record of waste disposal locations and quantities must be submitted to the Regional Administrator

TABLE 11.1 Examples of Records Maintained at a Land Treatment Facility

Initial Site and Waste Assessment Records

- USGS maps and other topography information
- Hydrogeological surveys
- Initial waste and soil analyses
- Climate information

Inspections of Facility (Signed and Dated)

- Inspections of the treatment area for excessive moisture
- Inspections of the security of the facility
- Inspection of berms and run-off and run-on management systems
- Inspection of road conditions

Waste Analysis Records

- Waste analysis plan
- Waste analysis records
- Documentation of changes in waste composition

Monitoring Data

- Data from groundwater monitoring
- Data from soils monitoring
- Data from retention pond monitoring
- Data from soil pore monitoring
- Data from other monitoring
- Documentation of statistically significant changes below the treatment zone

Records of All Waste Applications

- Description of wastes applied to the facility
- Records of dates, amounts, waste types, and rates of application

Records of Climate Parameters during Operation

- Rainfall records
- Records of unusual climatic events during operation

Records of Vegetation Planting

- Records of vegetative planting including dates, crop types, and area covered
- Records of fertilization of crops, including fertilizer type and amounts
- Records of crop growth

Maintenance Records

- Records of maintenance schedules and activities
- Records of decontamination of equipment

Records of Run-Off Management

- NPDES permit and monitoring reports

TABLE 11.1 *Continued*

- Documentation of breaches of run-off retention
- Documentation of water depth in retention basins after rainfull events
- Records of dates and times when the retention basins were emptied
- Other documents required in the permit

Records of Treatment Zone Activities

- Records of tilling activities
- Documentation of control methods to demonstrate that wastes were not applied below the treatment zone lower boundary
- Records of measures to control soil pH
- Records of measures to control the moisture content of the soil

Manifest Documents and Records

- For wastes accompanied by a manifest, records of where the waste was applied, cross-referenced with the manifest number
- For wastes accompanied by a manifest, records of quantities and application dates
- Copies of the manifest and certifications by the generator that the waste is not land-restricted

Accidents or Incidents

- Records of spills during the operation of the facility
- Records of personnel injuries
- Records of breaches of security
- Records of other unplanned events during the operation of the facility

Source: Information from DOE (1987).

and local land authority upon closure of the facility. In addition, records pertaining to corrective action, as applicable, should be kept and will be required upon closure.

Applications

Use of land treatment has been significantly curtailed for many hazardous waste sludges and other hazardous wastes as a result of land disposal restrictions. (Land disposal restrictions are discussed later in the chapter.) These restrictions included listed metal-bearing wastes from petroleum refineries, which once were primary candidates for disposal using this technology. Additionally, the owner or operator must demonstrate, prior to application of a waste, that the hazardous constituents can be completely degraded, transformed or immobilized in the treatment zone. Approvals for adequate demonstration have been difficult to obtain. Land treatment is now typically used for nonhazardous waste.

Closure

The cumulative amount of any waste that can be applied to a given parcel of land usually is predetermined by the concentration of metals in the waste and the acceptable contamination levels for the metals. For many wastes, metals are present only in trace amounts, which allows the facility to have long-term use. Once a limit for a specific metal is reached, however, the land treatment facility must be closed.

During the closure period, the owner or operator of the facility must continue all operations necessary to maximize degradation, transformation, or immobilization of the hazardous constituents within the treatment zone. When planted, the vegetative cover cannot substantially impede continued degradation of the waste. The owner or operator must maintain the run-off management system, control wind dispersal of hazardous waste, and comply with food-chain crop regulations. Additionally, the owner or operator must continue to monitor the unsaturated zone, except that soil-pore liquid monitoring may be terminated 90 days after the last application of waste to the treatment zone. These requirements do not apply if the owner or operator can demonstrate that the level of hazardous constituents in the treatment zone soil does not exceed the background value of those constituents by an amount that is statistically significant. Post-closure requirements are similar.

LANDFILL TECHNOLOGY

Technology standards for hazardous waste landfill design were promulgated as a result of the Hazardous and Solid Waste Amendments (HSWA) of 1984. This set of regulations covers requirements for landfill liners, leachate collection and removal systems, and leak detection systems. These regulations are detailed in 40 CFR 264 and 265 Subpart N, 40 CFR 267 Subpart C, and 40 CFR Part 270.

Liners for Landfills

There are several requirements that must be followed in the design of liners for new landfills. Key to landfill system design is the requirement for all new landfill systems to have double liners. Information pertaining to liners and other aspects of hazardous waste landfill systems is published by EPA as a seminar publication entitled "Requirements for Hazardous Waste Landfill Design, Construction, and Closure," and in EPA SW-869, "Lining of Waste Impoundment and Disposal Facilities" (see references).

The top liner, or primary liner, must be designed and constructed of materials that prevent the migration of hazardous constituents into the liner during the active life of the facility and during post-closure care. These liners typically consist of a flexible membrane liner (FML). The bottom liner, or secondary liner, must be a composite liner, with two components. These components consist of a FML for the upper component and compacted clay for the lower component. The upper component, like the top liner, must prevent migration of hazardous constituents into this component during the active life of the facility and during post-closure. The lower component must minimize the migration of the hazardous constituents if a breach in the upper component were to occur. An example of an acceptable liner configuration, with other system components, is presented in Figure 11.3.

Flexible Membrane Liners (FMLs). FMLs are liners made of polymers such as thermoplastics, crystalline thermoplastics, thermoplastic elastomers, and elastomers. These liners typically are 60 to 100 mils in thickness for most landfill applications (EPA 1989). Selected examples of polymers and liner materials are presented in Table 11.2.

Key considerations in selecting a FML include chemical compatibility of the FML with the waste leachate, stress-strain characteristics, survivability, and permeability.

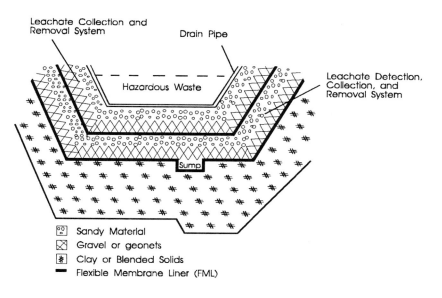

Source: Adapted from EPA (1989).

FIGURE 11.3 Example of a Landfill with Liners and a Leachate Management System.

TABLE 11.2   Selected Examples of Polymers and Liner Materials Used for Flexible Membrane Liners

Polymer	Liner Material
Thermoplastic	Polyvinyl chloride, nylon, polyester
Crystalline thermoplastic	High density polyethylene, linear low density polyethylene
Thermoplastic elastomers	Chlorinated polyethylene, chlorylsulfonated polyethylene
Elastomers	Neoprene, ethylene propylene diene monomer, urethane

Source: Information from EPA (1989).

Chemical Compatibility of the FML with the Waste Leachate.   Chemical degradation of a FML can occur if the leachate is incompatible with the liner material in any of several ways. High or low pH may negatively affect material, and high temperatures may accelerate degradation. Additionally, chemical constituents in the leachate may cause damage to the material. EPA Method 9090, which is used to assess chemical compatibility with the leachate, must be performed on the FML prior to installation. The test requires the material to be immersed in the site-specific chemical environment for 120 days at two different temperatures.

Stress-Strain Characteristics.   Stress-strain characteristics include tensile strength of the material, thermal expansion characteristics, and uniaxial and biaxial stress-strain performance. Depending on the application (i.e., side slopes vs. landfill bottom), some materials will not be suitable for installation because of potential stress-strain failure, even if chemical compatibility is acceptable.

Survivability.   Survivability of unexposed polymeric geomembranes can be determined using various test methods. Tests that indicate physical properties such as puncture strength, tear resistance, seam strength, and brittleness can be used to measure potential survivability of the FML in the field. Other characteristics that could affect survivability include environmental stress cracking, tensile properties at elevated temperatures, and dimensional stability.

Permeability.   Permeability of a FML can be evaluated using ASTM E 96–80—Test Methods for Water Vapor Transmission. A permeation rate for the liner can be determined from the test and evaluated for acceptability in field applications.

Quality Control during Placement of FMLs.   Quality control during placement of the FML is as important as material selection. Quality assurance factors during the placement of FMLs include ensuring that the material is not flawed, bedding considerations, seaming, temperature considerations, and in-situ testing.

Ensuring the FML Material is not Flawed.   All FML material should be inspected carefully before installation to ensure that damage has not occurred from shipping, storage, or placement of the liner material. Some high crystalline FMLs should not be folded when shipped, so inspections for these types of materials should include signs of stress failure such as white lines on the fabric.

Bedding Considerations.   Bedding preparations should be made prior to placing the membrane. These include subsurface compaction and treatment with compatible herbicides (if necessary), top layer soil compaction, and ensuring that the surface is free of rocks, roots, and water.

Seaming.   Seams are a likely place for liner leakage. Seaming and seam repair should be performed by a qualified seamer. Visual inspection should be made of each seam.

Temperature Considerations.   During installation and particularly during seaming, materials should not be exposed to rain or dust. If the temperature drops below 50°F, preheaters with chambers around them may be used to keep the FML warm.

In-Situ and Other Testing.   In-situ tests on seams should be performed throughout the installation process. These include nondestructive seam testing, mechanical point stress testing, and ultrasonic testing. Membrane property tests should be performed on samples of all batches of material that are used for the FML.

Compacted Soil or Clay Component of the Composite Liner.   Compacted soils or clays used for the lower component of composite liners must be at least 3 feet thick and have a hydraulic conductivity less than or equal to 1×10^{-7} centimeters per second (cm/sec). To achieve this, the soil material must have at least 20% fines and a plasticity index (PI) of greater than 10% (EPA 1989). Some soils with extremely high PIs in the range of 30% to 40% and above should be avoided due to stickiness, which makes them difficult to work with in the field. Blended soils are used when native soils cannot offer acceptable hydraulic conductivity when compacted. Sodium bentonite and calcium bentonite are common additives to native silts. Other additives include lime or cement.

Problems that can arise during the installation and operation of clay liners include:

- Uneven soil compaction resulting in varying hydraulic conductivity rates and PIs
- Varied materials within the soil structure, which result in varying hydraulic conductivity rates and PIs
- Potential channeling throughout the system because of uneven compaction or varied soil structure
- Uneven confining stress within the system resulting in uneven hydraulic conductivity within the system
- Attack by waste leachate acids and bases resulting in higher hydraulic conductivity within the system

As a result of these problems, EPA allows the use of clay liners as the lower component in a composite liner, but not as a stand alone liner in place of a composite liner.

Quality Control during Construction of Clay Liners.   Since the liner system is one of the most important aspects of the landfill facility, care must be taken to assure quality control during construction of clay liners. This topic is discussed in detail in the EPA's landfill seminar document (EPA 1989). Several factors can influence the quality of compacted soil or clay liners and must be managed carefully. These include soil water content, method of compaction, bonding between lifts and in-situ testing.

Soil Water Content.   The liner should be constructed when the water content of the soil is wetter than the optimum value for achieving the smallest hydraulic conductivity. If the soil is drier than optimum, hard clods are formed, which are difficult to break at compaction.

Method of Compaction.   "Kneading" compaction, wherein the soil is kneaded or remolded with a footed roller, is the best compaction method for breaking up clods. Heavy equipment is preferred to ensure maximum density of the soil liner.

Bonding between Lifts.   A fully penetrating footed roller is the preferred equipment to ensure that soil is compacted evenly between lifts. This is necessary to prevent defects from creating a hydraulic continuity from lift to lift.

In-Situ Testing.   Tests of prepared and compacted soils should be made continually during the construction process. Tests should include assessment of soil moisture, soil density, clod size, and hydraulic conductivity.

Alternative Barrier Materials.   Alternative barriers to FMLs and clay liners currently are being developed, although none are approved yet by EPA for use in hazardous waste landfills as a replacement to clay liners or FMLs (Grube and Daniel 1991). These alternative barriers include manufactured low-permeability, blanket-like products that contain a thin layer of bentonite clay between two layers of geotextiles. Other alternative barrier products include spray-on concretes, grout formulas, asphalt coatings, resins, epoxies, latexes, and other plastic formulations.

These materials are advertised as inexpensive, effective, and easy to install. Data pertaining to performance for various field applications, however, is insufficient to determine reliability. Current uses for these materials include use as a back-up liner to a FML and use as a temporary cap for a hazardous waste landfill or as a permanent cap for a nonhazardous waste landfill.

Leachate Management Systems

The leachate collection and removal system in a landfill, like the liner, is critical to the operation of the unit. Since all landfills have the potential for leakage—first through the landfill cap and subsequently through the top liner—there must be an adequate means of collecting and removing contaminated liquid as it passes through the facility. The leachate management system consists of the primary leachate collection removal system, which is associated with the top liner, and the secondary leachate detection, collection, and removal systems, which is associated with the bottom composite liner.

Leachate is formed as rainwater and other precipitation penetrates a landfill and combines with the liquids in the wastes. Typically, the rainwater and waste liquids also will leach constituents from solids, which adds to the leachate mixture. As the leachate flows to the bottom of the landfill the liquid becomes increasingly contaminated with constituents found in the wastes. Some of these wastes may be acidic or basic or have constituents that can cause deterioration to certain materials. Thus, leachate collection and removal systems must be designed and constructed to withstand chemical attack from the leachate as well as successfully remove the leachate for proper treatment or disposal.

The primary leachate collection and removal system, located below the waste and above the top liner, must be designed to ensure that the leachate depth over the liner does not exceed 30 centimeters or 1 foot. The secondary leachate management system is installed immediately above the composite liner and is used to detect and collect any leachate that percolates through the top liner. An additional part of the system diverts surface water away from the facility's cap.

Drainage materials with a hydraulic conductivity of 1 cm/sec or more are required in the primary system to allow the liquid to flow unimpeded into perforated drain pipes. Common materials used for this application include gravel that is ¼ to ½ inch in diameter or synthetic materials, called geonets. Above the drainage system is a filtration system made of synthetic or sandy materials to protect against invading fine particles.

For adequate drainage, the primary leachate collection and removal system should have a slope greater than 1%. The perforated drain pipe should be made of a material that can withstand chemical attack as well as pressure from the weight of the landfill. The liquid is routed from the collection zone to a sump for removal. An example of a cross-sectional view of the primary leachate system is presented in Figure 11.4.

The secondary system should be designed to indicate that leachate has passed through the top liner, and it also should have the capability of removing this liquid. If the system is designed properly, detection of leachate should occur within 24 hours (EPA 1989). A common design of this system includes a large diameter pipe, which is installed between the primary and secondary liners. A pump can be submersed through the pipe for liquid detection and removal.

Landfill Closure

Closure and post-closure care of a hazardous waste landfill are regulated under 40 CFR §§264.310. Closure of a landfill typically includes capping the facility, providing a run-on and run-off management system, the installation of a gas control system, and installation of a biotic barrier (EPA 1989). Caps are normally composite caps, and consist of a flexible membrane cap placed over a low permeable clay cap. The run-on and run-off management system is located above the cap and prevents, to the extent possible, infiltration of surface water into the landfill.

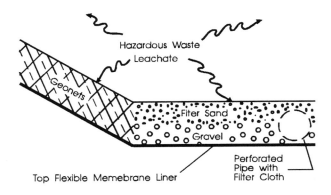

Source: Adapted from EPA (1989).

FIGURE 11.4 Cross-Sectional View of a Primary Leachate Collection System.

A gas control system can be installed directly beneath the low permeability clay cap in order to detect and vent gas that may be generated in hazardous waste landfills. Care must be made when designing and installing the gas vent system to ensure that surface water cannot drain directly into the landfill.

A biotic barrier—usually a gravel or rock layer—is installed around the cap to keep burrowing animals out of the landfill. If this barrier is applied on top of the cap, a protective layer of soil should be installed on top of the flexible membrane cap before the gravel or rock is poured to prevent tearing of the membrane.

Once a permitted hazardous waste landfill facility is closed, groundwater monitoring and other post-closure care requirements must be performed for 30 years, as required under 40 CFR Part 264 Subpart G. Leachate collection and removal systems and run-on and run-off management systems must be maintained in good order. Additionally, cap integrity must be verified periodically.

UNDERGROUND INJECTION CONTROL SYSTEMS

Underground injection control (UIC) systems, also termed deep-well injection systems, provide a method of waste disposal wherein a pretreated aqueous waste stream is pumped down a well shaft into a geologically confined zone below any potential drinking water aquifer. Chemical and other reactions that transform the waste into a less hazardous state take place within the injection zone. This type of disposal system normally is constructed and permitted at an industrial site, although some commercial facilities offer this disposal option. Because of geological or other reasons, some states have banned underground injection of hazardous waste. Hazardous waste UIC systems are regulated under RCRA and SDWA. Regulations are published under 40 CFR 267 Subpart G (RCRA) and 40 CFR 146 and 148 (SWDA).

The owner or operator of a UIC system cannot dispose of land restricted wastes in the UIC system without an exemption from the EPA administrator or delegated state authority. (Definitions of land restricted wastes and other information about land disposal restrictions are included in the next section.) In order to be granted this exemption, the UIC system must meet stringent criteria defined under 40 CFR Part 148, which includes a no migration standard. Under this standard, injection fluids cannot migrate upward out of the injection zone or laterally within the injection zone to a point of discharge or interface with an underground source of drinking water for 10,000 years. A second standard requires that before the injected fluids migrate out of the injection zone or to a point of discharge or interface with an underground source of drinking water, the fluid must no longer be hazardous because of attenuation, transformation, or immobilization of hazardous constituents within the injection zone by hydrolysis, chemical interaction, or other means. Waste treatment methods, a monitoring plan, and other information is required in support of the petition for exemption.

A reference guide published by EPA entitled "Assessing the Geochemical Fate of Deep-Well-Injected Hazardous Waste" (see references) provides information about injection well design, wastes that are compatible with this disposal method, the potential geochemical fate of these wastes, and case studies of underground injection of industrial waste. The guide also gives an overview of sources, amounts, and composition of wastes disposed of in a UIC system, and geographic distribution of hazardous waste injection wells.

Design Elements

A typical UIC system is defined by EPA (EPA/625/6-89/025a 1990) and consists of an aboveground pretreatment system, the well hole, and a monitoring system. The aboveground pretreatment system will incorporate elements such as an oil/water separator, a clarifier, a holding tank, and an injection pump. An example of this system is depicted in Figure 11.5.

The well hole always is lined with casing to prevent the hole from caving in and to prevent any waste from migrating into an overlying aquifer before it has reached the injection zone. An outer casing is placed far enough down the hole to seal the well hole below any potential drinking water sources. An inner casing is installed that extends to the injection zone. Tubing is placed inside the inner casing to serve as a conduit for the injected wastes. The annular space between the tubing and inner casing is plugged with packers at or near the end of injection tubing. Cement is used to seal the space between the outer and inner casing.

Typically, injection wells are completed using a filter pack and perforations. Open hole completion can be used in some competent formations. More information pertaining to well construction is presented in the next

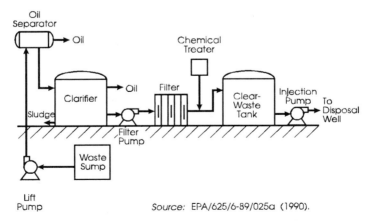

Source: EPA/625/6-89/025a (1990).

FIGURE 11.5 Example of an Aboveground Pretreatment System Used with an Underground Injection System.

chapter, which discusses well systems for groundwater monitoring and other aspects of groundwater assessment. Mechanical integrity testing is required to ensure the well has been constructed and is operating without leakage or upward movement of fluid through vertical channels adjacent to the well bore. The well should be tested at the time of installation and at a specified frequency after operation begins. EPA has published guidance for testing the wells in the manual "Injection Well Mechanical Integrity Testing" (see references). Types of well testing that are applicable to wells cased with metal or polyvinyl chloride (PVC) include:

- Pressure test—This test can detect the presence but not the location of leaks in metal or PVC casing.
- Monitoring of annulus pressure—This test can detect the presence but not the location of leaks in metal or PVC casing or tubing.
- Temperature logs, noise logs, or radioactive tracer logs—Any of these tests can detect the presence and location of leaks in metal or PVC casing or tubing. These tests also can detect the presence and location of fluid moving behind the casing. Some limitations apply to radioactive tracer logs.
- Cement bond log—This test can detect the presence and location of fluid moving behind metal or PVC casing. It cannot detect the presence of leaks in the casing or tubing.
- Casing condition log—This test provides information on corrosion spots and holes and can indicate potential failure for metal casings.

Other tests EPA plans to evaluate include a helium leak test, a "mule tail" test, and others (EPA/625/9-89/007 1990).

Monitoring Wells

Monitoring wells drilled into the injection zone at selected distances and directions from the injection well allow direct observation of formation water characteristics and can provide data about waste transformations as the waste front reaches the monitoring well. Monitoring wells placed near the injection well in the aquifer above the confining layer allow for detection of upward migration of wastes caused by failure of the casing or confining-layer. In addition to or in lieu of data collection via monitoring wells, backflushing of injected wastes can be used to sample and characterize injected waste after a prespecified residence time.

Applications

UIC systems are used for disposal of aqueous solutions containing acids, heavy metals, organics, and inorganics. The disposal method is used most extensively by the organic chemical, petrochemical, and chemical products industries, although other industries such as aerospace and related industries also use this disposal method.

Closure of a UIC System

Closure of a UIC system includes plugging and sealing the well hole and decontamination and proper closure of all surface equipment. Monitoring wells in the injection zone and in the aquifer above the confining layer should continue to be tested for waste characterization purposes as well as for evidence of migration of waste out of the confined zone.

LAND DISPOSAL RESTRICTIONS

Land disposal restriction regulations, promulgated as a result of HSWA and defined under 40 CFR Part 268, placed significant restrictions on land disposal of certain wastes, including sludges, solvents, toxic chemicals, and wastewaters. The definition of land disposal includes placement of any hazardous waste in landfills, land treatment facilities, injection wells, salt domes or salt bed formations, underground mines or caves, and other waste management units such as surface impoundments and waste piles intended for disposal purposes.

Maximum concentration limits or treatment technology-based standards have been established for land restricted wastes. If a waste meets the established concentration limits (in the waste extract or in the waste itself, depending on the waste), or if it meets treatment technology-based standards, then the waste may be land disposed. Dilution of a waste stream as a substitute for treatment is prohibited.

Wastes regulated under the land disposal restrictions include wastewaters and nonwastewaters. Wastewaters contain less than 1% by weight of total suspended solids and total organic carbon, with some exceptions. Concentration limits and acceptable technologies to treat the same constituent in the two forms will vary. Selected examples of concentration limits for land restricted wastes in the form of wastewaters and nonwastewaters are presented in Table 11.3. Selected examples of technologies that have been established as acceptable for treatment of restricted wastes are presented in Table 11.4.

Acceptable concentrations in the waste extract are defined under 40 CFR §268.41, Table CCWE, and acceptable concentrations in the waste itself are defined under 40 CFR §268.43 Table CCW. A complete listing of technologies acceptable for treatment of restricted wastes is provided under 40 CFR §268.42, Table 2. These tables and other information about land disposal restrictions—including testing, recordkeeping, certification statements, and other administrative requirements—can be found in EPA's document entitled "Land Disposal Restriction: Summary of Requirements" (see references).

TABLE 11.3   Selected Examples of Concentration Limits for Land Restricted Chemicals in the Form of Wastewaters and Nonwastewaters

Constituent	Wastewaters (mg/l)	Nonwastewaters (mg/kg)
Acetone (from waste codes F001–F005)	0.05*	0.59*
Chlordane—alpha and gamma (waste code U036)	0.0033	0.13
Chloroform (from waste codes K009 and K010)	0.10	6.0
Total Cyanides (from waste codes F011 and F012)	1.9	110
Cyclohexane (waste code U057)	0.36	NA**
Dimethyl phthalate (waste code U102)	0.54	28
Mercury (from waste code K071)	0.030	0.025*
Toluene (waste code U220)	0.080	28
Xylenes (waste code U239)	0.32	28

* Treatment standard expressed as concentration in the waste extract.
** Technology-based standard defined under 40 CFR §268.42, Table 2 applies.
Source: Information from 40 CFR §268.41, Table CCWE and §268.43, Table CCW.

DISPOSAL OF RADIOACTIVE WASTES

Low-Level Radioactive Wastes

Disposal of low-level waste (LLW) is regulated under 10 CFR Part 61. Requirements pertaining to near-surface disposal, including site suitability requirements, site design requirements, and other requirements, are detailed in 10 CFR 61 Subpart D. Site closure and stabilization requirements are found in 10 CFR §§61.27-28, 61.52, and 61.53.

Site Selection and Design.   Suitability of the site for LLW disposal is determined by the site's natural characteristics such as hydrogeology, climate, soil chemistry, and other factors (NRC 1989). Technical evaluation of the site includes:

- Evaluation of ability of the site to maximize long-term (over 500 years) stability and isolation of waste
- Evaluation of ability of the site to be characterized, modeled, and monitored

TABLE 11.4 **Selected Examples of Technologies Acceptable for Treatment of Restricted Wastes**

EPA Code	Technology Description	Waste Applications
CHOXD fb CRBN	Chemical or electrolytic oxidation followed by carbon adsorption	Wastewaters containing allyl alcohol, fluoro-acetamide, acrylic acid, cumene, formaldehyde, others
INCIN	Incineration	Concentrated wastes (nonwastewaters) containing cyclohexane, phosphine, allyl alcohol, fluoroacetamide, acrolein, strychnine and salts, pentachloroethane, others
FSUBS	Fuel substitution	Nonwastewaters of high TOC ignitable liquids, ignitable regulated residues and heavy ends, nitroglycerin, acrylic acid, cyclohexane, allyl alcohol, cumene, others
HLVIT	High level mixed radioactive waste vitrification	Radioactive high level wastes generated during the reprocessing of fuel rods

Source: Information from 40 CFR §268.42 Table 2.

- Evaluation of the location of the site to determine sufficient distance from projected population growth areas, other future developments, or known natural resources
- Evaluation of the topography to ascertain if it is free from flooding or frequent ponding
- Evaluation of the location of the site to determine if there is sufficient distance above groundwater to prevent any possible groundwater intrusion into the bottom of the disposal unit
- Evaluation of the site's potential for faulting, folding, or seismic and/or volcanic activity
- Evaluation of the ability of the site to maintain a comprehensive environmental monitoring program and to meet all performance criteria

After the selection criteria are met, the LLW disposal facility must be designed to meet basic criteria. First, the facility should be designed to minimize water contact with waste during storage, disposal, and after disposal.

Additionally, the design should minimize the need for any active maintenance after closure to ensure long-term isolation of the waste. Finally, the design should complement the site's natural characteristics and provide for erosion control and run-on and run-off management.

Environmental Monitoring and Other Management Practices.   As with other disposal facilities discussed in this chapter, LLW disposal facilities must be monitored continually throughout the life of the facility and after closure to ensure that the facility poses no threat to human health and the environment. As specified in the regulations, monitoring systems must be capable of providing early warning of releases of radionuclides before they leave the site boundaries. Typically, upgradient and downgradient groundwater monitoring wells are used to collect data, and other monitoring may be required.

Other management practices include intruder protection for some Class B and all Class C facilities. For facilities that accept Class B waste, intruder protection in the form of waste stabilization is sufficient. For facilities accepting Class C wastes, waste stabilization and deeper disposal or barriers are required. Class A waste does not require stabilization since it is cost prohibitive and impractical to require small generators such as medical and university research centers to meet these requirements. NRC has ruled that segregation of Class A from Classes B and C is sufficient (NRC 1989).

Closure of LLW Disposal Facilities.   The following standards must be met at the time of LLW disposal site closure and during post-closure:

- The site must be closed according to an approved plan.
- All Class A wastes must be segregated from Class B and C wastes.
- Class C wastes must be disposed of so that waste containers are not less than five meters below the top of the disposal unit covers.
- All waste containers must be backfilled properly with soil or other materials to reduce the possibility of cracking of the cap or barrier due to subsidence.
- The boundaries of the disposal unit must be surveyed, marked, and recorded on a map.
- Buffer zones around the disposal unit and beneath the disposal zone must be provided and maintained.

Post-closure observation and maintenance must be performed for a period of time to assure that the site is performing as expected. Transfer of control to a governmental agency can occur after the period of observation, if no problems are encountered.

High-Level Radioactive Wastes

Disposal of high-level radioactive waste (HLW) is regulated under 10 CFR Part 60. This type of waste requires permanent isolation in a geologic repository. NRC issues licenses to DOE to receive and possess source, spent nuclear, and by-product materials at a geologic repository operations area sited, constructed, or operated in accordance with the Nuclear Waste Policy Act of 1982.

Technical Performance Criteria.   Technical performance criteria for HLW repositories are defined in the regulations. All geologic repository operations areas must be designed so that, until permanent closure has been completed, radiation exposure, radiation levels, and release of radioactive materials will meet exposure standards defined by NRC and EPA. Additionally, waste must be retrievable for up to 50 years. Upon permanent closure, engineered barriers must be designed to effect substantially complete containment. Any release of radionuclides must be small and occur over long periods of time. Containment of a HLW within waste packages must be substantially complete for a period of 300 to 1,000 years after permanent closure of the facility.

Design Criteria.   The following design criteria for the geologic repository operations area must be met before a license is issued to the facility (ORAU 1988):

- A means to limit concentrations of radioactive material in air must be established.
- A means to limit time required to perform work in the vicinity of radioactive materials must be developed.
- Suitable shielding must be designed.
- A means to control access to high radiation areas or airborne radioactivity areas must be established.
- Surface facilities must be designed to control the release of radioactive materials in effluents within the limits specified by NRC and EPA.
- HLW waste packages must be designed so that the chemical, physical, and nuclear properties of the waste package and its interaction with the site environment do not compromise the performance of the waste package or the underground facility or geologic setting.
- A geologic setting must exhibit an appropriate combination of favorable conditions to reasonably assure that the performance objectives relating to isolation of the waste will be met.

CONCLUSION

Disposal of hazardous waste is used when other preferred alternatives are not available. Disposal technologies include land treatment, landfill, and underground injection, and others. Facilities using disposal technologies are extensively regulated by EPA. Minimum technology standards have been developed, monitoring of the facility during its active life and during post-closure is required, and requirements for closing the facility have been established. Additionally, EPA has published land disposal restrictions in order to mitigate risk from disposal practices. Likewise, NCR has extensively regulated disposal facilities for radioactive waste.

The next section of the book discusses methods for contamination assessment and remediation of soils and groundwater for areas where land disposal facilities and other hazardous materials and hazardous waste management units pose a threat to the environment. In addition, the next section discusses methods of pollution prevention—which are preferred from an environmental and resource management standpoint over treatment or disposal. Also included is a methodology for performing a facility assessment, which can help an owner or operator of a facility find and fix problems before they become threats to the environment, as well as ensure that the facility is in compliance with administrative requirements.

REFERENCES

Department of Energy, "Hazardous Waste Land Treatment: A Technology and Regulatory Assessment," DE88 005571, prepared by M. Overcash, K.W. Brown, and G.B. Evans, Jr. for Energy and Environmental Systems Division, Argonne National Laboratory, Argonne, IL, 1987.

Grube, Walter E., Jr., and David E. Daniel, "Alternative Barrier Technology for Landfill Liner and Cover Systems," *Proceedings of the 84th Annual Meeting of the Air and Waste Management Association*, Vancouver, June 1991.

Nuclear Regulatory Commission, "Regulating the Disposal of Low-Level Radioactive Waste: A Guide to the Nuclear Regulatory Commission's 10 CFR Part 61," NUREG/BR-0121, Office of Nuclear Material Safety and Safeguards, Division of Low-Level Waste Management and Decommissioning, Washington, D.C., 1989.

Oak Ridge Associated Universities, "A Compendium of Major U.S. Radiation Protection Standards and Guides: Legal and Technical Facts," prepared by W.A. Mills et al., Oak Ridge TN, 1988.

U.S. Environmental Protection Agency, "Assessing the Geochemical Fate of Deep-Well-Injected Hazardous Waste," EPA/625/6-89/025a, prepared by Eastern Research Group, Inc., for Center for Environmental Research Information, Cincinnati, OH, 1990.

U.S. Environmental Protection Agency, "Hazardous Waste Land Treatment," SW-874, Office of Solid Waste and Emergency Response, Washington, D.C., 1983.

U.S. Environmental Protection Agency, "Injection Well Mechanical Integrity Testing," EPA/625/9-89/007, prepared by Jerry T. Thornhill and Bobby G. Benefield, Robert S. Kerr Laboratory, Ada, OK, 1990.

U.S. Environmental Protection Agency, "Land Disposal Restrictions: Summary of Requirements," OSWER 9934.0-1A, Office of Waste Programs Enforcement, Solid Waste and Emergency Response, Washington, D.C., 1991.

U.S. Environmental Protection Agency, "Lining of Waste Impoundment and Disposal Facilities," EPA SW-869, Office of Solid Waste and Emergency Response, U.S. Environmental Protection Agency, Washington, D.C., 1983.

U.S. Environmental Protection Agency, "Requirements for Hazardous Waste Landfill Design, Construction, and Closure," EPA/625/4-89/022, prepared by Eastern Research Group, Inc., for U.S. Environmental Protection Agency, Cincinnati, OH, 1989.

U.S. Environmental Protection Agency, "Test Methods for Evaluating Solid Waste: Physical/Chemical Methods," SW–846, Office of Solid Waste and Emergency Response, Washington, D.C., 1986.

BIBLIOGRAPHY

API (1984), "The Land Treatability of Appendix VIII Constituents Present in Petroleum Industry Wastes," API Publication 4379, American Petroleum Institute, Washington, D.C.

―――― (1987), "The Land Treatability of Appendix VIII Constituents Present in Petroleum Refinery Wastes: Laboratory and Modeling Studies," API Publication 4455, American Petroleum Institute, Washington, D.C.

ASTM (1985), "Standard Test Method for Classification of Soils for Engineering Purposes," ASTM D 2487–85, Annual Book of ASTM Standards, Vol. 4.08, Philadelphia.

―――― (1991), *ASTM Standards on Geosynthetics*, 2nd ed., American Society of Testing and Materials, Philadelphia.

Brown, K.W., G.B. Evans, Jr., and B.D. Frentrup, eds. (1983), *Hazardous Waste Land Treatment*, Butterworth Publishers, Woburn, MA.

DOE (1990), "Database for Compliance with Land Disposal Restrictions," DE91 002809/XAB, prepared by M.W. McCoy, Battelle Pacific Northwest Laboratory, Department of Energy, Richland, WA.

Fogel, Samuel (1987), "Feasibility of Coal Tar Biodegradation by Land Treatment," Cambridge Analytical Associates, Inc., Boston, MA.

Goldman, L.J., et al. (1990), *Clay Liners for Waste Management Facilities*, Noyes Data Corporation, Park Ridge, NJ.

Haxo, Henry E., Jr., et al. (1985), *Liner Materials for Hazardous and Toxic Wastes and Municipal Solid Waste Leachate*, Noyes Data Corporation, Park Ridge, NJ.

Loehr, R.C., and J.F. Malina, eds. (1986), *Land Treatment: A Hazardous Waste Management Alternative*, Center for Research in Water Resources, University of Texas, Austin, TX.

Overcash, M.R., and D. Pal (1979), *Design of Land Treatment Systems for Industrial Wastes—Theory and Practice*, Ann Arbor Science Publishers, Inc., Ann Arbor, MI.

U.S. EPA (1980), "Land Disposal of Hexachlorobenzene Waste: Controlling Vapor Movement in Soils," EPA/600-2-80/119, prepared by W.J. Farmer et al., for Environmental Research Laboratory, U.S. Environmental Protection Agency, Cincinnati, OH.

―――― (1982), "Handbook for Remedial Action at Waste Disposal Sites," EPA-625/6-82/006, U.S. Environmental Protection Agency, Cincinnati, OH.

―――― (1983), "Hazardous Waste Materials: Hazardous Effects and Disposal Methods," Vol. 2, prepared by Allen Booz & Hamilton, Inc., for the U.S. Environmental Protection Agency, Washington, D.C.

———— (1984), "Use of Short-Term Bioassay to Evaluate Environmental Impact of Land Treatment of Hazardous Industrial Wastes, EPA-600/S2/84-135, prepared by K.W. Brown and K.C. Donnelly for the U.S. Environmental Protection Agency, Washington, D.C.

———— (1985), "Draft Minimum Technology Guidance on Double Liner Systems for Landfills and Surface Impoundments: Design, Construction, and Operation," EPA 530 SW/84-014, U.S. Environmental Protection Agency, Washington, D.C.

———— (1986), "Design, Construction, and Evaluation of Clay Liners for Waste Management Facilities," EPA 530/SW 86-007, U.S. Environmental Protection Agency, Cincinnati, OH.

———— (1986), "Field and Laboratory Evaluation of Petroleum Land Treatment System Closure," EPA-600/2-85/134, prepared by M.R. Overcash et al., for the U.S. Environmental Protection Agency, Washington, D.C.

———— (1986), "Permit Guidance Manual on Hazardous Waste Land Treatment Demonstrations, EPA/530/SW-84/015, U.S. Environmental Protection Agency, Washington, D.C.

———— (1986), "Permit Guidance Manual on Unsaturated Zone Monitoring for Hazardous Waste Land Treatment Units," EPA/530/SW-86/040, U.S. Environmental Protection Agency, Washington, D.C.

———— (1986), "Saturated Hydraulic Conductivity, Saturated Leachate Conductivity, and Intrinsic Permeability," EPA Method 9100, U.S. Environmental Protection Agency, Washington, D.C.

———— (1986), "Technical Guidance Document: Construction Quality Assurance for Hazardous Waste Land Disposal Facilities," EPA 530/SW 86-031, U.S. Environmental Protection Agency, Washington, D.C.

———— (1986), "Waste/Soil Treatability Studies for Four Complex Industrial Wastes: Methodologies and Results, Vol. 2, Waste Loading Impacts on Soil degradation Transformation," EPA/600/6086/003b, prepared by R.C. Sims et al., for the Robert S. Kerr Environmental Research Laboratory, U.S. Environmental Protection Agency, Ada, OK.

———— (1987), "Background Document on Proposed Liner and Leak Detection Rule," EPA 530/SW 87-015, U.S. Environmental Protection Agency, Washington, D.C.

———— (1988), "Draft Background Document on the Final Double liner and Leachate Collection System Rule," prepared by EMCON Associates for the Office of Solid Waste, U.S. Environmental Protection Agency, Washington, D.C.

———— (1988), "Model RCRA Permit for Hazardous Waste Management Facilities (Draft)," EPA/530/SW-90/049, Office of Solid Waste, U.S. Environmental Protection Agency, Washington, D.C.

———— (1990), *How to Meet Requirements for Hazardous Waste Landfill Design, Construction and Closure*, Noyes Data Corporation, Park Ridge, NJ.

———— (1990), "Mobility and Degradation of Residues at Hazardous Waste Land Treatment Sites at Closure," EPA/600/2-90/018, prepared by Raymond Loehr et al., for Robert S. Kerr Environmental Research Laboratory, Office of Research and Development, Ada, OK.

———— (1991), "Status of Land Treatment as a Hazardous Waste Management Alternative in the United States," prepared by J. Matthews et al., Risk Reduction Engineering Laboratory, Office of Research and Development, U.S. Environmental Protection Agency, Cincinnati, OH.

SECTION IV

ASSESSING AND MANAGING ENVIRONMENTAL CONTAMINATION

12

GROUNDWATER ASSESSMENT AND REMEDIATION

A basic understanding of pollution problems at an industrial or waste management facility can be obtained from an assessment of the groundwater. Sometimes, activities from past operations have caused contamination, which is evident only years later when an environmental investigation takes place. At present, more than 30,000 sites have been identified as potentially being contaminated according to a DOE overview on uncontrolled hazardous waste site cleanup programs (DOE, DE90 002215 1990). This chapter discusses methodologies for determining pollution in groundwater. Additionally, the topic of groundwater remediation is discussed.

TRANSPORT OF CONTAMINANTS IN THE SUBSURFACE

Understanding transport phenomena of contaminants as they migrate into and throughout the subsurface is an important part of groundwater assessment and remediation. Transport characteristics often are used as the basis for groundwater monitoring system design and for the selection of remediation methods. This subject is explored in detail in EPA's seminar publication entitled "Transport and Fate of Contaminants in the Subsurface" (see references). Included in the publication is a discussion of the advection-dispersion theory for evaluating contaminant transport in a saturated, porous media, as well as diffusive transport theory for low-permeability materials. Transport parameters used in most water models include hydraulic conductivity, bulk density, porosity, volumetric water content, and average linear velocity. These parameters can be derived by fairly simple field tests.

DESIGNING A GROUNDWATER MONITORING SYSTEM

The primary reason for installing a groundwater monitoring system is to provide access to groundwater in order to determine in situ groundwater conditions. If the system is installed properly, samples collected and analyzed from the wells will provide a means for evaluating groundwater contamination. An excellent guidance document on the subject of design and installation of groundwater monitoring systems is EPA's "Handbook of Suggested Practices for the Design and Installation of Ground-Water Monitoring Wells" (see references).

Factors to Consider

Several factors influence the design and installation of an effective groundwater monitoring system. These include:

- Geologic and hydrogeologic conditions—Geologic and hydrogeologic conditions directly affect the occurrence and movement of groundwater and contaminant transport. Considerations include soil type, soil permeability, water table height and seasonal fluctuations, characteristics of aquifers, groundwater direction, and others. In addition, the presence of consolidated or unconsolidated formations must be understood in order to properly evaluate and design water intake methods.
- Facility characteristics—Different types of facilities (i.e., disposal facilities, such as hazardous waste landfills and underground injection control systems, waste treatment facilities, underground storage facilities, etc.) will require different monitoring system designs. Waste characteristics, such as solubility and miscibility of wastes managed at the facility, also should be considered.
- Other site-specific information—Other information about the site should be taken into account, such as age of tanks, activities at the site, location of previously existing pumping wells, exact property boundaries, and other data that may be available.
- Equipment to be used in the well—The well must be designed to accommodate certain equipment such as water level measuring devices and groundwater sampling devices.

Drilling

Drilling methods should be adequate for the geologic conditions. For wells in stable unconsolidated materials under 150 feet deep, solid flight augers can be used. For deeper wells in consolidated materials, rotary drilling tool typically will be needed. Soil samples should be taken during well drilling and analyzed for contamination.

Design Components of Monitoring Wells

Several components of the monitoring well must be designed properly to provide an adequate sampling well. These include the well casing, well intake, annular seals, and surface completion. An example of a completed monitoring well with these design components is presented in Figure 12.1.

Well Casing.   The well casing should be of adequate strength to resist borehole pressures (collapse forces) and compressive and tensile stresses exerted during installation. In addition, the casing should be made of materials that will resist chemical attack and do not themselves have the potential to leach constituents into the groundwater. Selected examples of casing materials commonly used for groundwater monitoring installations are presented in Table 12.1.

Compatibility of the material with the expected groundwater constituents should be investigated thoroughly before well installation, since some materials have limited use. For instance, stainless steel can be used only in noncorrosive conditions, and thermoplastics are not recommended for use where potentially sorbing organics are of concern. Additionally, studies have indicated that cuts and abrasions on stainless steel and, to a lesser degree, PVC casing during installation are readily susceptible to surface oxidation and provide active sites for sorption of metal impurities such as lead, cadmium, and other heavy metals (U.S. Army Corps of Engineers 1989).

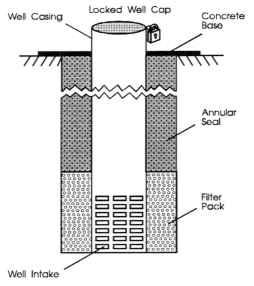

FIGURE 12.1   Example of a Groundwater Monitoring Well.

TABLE 12.1 Selected Examples of Materials Used for Well Casing

Material Category	Material Type
Fluoropolymer materials	Polytetrafluoroethylene (PTFE), tetrafluoroethylene (TFE), fluorinated ethylene propylene (FEP), perfluoroalkoxy (PFA), polyvinylidene fluoride (PVDF)
Metallic materials	Carbon steel, low-carbon steel, galvanized steel, AISI 304 and 316 stainless steels
Thermoplastic materials	Polyvinyl chloride (PVC), acrylonitrile butadiene styrene (ABS)
Fiberglass-reinforced materials*	Fiberglass-reinforced epoxy (FRE), fiberglass-reinforced plastic (FRP)

* These materials have not yet been used in general application across the country, and very little data is available on characteristics and performances.
Source: Information from EPA (1990).

Materials used in joining sections of casing should be of adequate strength and should be chemically compatible with the application. When possible, mechanical joints, such as threaded joints, are recommended. A fluoropolymer tape can be wrapped around the threads of the joint for water-tightness. Other joints found in the field include welded joints for metal applications and solvent-cemented joints for thermoplastic casings. Thermoplastic casings with solvent joints generally are not preferred for groundwater monitoring system installations. Solvent-cemented joints are never appropriate for fluoropolymer applications since these materials are inert to chemical attach and a bond cannot be formed.

Well Intake.   The design and installation of the monitoring well intake is critical to proper functioning of the well. In unconsolidated or poorly consolidated materials, the intake design includes a natural or artificial filter pack of granular materials to maximize well development and prevent plugging. Materials in natural filter packs that are effective include gravels and coarse sands. Materials normally used in artificial filter packs include quartz sand or glass beads. The filter pack should extend from the bottom of the well intake to approximately 2 to 5 feet above the well intake, unless this extension would provide a cross connection into an overlying water-bearing zone.

Another key aspect of well intake design is the opening or slot size. The slot size is selected based on the uniformity coefficient of the formation material for naturally packed wells or the size of the materials in the filter pack for artificially packed wells. Well intakes are typically 2 to 10 feet in length (EPA 1990).

The intake should be placed in the aquifer of interest for evaluation of that specific aquifer. If several aquifers are to be evaluated, separate wells should be constructed with the intakes placed at proper depths for sampling discrete water-bearing zones.

Annular Seals. Once the casing is placed in the borehole and the well intake is protected with filter packs, the annular space above the filter pack must be sealed. This will prevent vertical migration of the groundwater and will provide protection from surface water run-off into the well. In addition, the seal allows for the sampling of discrete zones. The annular seal is made with a stable, low permeable material such as bentonite or cement.

Surface Completion. There are two types of surface completions for monitoring wells—aboveground completion and flush-to-the-ground completion. In both types of completions, there is a surface seal of cement or concrete placed around the well casing for added protection to the well. In the aboveground completion, a protective casing 18 to 24 inches high is installed in the cement or concrete surface seal while it is wet and uncured. This casing should be fitted with a lockable cap for well protection and should be vented and have a drain hole. In the flush-to-the-ground completion, a lockable protective cover such as a meter box should be installed around the well casing. As in the aboveground completion, the structure is placed in the surface seal while it is still wet. This type of completion is susceptible to invasion by surface run-off if watertight seals are not maintained.

Monitoring Releases

General groundwater monitoring requirements for detection of releases from hazardous waste management units, including detection monitoring, compliance monitoring, and corrective action, are detailed in 40 CFR §§264.90–264.101. Key to a valid monitoring program is the proper placement of the monitoring wells for statistical evaluation of contamination in the uppermost aquifer. Guidance for installing a groundwater monitoring system and evaluating potential contamination are presented in several EPA guidance documents including:

- SW-611, "Procedures Manual for Groundwater Monitoring at Solid Waste Disposal Facilities"
- OSWER 9950.1, "RCRA Ground Water Monitoring Technical Enforcement Guidance Document"
- EPA/600/2-85/104, "A Practical Guide for Ground Water Sampling"

The groundwater monitoring system must consist of a sufficient number of wells, installed at appropriate locations and depths, to yield representative background and point-of-compliance groundwater samples from the uppermost aquifer. Upgradient wells should be installed in positions that will yield samples of groundwater not affected by the hazardous waste management unit. Downgradient well locations should be selected to ensure that potential pathways of contaminant migration are sampled. If several water-bearing zones are to monitored, this can be accomplished through well clusters, multiport well systems, nested piezometers (for shallow depths), and other groundwater profile systems.

As required under RCRA regulations, upgradient wells and downgradient wells must be compared statistically to demonstrate that waste management units have not contaminated the groundwater. For permitted facilities, statistical evaluation of groundwater contamination can be performed using one of four statistical methodologies. These include parametric analysis of variance (ANOVA), nonparametric ANOVA based on ranks, tolerance or prediction interval analysis, or control chart analysis. RCRA-permitted facilities may use an alternate method if it meets the same performance standards required of the four delineated methods. Interim status facilities (those that have applied for, but have not received a RCRA permit) must make their statistical demonstration using the Student's t-test. EPA-approved statistical methods are delineated in EPA's guidance document "Statistical Analysis of Ground-Water Monitoring Data at RCRA Facilities: Interim Final Guidance" (see references).

ANALYTICAL METHODS

Acceptable analytical methods for evaluating groundwater are documented in EPA's SW–846, "Test Methods for Evaluating Solid Waste Physical/Chemical Methods" (see references). Two quantitation limits are defined in the manual—the method quantitation limit (MQL), which is the minimum concentration of a substance that can be measured using the method, and the practical quantitation limit (PQL), which is the lowest level that can be reliably achieved within specified limits of precision and accuracy during routine laboratory operating conditions. The PQL generally is accepted as adequate for groundwater evaluation.

Laboratory Analysis

Most commonly, sample analysis for groundwater is performed in a laboratory. Typical analytical methods used in groundwater analysis for organics and pesticides include gas chromatographic (GC) methods and gas chromatographic/mass spectroscopic (GC/MS) methods. Hydrocarbons are analyzed using GC and high performance liquid chromatographic methods. Metals (except for hexavalent chromium and mercury) generally are evaluated using atomic absorption (AA) methods. Other constituents are evaluated using colorimetric, spectrophotometric, and other test methods.

Selected examples of types contaminants commonly found in groundwater, appropriate method number, and description of method are listed in Table 12.2. Analytical methods and PQLs for RCRA-regulated groundwater constituents are identified in 40 CFR Part 264, Appendix IX.

TABLE 12.2   Selected Examples of Contaminants Commonly Found in Groundwater and Appropriate Analytical Methods

Contaminant Type	Method Number	Method Description
Metals: aluminum, antimony, arsenic, barium, beryllium, cadmium, calcium, chromium	7020-7191	Atomic absorption (AA), direct aspiration or furnace technique (method 7061 for arsenic uses AA, gaseous hydride)
Hexavalent chromium	7195 7196 7197 7198	Coprecipitation Colorimetric Chelation/extraction Differential pulse polarography
Metals: cobalt, copper, iron, lead, lithium, magnesium, manganese	7200–7461	AA, direct aspiration or furnace technique
Mercury	7470–7471	Manual cold-vapor technique
Metals: molybdenum, nickel, osmium, potassium, selenium, silver, sodium, strontium, thallium, tin, vanadium, zinc	7480–7950	AA, direct aspiration or furnace technique
Halogenated volatile organics	8010	Gas chromatography (GC)
Nonhalogenated volatile organics	8015	GC
Aromatic volatile organics	8020	GC
Volatile organics	8240 8260 8021	Gas chromatography/mass spectrometry (GC/MS) GC/MS, capillary column technique GC with photoionization and electrolytic conductivity detectors in series

continued

TABLE 12.2 *Continued*

Contaminant Type	Method Number	Method Description
Semivolatile organics	8250	GC/MS, packed column technique
	8270	GC/MS, capillary column technique
Polynuclear aromatic hydrocarbons	8100	GC
	8310	High performance liquid chromatographic methods
Chlorinated hydrocarbons	8120	GC
Organophosphorus pesticides	8140	GC
	8141	GC, capillary column technique
Total cyanide	9010	Colorimetric, manual
	9012	Colorimetric, automated UV
Sulfate	9035	Colorimetric, automated chloranilate
	9036	Colorimetric, automated methylthymol blue, AA II
Phenolics	9065	Spectrophotometric
	9066	Colorimetric
Total recoverable oil and grease	9070	Gravimetric, separatory funnel extraction
Radium-228	9320	Coprecipitation, beta counter

Source: Information from EPA, SW-846 (1986).

In Situ Analytical Methods

As a result of Superfund and other remediation activities, analytical methods for groundwater evaluation that can be used directly in the field are being investigated and, when possible, employed. Some of these methods that have proved successful in pilot or field tests include:

- Optical detection of organic nitro compounds (experimental)—This method uses a membrane for in situ optical detection of organic nitro compounds based on fluorescence quenching. A fluorophor is incorporated into a plasticized membrane, which preconcentrates organic nitro compounds from the groundwater. This leads to fluorescence quenching. The membrane can be used to sense explosives that are present at the parts per million (ppm) level or higher (U.S. Army Corps of Engineers 1991).

- Optical detection of polynitroaromatic hydrocarbons—This method uses a membrane that reacts with polynitroaromatic hydrocarbons to form a brown product when placed in groundwater contaminated with these compounds. This brown product can be remotely detected in the parts per billion (ppb) range with fiber optics, as a one-time reading (U.S. Army Corps of Engineers 1991).
- Synchronous excitation (SE) fluorescence spectroscopy—This method is an advanced type of ultraviolet fluorescence spectroscopy. The method is different from the conventional single wavelength excitation (SWE) technique in that the SE procedure can avoid scatter distortion in situ that the SWE cannot. Examples of contaminants that can be quantified in the ppb (and sometimes parts per trillion) range include aniline, o-cresol, naphthalene, phenol, toluene, and xylene (EPA 1988).
- Fiber-optic sensor—This method uses a differential-absorption fiber optic sensor to monitor certain volatile organochlorides such as trichloroethylene and chloroform. Optical fibers are sealed into a capillary tube containing a reagent. A porous teflon membrane at one end allows the target molecules to enter the tube and mix with the reagent, which turns color in the presence of these constituents. This color results in a decreased transmission of light, which the optical fibers measure (DOE 1991).

Other methods for monitoring changes in the groundwater are being investigated by Lawrence Livermore National Laboratory's Environmental Technology Program and include high-frequency electromagnetic measurements and electrical resistance tomography (DOE, DE91 009070 1990).

GROUNDWATER REMEDIATION

Over the last decade, groundwater remediation has been a major focus of engineers, scientists, and practitioners working in the environmental field. As such, numerous technologies have been developed and utilized to expedite cleanup of contaminated groundwater. In addition, the topic of cleanup standards often has been a controversial subject addressed by regulators, industry officials, and the public. As a result, several methods for determining cleanup standards have been developed that are deemed acceptable, depending on the contamination type, hydrogeology of the area, and other factors.

Technologies

Because of the quantities of groundwater involved with any remediation project, treatment technologies generally are limited to those that can be performed at the remediation site. Considerations that should be taken into account when selecting a technology include source of contamination, types of contaminants, hydrogeologic features, groundwater recharge zones, and other factors. Groundwater treatment technologies that have proved successful for some applications include:

Extraction/Treatment (also termed Pump and Treat)
- Process description—Contaminated groundwater is extracted out of the ground with a system of pumping wells and is treated with an air stripper, carbon adsorption unit, or other appropriate treatment unit. Remediated water is discharged to a stream, a groundwater recharge zone, or the sanitary sewer. This is the most commonly used remediation technology.
- Applications—This technology is most appropriate for water soluble compounds. This technology generally is not suitable for free phase contaminants in permeable aquifers.
- Advantages—The technology is simple to operate and maintain and has been successful in many field applications over the years.
- Disadvantages—Complete remediation using this method is difficult to achieve due to tailing effects after initial high concentrations are remediated. Also, the volume of water that must be pumped to achieve acceptable cleanup levels can be excessive, and wells may become dry or perform poorly before remediation is completed.

Biological Barriers or Filters
- Process description—Microbial organisms attach to a surface or a liquid/liquid interface and form a biofilm, which allows for uptake and decomposition of contaminants. The process is useful for groundwaters in the saturation zone and for aquifer systems (Bellamy and deLint 1991 and DOE, DE91 009070 1990).
- Applications—Contaminants that can be treated using this technology include organic compounds and monoaromatic hydrocarbons, benzene, toluene, ethylbenzene and xylenes (termed BTX as a group).
- Advantages—Remediation can be performed in situ.
- Disadvantages—Biofouling can occur in some situations if pH, temperature, oxygen and/or the availability of electron acceptors cannot be controlled adequately.

In Situ Aeration
- Process description—Air is injected above or below the water table for the purpose of volatilizing organics. The air is injected through injection wells or horizontal trenches in the contaminated zone (Adomait and Whiffin 1991).
- Applications—This technology is suitable for volatile compounds.
- Advantages—This process is inexpensive and does not have the potential to create break-down products.

- Disadvantages—The amount of material removed decreases substantially as the concentration levels decrease.

Other innovative groundwater treatment methods that are being evaluated include in situ steam extraction and accelerated biodegradation through nutrient dosing.

Cleanup Standards

Remedial action objectives for contaminated groundwaters normally are set through numerical standards, which must be achieved in order to terminate the remediation effort. Regulatory authorities and industry negotiate these standards, with input from the public. These cleanup standards can be based on several criteria, including natural water quality or background concentration levels, alternate concentration levels, maximum concentration levels, and practical quantitation limits.

Natural Water Quality or Background Concentration Levels.   This standard requires the return of the aquifer to a precontamination, or pristine, state. An upstream, noncontaminated segment of the aquifer is used as the basis for evaluating concentration levels that are deemed natural in the water. This standard is used for RCRA-defined constituents and other constituents found in groundwater. This type of standard is the most difficult to meet, and some technologies such as the extraction/treatment method of remediation typically cannot achieve this standard.

Alternate Concentration Levels (ACLs).   Alternate concentration levels typically are higher than natural water quality, or background, levels. ACLs may be established as cleanup standards for RCRA-defined constituents when the constituents do not pose a substantial present or potential hazard to human health and the environment at levels that are higher than background.

Maximum Concentration Levels (MCLs).   Maximum concentration levels are concentrations equivalent to drinking water standards. MCLs often are used as the basis for cleanup standards for constituents regulated under the Safe Drinking Water Act.

Practical Quantitation Limits (PQLs).   Practical quantitation limits may be used as the basis for cleanup standards in certain situations where health-based or water-quality based standards do not exist. This type of standard requires the level of contamination to be below the PQL before remediation is terminated.

Other Factors.   Other factors are reviewed when determining cleanup standards. These are detailed in 40 CFR §264.94 and are summarized in Table 12.3.

TABLE 12.3   Factors Considered when Determining Cleanup Standards for Contaminated Groundwater

Category	Factors Considered
Regulated waste unit	Physical and chemical characteristics of waste; volume of waste; potential of waste for migration
Hydrogeological characteristics	Hydrogeological characteristic of facility and surrounding area; quantity of groundwater; direction of groundwater flow; hydraulic connections from surface to groundwater; patterns of rainfall in the region; existing quality of surface water and groundwater in nearby areas
Land use	Location of underground drinking water sources; proximity and withdrawal rates of all groundwater users; current and future uses of groundwater in the area
Potential exposure	Potential for human health risks; potential for damage to wildlife, crops, vegetation, and physical structures; persistence and permanence of the potential adverse effects

Source: Information from 40 CFR §264.94.

Cleanup Verification.   After cleanup standards are met and remediation actions are terminated, there might be a verification period of several years to ensure that contamination levels do not exceed the standard at a later date. In some cases, rainfall events may continue to wash contaminants trapped in the unsaturated zone into the aquifer, and remediation may need to be resumed after a period of time.

CONCLUSION

The discovery of widespread groundwater contamination in the late 1970s and early 1980s, which resulted from releases from hazardous waste management units, made groundwater evaluation and cleanup a top national focus. Much information in the form of technology guidance documents, hydrogeologic transport models, analytical standards, and other books and documents pertaining to the subject have been published in the last decade. Groundwater cleanup efforts have been underway for over a decade and many facilities are now asking the question "how clean is clean"? This question of how much remediation is needed to achieve an acceptable level of cleanup is being resolved through cleanup standards, which are negotiated between the regulatory authorities and industry, with input from the public.

Concurrent with the focus on groundwater came a focus on soil contamination, as Superfund efforts revealed that, like groundwater, soils associated with hazardous waste management units were contaminated. The subject of soils assessment and cleanup is discussed in the next chapter. The chapter includes methods of collecting soil samples for analysis, acceptable analytical methods including in situ analysis, and methods of remediation applicable to contaminated soils.

REFERENCES

Adomait, Martin, and Brian Whiffin, "In Situ Volatilization Technologies: R, D & D Scoping Study," in *Proceedings of the 1st Annual Soil and Groundwater Remediation R, D & D Symposium,* Ottawa, Ontario, January 1991.

Bellamy, Kevin L., and Nicole de Lint, "In-Situ Intercedent Biological Barriers for the Containment and Remediation of Contaminated Groundwater," in *Proceedings of the 1st Annual Soil and Groundwater Remediation R, D & D Symposium,* Ottawa, Ontario, January 1991.

Department of Energy, "Environmental Technology Program: Annual Report," UCRL-LR-105199, DE91 009070, Lawrence Livermore National Laboratory, Livermore, CA, 1990.

Department of Energy, "Preliminary Field Demonstration of a Fiber-Optic TCE Sensor," DE91 008453, prepared by S.M. Angel et al., Lawrence Livermore National Laboratory, Livermore, CA, 1991.

Department of Energy, "Uncontrolled Hazardous Waste Site Cleanup Programs in the U.S.—An Overview," DE90 002215, prepared by Shen-yann Chiu and Steve Y. Tsai, Argonne National Laboratory, Argonne, IL, 1990.

U.S. Army Corps of Engineers, "A Membrane for In Situ Optical Detection of Organic Nitro Compounds Based on Fluorescence Quenching," AD-A244 261, prepared by Rudolf W. Seitz, Chen Jian, and Donald C. Sundberg for U.S. Army Toxic and Hazardous Materials Agency, Aberdeen Proving Ground, MD, January 1991.

U.S. Army Corps of Engineers, "Influence of Well Casing Composition on Trace Metals in Ground Water," AD-A208 109, prepared by Alan D. Hewitt, U.S. Army Cold Regions Research and Engineering Laboratory, Hanover, NH, for U.S. Army Toxic and Hazardous Materials Agency, Aberdeen Proving Ground, MD, April 1989.

U.S. Environmental Protection Agency, "Development of an Environmental Monitoring Technique Using Synchronous Excitation (SE) Fluorescence Spectroscopy," EPA/600/D-90/060, prepared by Dennis Stainken and Uwe Frank for Risk Reduction Engineering Laboratory, Office of Research and Development, Cincinnati, OH, 1988.

U.S. Environmental Protection Agency, "Handbook of Suggested Practices for the Design and Installation of Ground-Water Monitoring Wells," EPA/600/4-89/034, Environmental Monitoring Systems Laboratory, Las Vegas, NV, 1990.

U.S. Environmental Protection Agency, "A Practical Guide for Ground Water Sampling, EPA/600/2-85/104, Environmental Research Laboratory, Ada, OK, 1985.

U.S. Environmental Protection Agency, "Procedures Manual for Ground Water Monitoring at Solid Waste Disposal Facilities," SW-611, Office of Solid Waste and Emergency Response, Washington, D.C., 1980.

U.S. Environmental Protection Agency, " RCRA Ground Water Monitoring Technical Enforcement Guidance Document," OSWER 9950.1, Office of Solid Waste and Emergency Response, Washington, D.C., 1986.

U.S. Environmental Protection Agency, "Statistical Analysis of Ground-Water Monitoring Data at RCRA Facilities: Interim Final Guidance," Office of Solid Waste Management Division, Washington, D.C., 1989.

U.S. Environmental Protection Agency, "Test Methods for Evaluating Solid Waste: Physical/Chemical Methods," 3rd ed., SW-846, Office of Solid Waste and Emergency Response, Washington, D.C., 1986.

U.S. Environmental Protection Agency, "Transport and Fate of Contaminants in the Subsurface," EPA/625/4-89/019, Center for Environmental Research Information and Robert S. Kerr Environmental Research Laboratory, Cincinnati, OH, 1989.

BIBLIOGRAPHY

Aller, Linda et al. (1989), *Handbook of Suggested Practices for the Design and Installation of Ground-Water Monitoring Wells*, National Water Well Association, Dublin, OH.

Aral (1990), *Ground Water Modeling in Multilayer Aquifers: Steady Flow*, Lewis Publishers, Boca Raton, FL.

ASTM (1990), *Ground Water and Vadose Zone Monitoring*, ASTM Special Technical Publication 1053, D.M. Nielsen and A.I. Johnson, eds., American Society of Testing and Materials, Philadelphia.

ASTM (1988), *Ground-Water Contamination: Field Methods*, Special Technical Publication 963, Collins and Johnson, eds., American Society of Testing and Materials, Philadelphia.

ASTM (1991), Monitoring Water in the 1990's: Meeting New Challenges, Special Technical Publication 1102, Hall and Glysson, eds., American Society of Testing and Materials, Philadelphia.

Canter, Larry W. et al. (1987), *Ground Water Quality Protection*, Lewis Publishers, Boca Raton, FL.

DOE (1988), "Recommendations for Holding Times of Environmental Samples," DE88 011831, prepared by M.P. Maskarinec, L.H. Johnson, and S.K. Holladay, Oak Ridge National Laboratory, Department of Energy, Oak Ridge, TN.

_____ (1990), "Performance Evaluation of a Groundwater and Soil Gas Remedial Action," DE90 017659, prepared by Mary C. Hansen and Suzanne L. Hartnett, Argonne National Laboratory, U.S. Department of Energy, Argonne, IL.

_____ (1991), "Bioremediation of Hanford Groundwater," DE92 000799, prepared by T.M. Brouns et al., Pacific Northwest Laboratory, Department of Energy, Richland, WA.

_____ (1991), "Evaluation of a Multiport Groundwater Monitoring System," DE91 011073, prepared by T.J. Gilmore et al., Pacific Northwest Laboratory for Department of Energy, Richland, Washington, D.C.

———— (1991), "Statistical Approach on RCRA Groundwater Monitoring Projects at the Hanford Site," DE91 014682, prepared by C.J. Chou for the U.S. Department of Energy, Washington, D.C.

DOE-RL (1991), "Annual Report for RCRA Groundwater Monitoring Projects at Hanford Facilities for 1990," DOE/RL-91-03, U.S. Department of Energy-Richland Operations Office, Richland, Washington.

GRI (1990), "Laboratory and Pilot-Scale Evaluations of Physical/Chemical Treatment Technologies for MGP Site Groundwaters," prepared by Remediation Technologies, Inc., for Gas Research Institute, Chicago, IL.

Illinois State Water Survey (1983), "A Guide to the Selection of Materials for Monitoring Well Construction and Ground-Water Sampling," SWS Contract Report 327, prepared by M.J. Barcelona, J.P Gibb, and R. Miller, Illinois State Water Survey, Champaign, IL.

———— (1985), "Practical Guide for Ground-Water Sampling," SWS Contract Report 374, prepared by M.J. Barcelona et al., Illinois State Water Survey, Champaign, IL.

———— (1988), "In Situ Aquifer Reclamation by Chemical Means: A Feasibility Study," HWRIC-RR/028, prepared by G.R. Peyton, M.H. Lefaivre, and M.A. Smith, Hazardous Waste Research and Information Center, Illinois State Water Survey, Champaign, IL.

Matthess, G. (1982), *The Properties of Groundwater*, Wiley-Interscience, Toronto.

Nielsen, David M. (1991), *Practical Handbook of Ground-Water Monitoring*, Lewis Publishers, Boca Raton, FL.

O'Brien & Gere Engineers, Inc. (1988), *Hazardous Waste Site Remediation*, Van Nostrand Reinhold, New York.

U.S. Army (1987). "In-Situ Groundwater Treatment Technology Using Biodegradation," AD-A244 079, prepared by Edward J. Bower and Gordon D. Cobb for U.S. Army Toxic and Hazardous Materials Agency. U.S. Army, Aberdeen Proving Ground, MD.

———— (1988), "Monitoring Well Installation and Groundwater Sampling and Analysis Plan," prepared by Donahue and Associates, Inc., for U.S. Army Training Reserve-84th Division, Milwaukee, WI.

———— (1989), "Single Fiber Measurements for Remote Optical Detection of TNT," CETHA-TE-CR-89102, U.S. Army Toxic and Hazardous Materials Agency, U.S. Army, Washington, D.C.

U.S. EPA (1983), *Characterization of Hazardous Waste Sites—A Methods Manual*, Vol. 2, "Available Sampling Methods," EPA/600/4-83/040, U.S. Environmental Protection Agency, Washington, D.C.

———— (1985), "Modeling Remedial Actions at Uncontrolled Hazardous Waste Sites," EPA/540/2-85/001, U.S. Environmental Protection Agency, Cincinnati, OH.

———— (1985), "Remedial Action at Waste Disposal Sites (Revised)," EPA/625/6-85/006, Hazardous Waste Engineering Research Laboratory, Cincinnati, OH, and Office of Emergency and Remedial Response, U.S. Environmental Protection Agency, Washington, D.C.

———— (1986), "Testing and Evaluation of Permeable Materials for Removing Pollutants from Leachates at Remedial Action Sites," EPA/600/2-86/074, U.S. Environmental Protection Agency, Cincinnati, OH.

_____ (1987), "A Handbook on Treatment of Hazardous Waste Leachate," EPA/600/8-87/006, U.S. Environmental Protection Agency, Washington, D.C.

_____ (1988), "Comparison of Water Samples from PTFE, PVC, and SS Monitoring Wells," EPA/600/X-88/091, prepared by M.J. Barcelona, G.K. George, and M.R. Schock, Environmental Monitoring Systems Laboratory, U.S. Environmental Protection Agency, Las Vegas, NV.

_____ (1988), "Ground-Water Modeling: an Overview and Status Report," EPA/600/2-89/028, U.S. Environmental Protection Agency, Cincinnati, OH.

_____ (1988), "Guidance for Conducting Remedial Investigation and Feasibility Studies Under CERCLA," EPA/540/G-89/004 and OSWER-9355.3-01, Office of Emergency and Remedial Response, U.S. Environmental Protection Agency, Washington, D.C.

_____ (1988), "Guidance on Remedial Actions for Contaminated Groundwater at Superfund Sites," EPA/540/G-88/003, U.S. Environmental Protection Agency, Washington, D.C.

_____ (1988), "Treatment of Hazardous Landfill Leachates and Contaminated Groundwater," EPA/600/2-88/064, prepared by Robert C. Ahlert and David S. Kosson, Rutgers University for Risk Reduction Engineering Laboratory, U.S. Environmental Protection Agency, Cincinnati, OH.

_____ (1989), "Evaluation of Ground-Water Extraction Remedies, Vol. 1, Summary Report," EPA/540/2-89/054, U.S. Environmental Protection Agency, Washington, D.C.

_____ (1989), "Guide on Remedial Actions for Contaminated Groundwater," EPA/9283-1-02FS, U.S. Environmental Protection Agency, Washington, D.C.

_____ (1989), "Performance of Pump-and-Treat Remediations," EPA/540/4-89/005, prepared by J.F. Keely, U.S. Environmental Protection Agency, Cincinnati, OH.

_____ (1989), "Transport Processes Involving Organic Chemicals," EPA/600/D-89/161, Robert S. Kerr Environmental Research Laboratory, U.S. Environmental Protection Agency, Ada, OK.

_____ (1990), "Basics of Pump-and-Treat Ground-water Remediation Technology," EPA/600/8-90/003, U.S. Environmental Protection Agency, Cincinnati, OH.

_____ (1990), "Cleanup of Underground Storage Tank Releases Using Pump and Treat Methods," EPA/101/F-90/037, prepared by Mary Ann Susavidge, Drexel University, Philadelphia for Office of Cooperative Environmental Management, U.S. Environmental Protection Agency, Washington, D.C.

_____ (1990), "Emerging Technologies: Bio-recovery Systems Removal and Recovery of Metals Ion from Groundwater," EPA/540/5-90/005a, Risk Reduction Engineering Laboratory, Office of Research and Development, U.S. Environmental Protection Agency, Cincinnati, OH.

_____ (1990), "A Guide to Pump and Treat Groundwater Remediation Technology," EPA/540/2-90/018, Office of Solid Waste and Emergency Response, U.S. Environmental Protection Agency, Washington, D.C.

_____ (1990), "Innovative Operational Treatment Technologies for Application to Superfund Site: Nine Case Studies," U.S. Office of Solid Waste and Emergency Response, U.S. Environmental Protection Agency, Washington, D.C.

_____ (1991), "Compendium of ERT Groundwater Sampling Procedures," OSWER Directive 9360.4-06, Environmental Response Team, Office of Emergency and Remedial Response, U.S. Environmental Protection Agency, Washington, D.C.

_____ (1991), "Innovative Treatment Technologies: Semi-Annual Status Report," Office of Solid Waste and Emergency Response Technology Innovation Office, U.S. Environmental Protection Agency, Washington, D.C.

_____ (1991), "In Situ Steam Extraction Treatment," EPA/540/2-91/005, prepared by Science Applications International Corporation, Cincinnati, OH, for U.S. Environmental Protection Agency, Washington, D.C.

Van Der Leeden, Fritz (1992), *Geraghty & Miller's Groundwater Bibliography*, 5th ed., Lewis Publishers, Boca Raton, FL.

Walton (1989), *Numerical Groundwater Modeling—Flow and Contamination Migration*, Lewis Publishers, Boca Raton, FL.

Wanielista, Martin, P. (1990), *Hydrology and Water Quantity Control*, John Wiley & Sons, New York.

13

SOILS ASSESSMENT
AND REMEDIATION

A soils assessment typically is required during the closure of a regulated waste storage, treatment, or disposal facility in order to demonstrate that the area has been decontaminated to an extent that it poses no threat to human health or the environment. A soils assessment also may be performed to assure adequate cleanup of a spill or to verify that activities involving hazardous materials and wastes have not affected the surrounding area adversely. If the assessment reveals that there is contamination, some type of remediation normally is required. This chapter discusses a methodology for performing a soils assessment and describes remediation technologies used for cleanup of contaminated soils.

FACILITY HISTORY

In order to adequately plan a soils assessment strategy, a facility history must be developed. This history is used to determine areas of potential contamination so that the number of samples and the money spent on sampling and analysis can be limited as much as possible. Examples of items that might be included in the facility history are shown in Table 13.1. Using the facility history, areas that are potentially contaminated can be defined.

DETERMINATION OF SAMPLING POINTS

Once the areas of potential contamination are defined, sampling points and sampling techniques can be determined. A useful guidance document for this process is EPA's SW-846, "Test Methods for Evaluating Solid Waste" (see

TABLE 13.1   Examples of Items to Include in a Facility History

• Facility age	• Description of major upgrades
• Materials/wastes stored at the facility over time	• List of spills, location, quantities, and cleanup techniques
• Description of facility's chemical operations, present and past	• List of chemical operations performed by vendors, present and past
• RCRA permit	
• Air emissions permit and monitoring data	• RCRA and other closures
• Wastewater discharge permit and reports	• Air modeling data (plume assessments)
	• Inspection logs for docks, tanks, and containment structures

references). A three-dimensional grid can be superimposed over the area of investigation. Samples can be taken at varying locations and depths within the grid to assess the extent and depth of contamination, if any. Typically, a circular pattern of sampling is used, with samples being concentrated near expected contamination.

In order to effectively interpret data gathered from the analysis of the samples, background samples must be collected. These background samples will indicate what constituents are naturally occurring in the soil. The background samples should come from areas of the same or similar soil classification as the area under investigation. This determination can be made using the visual-manual procedure outlined by the American Society of Testing and Materials (ASTM) in ASTM D 2488-90. In addition to having the same or similar classification, the areas must have been secluded from chemical and waste traffic, manufacturing, and/or other chemical-use activities to provide effective background data.

SAMPLE COLLECTION METHODS

There are several methods used for soil sample collection that are recognized throughout industry. These include:

- Collection using a trowel
- Collection using an auger (screw-type, barrel, or hollow-stem)
- Collection using a thin-walled tube
- Collection using a split-spoon or split barrel

Numerous ASTM standards for soil collection and classification are documented in Volume 4.08 of the *Annual Book of Standards* (see references). Standards and practices pertaining to this topic are delineated in Appendix C. Presented in Table 13.2 is a brief description of the most frequently used sample collection methods, including advantages and disadvantages of each.

Whenever possible, liners should be used with sampling devices designed to incorporate liners to help prevent cross-contamination during sampling. Liners should be made of materials such as stainless steel, brass, or another material that prevents leaching of contaminants into the sample from the liner. In particular, flexible PVC and polyethylene liners are not recommended when sampling for certain volatile organics due to sorption.

In some applications, preparation methods are required, such as coring of concrete for sampling under a slab or boring of the soil in order to collect samples at progressively deeper locations. In these instances, drilling equipment typically is used, and drilling operations occur in tandem with sample collection.

In most applications, the most efficient preservation method for soil samples is refrigeration. Sample integrity can be maintained for several days and up to several weeks, depending on soil type and preservation temperature, with the optimal temperature range being $0\,^{\circ}C$ to $4\,^{\circ}C$ (EPA 1991). If the sample has to be transferred to a container, the time the sample is exposed to the atmosphere should be minimized and measures should be taken to ensure the sample strata is not disturbed.

FIELD CONSIDERATIONS

Field sampling is not always performed under optimum conditions, so adequate field equipment must be on hand to ensure proper sampling for varying sets of circumstances. Items that might be included in a field vehicle are presented Table 13.3.

The field log is an integral part of the sampling event and should be completed as accurately as possible. Items to be included in the field log are presented in Table 13.4. Once the log for an event is completed, it should be kept as part of the sampling record, and should be filed with sampling data and other information about the project.

ANALYTICAL METHODS

Standard Methods

Standard methods for soils analysis are documented in EPA's manual SW-846. The methods presented for analysis of soils are the same methods used for groundwater, which is described in Chapter 12. Preparation procedures typically will be different, however, since soils will require digestion and extraction before the analysis is performed. Because of this, practical quantitation limits (PQLs) of soil and groundwater may vary significantly from contaminant to contaminant, with quantitation limits of groundwater contaminants generally being lower.

TABLE 13.2 Soils Sample Collection Methods

Sampling Method	Description	Advantages/Disadvantages
Trowel	Scoop device used to collect samples from side or bottom of sample area	Inexpensive to use; applications are limited to shallow depths and nonvolatile materials; method tends to cause disturbance of sample
Screw-type auger	Hand-operated, screw type devices are used to collect samples at shallow to medium depths; machine-operated augers are used to collect samples at medium depths or greater	Can be hand- or power-operated; can be used in loose or dense soils; use is limited to nonvolatile materials; some potential for cross contamination between segments exists
Barrel auger	Hand-operated; some types can collect samples in muds or sands; can be used at shallow to medium depths	Can be used for nonvolatile materials; not recommended for sampling volatile materials
Hollow-stem auger	Power-operated; is used to collect core samples; can be used in most soils; can be used at shallow to medium depths	Can remove undisturbed samples from cohesive soils; can be used in conjunction with thin-walled tubes
Thin-walled tube	Light-weight pipe to encapsulate soil samples; can be used at medium depths or greater	Can be used for volatile materials; will preserve stratification of soils; cannot be used in dense material due to buckling problems; can be used in conjunction with extension rods
Split spoon or split barrel sampler	Heavy tubular device used in conjunction with a drilling rig; can be used at medium depths or greater	Tube can be lined for collection of volatile materials; can be used in very dense or rock-like materials; will preserve stratification of soils; is expensive due to rig charges

Source: Information from EPA/600/4-89/034 (1990).

TABLE 13.3 Items that Might be Included in a Field Vehicle During Sampling Events

Category	Equipment
Safety and personal equipment	Pylons, tape or rope for marking off area, *Do Not Enter* signs, oxygen meter and LEL meter for sampling in confined spaces, dust mask, water bottle, ear plugs or other hearing protection, safety glasses, hardhat, sunscreen, coveralls, other appropriate clothing
Recordkeeping items	Clipboard, all-weather paper, field notebook with all-weather paper, indelible ink pen and markers, other pens, scotch tape, watch, camera and film, compass, thermometer
Sampling items	Appropriate noncontaminated jars, plastic containers, lids, sample labels, indelible ink pens, airtight tape, thin-walled tubes with plastic end caps, hand augur, trowel, ice chest, ice, sampling equipment decontamination solutions, clean rags, chain of custody forms

TABLE 13.4 Items to Include in a Field Log

Personnel Data

- Name, company, and telephone number of person collecting the sample and recording the event
- Names of other field crew members, company affiliations, phone numbers
- Name of driller, company affiliation, phone number
- Name of other persons involved in the sampling activity
- Signature of person recording information in the log

Site Data

- Date and time of the start of the sampling event
- Time of various events during sampling, such as drilling activities, sample collection and preservation, and transport to lab
- Location of sampling point (include map or sketch, if possible)
- Weather conditions, including temperature, cloud conditions, humidity (if known), other conditions such as mist, rain, or sleet
- Out of the ordinary aspects, such as confined space, enclosed building, other

Sample Data

- Depth to sample point
- Soil color, soil type or classification, unusual characteristics of soil
- Sample equipment used and method of collection
- Preservation techniques utilized
- Other observations

In Situ Analysis

Soil-Gas Analysis.   Several analytical devices currently are marketed that can perform in situ soil-gas analysis. In situ analysis often is preferable to laboratory analysis since it is less time consuming and cuts down on sample handling and exposure of the sample to the atmosphere. In addition, in situ soils analysis allows for better selection of sampling locations, since concentrations can be mapped during the sampling event. Many of the procedures used in soil-gas analysis were originally developed for oil and gas exploration, and have been modified in recent years for investigations of hazardous waste sites. In general, in situ soil-gas analysis is used as a screening method.

Active in situ soil-gas analysis utilizes soil-gas cores or probes, which are 1 to 3 meters in depth, and a vacuum system such as a syringe to pull the vapor to the surface for immediate analysis. Probes can be inserted into the ground with hammers or hydraulic rams to reach depths of 3 to 5 feet. Numerous chemical detectors can be used in the field to give a quantitative measure of soil-gas contamination. These detectors are listed in Table 13.5.

In addition to active soil-gas analysis using detectors listed in the table, initial screening for contamination can be performed using an organic vapor analyzer (OVA). This type of device does not distinguish between hydrocarbons and chlorinated solvents. Thus, once contamination is identified, more sophisticated equipment must be used to speciate and quantify the con-

TABLE 13.5   Selected Detectors that Can be Used in Soil-Gas Analysis

Detection Method	Sensitivity	Applications
Flame ionization detector (FID)	Can detect 4 picograms/sec of carbon	Can detect total hydrocarbons
Electron capture detector (ECD)	Can detect 0.01 picograms/sec of material	Can detect halogenated hydrocarbons; more sensitive to iodine than chlorine or fluorine
Hall electrolytic conductivity detector (HECD)	Can detection 0.5 picograms/sec of chlorine	Can detect halogenated species, nitrogen-containing organics, and sulfur-containing organics
Photoionization detector (PID)	Can detect 10–100 picograms/sec of aromatic material	Can detect aromatic hydrocarbons
Flame photometric detector (FPD)	Can detect 1 picograms/sec of phosphorous or 100 picograms/sec of sulfur	Can detect sulfur and phosphorus compounds

Source: Information from EPA (1989).

taminants. Other hand-held instruments that can be used as screening devices include portable infrared spectrometers, ion mobility spectrometers, and fiber optical sensors. If budgets and trained personnel do not allow for active sampling, passive samplers made of an adsorbent such as activated charcoal can be buried in the soil and collected at a later time for analysis in the laboratory.

Surface Radiation Survey.   Portable beta/gamma detectors are used widely for screening contamination at a hazardous waste site containing radioactive wastes. Most of these detectors are easy to operate, with probes that are used to scan the surface and provide continuous readings while the operator walks in a predetermined pattern. When areas contaminated above background levels are found, they can be marked with a stake, and measured values can be recorded on the stake and in a log book. Using this method, the extent of contamination can be charted easily (DOE 1990).

IN SITU TREATMENT OF SOILS

If soils are determined to be contaminated, there are several methods for in situ treatment of soils, depending on soil type and waste characteristics. Extensive evaluation of in situ treatment technologies is presented by Chambers et al., in *In Situ Treatment of Hazardous Waste-Contaminated Soils* (see references). This book is a compilation of two EPA manuals: "Handbook on In Situ Treatment of Hazardous Waste-Contaminated Soils" and "Review of In-Place Treatment Techniques for Contaminated Surface Soils," Vol. 2. Additionally, numerous soil treatment technologies have been documented by EPA in engineering bulletins and other technical reports.

Treatment Technologies

Numerous in situ treatment technologies have proved effective for field use. These include soil flushing, solidification/stabilization, vitrification, chemical degradation, biodegradation, soil vapor extraction, photolysis, and in situ incineration.

Soil Flushing
- Process description—Soil flushing is accomplished by flooding the area to be remediated with an appropriate washing solution such as acidic or basic solutions, surfactants, or water. The elutriate is collected via shallow wells or drains and treated and/or reused.
- Applications—This technology is suitable for soluble compounds.
- Advantages—If successful, no additional treatment is needed. The technology can be applied successfully in permeable soils.
- Disadvantages—This technology is difficult to use in tight soils. Additionally, flushing may cause contamination to migrate to other soil areas or into the groundwater if the system is not installed properly.

Solidification/Stabilization

- Process description—Solidification techniques produce a solid from liquid or semi-liquid wastes, thus improving handling and physical characteristics through agents such as lime, kiln dust, and cement. Stabilization techniques reduce the hazard potential of a waste by reducing solubility, mobility, or toxicity through organic binding or thermoplastic microencapsulation. This technology is described more fully in Chapter 10.
- Applications—This technology is suitable for inorganics, low-level radioactive wastes, and has some limited use for organics.
- Advantages—Additives needed in the process are widely available and relatively inexpensive. Additionally, leaching is reduced and, under the right conditions, no other treatment is necessary.
- Disadvantages—In situ mixing can be difficult, and the volume of material will expand with the addition of solidifaction/stabilization agents.

In Situ Vitrification

- Process description—This process is a thermal treatment process that converts contaminated soils into inert glass or other crystalline material. Large electrodes are inserted into silicate-containing soils, and an electric current generates enough heat to pyrolyze contaminants.
- Applications—This technology is suitable for organic compounds that will either fuse or vaporize. The technology also proved effective for soils containing radionuclides and hazardous metals (DOE 1988).
- Advantages—Excavation is not necessary, thus minimizing worker's exposure to contaminants. Products from this process have a low leachability.
- Disadvantages—The process is very energy intensive. Specialized equipment and trained personnel are required, which increases the cost of this technology.

Chemical Degradation

- Process description—Wastes are transformed into less toxic, more elemental compounds through oxidation, reduction, and other chemical processes.
- Applications—In situ oxidation is suitable for pesticides and organic wastes that are water soluble. In situ reduction is suitable for metals and chlorinated organics.
- Advantages—A variety of chemicals such as pesticides, halogenated organics, and metals can be treated using this method. In addition, the achievable level of treatment potentially can be very high.
- Disadvantages—Treatment effectiveness is variable for different wastes and soil types, and some breakdown products may be more problematic or toxic than the parent compound. Additionally, some of the treatment reagents are costly.

Biodegradation

- Process description—Microbial organisms are used to convert waste to biomass and harmless by-products.
- Applications—This technology is suitable for oils and numerous organic compounds.
- Advantages—Achievable levels of treatment potentially can be very high, while costs of using this technology are relatively low. Additionally, soils can be treated without excavating large quantities of soil.
- Disadvantages—Treatment effectiveness is variable depending on waste and soil type, and some by-products may be water soluble or mobile in soil.

Soil Vapor Extraction

- Process description—Volatile organics are removed by extracting vapors from the unsaturated zone using a vacuum or other extraction mechanism. This process is best suited for fractured rock, or homogenous, high permeability soils.
- Applications—This technology is suitable for highly volatile compounds.
- Advantages—This technology provides permanent remediation once the material is extracted from the unsaturated zone.
- Disadvantages—Use of this technology is limited to permeable soils.

Photolysis (Photodegradation)

- Process description—This process breaks down chemicals with light energy such as ultraviolet light or incident solar radiation. Photo-enhancements can be used to increase treatment efficiencies.
- Applications—This technology is suitable for PCBs and specific organics such as acetaldehyde, dioxin, formaldehyde, and methylene chloride.
- Advantages—This technology does not require excavation.
- Disadvantages—The depth of treatment is limited to a thin upper layer of soil. Also, field data of expected breakdown products from the use of this technology is not fully developed.

In Situ Incineration

- Process description—This is a thermal process that uses a mobile rotary kiln incinerator system or other thermal system at the field site.
- Applications—This technology is suitable for PCB- contaminated soils and soils containing volatile and semi-volatile compounds.
- Advantages—High treatment efficiencies are achievable with this technology.
- Disadvantages—The process is expensive and energy intensive. Soils must be excavated to be treated.

Innovative Treatment Technologies

In Situ Treatment of Radionuclide-Contaminated Soils. In situ treatment of radionuclide-contaminated soils has been investigated, and several technologies proved effective during bench scale tests. These included vibratory grinding, scrubbing, screening and desliming, and flotation. For full-scale applications, attrition scrubbing at a high pH was deemed the most feasible (DOE 1988).

Other Innovative In Situ Treatment Technologies. Other in situ treatment technologies are still in the investigative stages. These include biochemical enhancement mechanisms such as cometabolism and cell-free enzymes, electrokinetics, low energy extraction, ozone treatment, steam heating and extraction, vacuum-induced venting for volatile organic compounds in saturated low-permeable soils, and field treatment with a bioreactor. Methods to control volatile materials during remediation activities also are being researched and include radio frequency heating and soil cooling.

CONCLUSION

Assessment of soils for contamination and the development of increasingly effective cleanup technologies are activities that will continue throughout this decade and into the next. Since contamination in soil typically leads to groundwater contamination, remediation of the unsaturated zone through techniques such as vapor extraction and other innovative methods can be of great benefit. Billions of dollars already have been spent on soil and groundwater cleanup activities in the United States under Superfund during the 1980s and 1990s. Thus, it is imperative that techniques be found to increase the effectiveness of these efforts.

The first two chapters of this section have reviewed how to assess and manage contamination of groundwater and soils. The next chapter discusses techniques for assessing atmospheric air quality. As a result of the passage of the Clean Air Act in 1990, new air regulations will be promulgated during the 1990s. These new regulations are expected to be comprehensive and encompass point source emissions, fugitive emissions, mobile sources, and regions that are not attaining established air quality standards.

REFERENCES

American Society for Testing and Materials, *Annual Book of ASTM Standards*, Vol. 4.08, "Soil and Rock; Dimension Stone; Geosynthetics," Philadelphia, PA., 1991.

Chambers, Catherine D. et al., *In Situ Treatment of Hazardous Waste-Contaminated Soils*, 2nd ed., Noyes Data Corporation, Park Ridge, NJ, 1991.

Department of Energy, "Treatment of Hazardous Metals by In Situ Vitrification," DE89 008247/XAB, Battelle Pacific Northwest Laboratory, Richland, WA, 1989.

Department of Energy, "Surface Radiation Survey and Soil Sampling of the 300-FF-1 Operable Unit, Hanford Site, Southeastern Washington: A Case Study," EMO-SA-5501, prepared by S.S. Teel and K.B. Olsen for Pacific Northwest Laboratory, Richland, WA, 1990.

Department of Energy, "Treatment of Radionuclide Contaminated Soils," RFP–4228, prepared by S.A. Pettis et al., for Rockwell International Aerospace Operations, Rocky Flats Plant, Golden, CO, 1988.

U.S. Environmental Protection Agency, "Handbook on In Situ Treatment of Hazardous Waste-Contaminated Soils," EPA/540/2-90/001, prepared by C. Chambers et al., Risk Reduction Engineering Laboratory, Office of Research and Development, Cincinnati, OH, 1990.

U.S. Environmental Protection Agency, "Handbook of Suggested Practices for the Design and Installation of Ground-Water Monitoring Wells," EPA/600/4-89/034, Environmental Monitoring Systems Laboratory, Las Vegas, NV, 1990.

U.S. Environmental Protection Agency, "Review of In-Place Treatment Techniques for Contaminated Surface Soils," Vol. 2, EPA/540/2-84/003b, prepared by Ronald C. Sims et al., Cincinnati, OH, 1984.

U.S. Environmental Protection Agency, "Soil-Gas and Geophysical Techniques for Detection of Subsurface Organic Contamination," ESL-TR-87-67, Environmental Monitoring Systems Laboratory, Las Vegas, NV, 1989.

U.S. Environmental Protection Agency, "Soil Sampling and Analysis for Volatile Organic Compounds," EPA/540/4-91/001, prepared by T.E. Lewis et al., for Environmental Monitoring Systems Laboratory, Las Vegas, NV, 1991.

U.S. Environmental Protection Agency, "Test Methods for Evaluating Solid Waste," SW-846, Office of Solid Waste Management and Emergency Response, Washington, D.C., 1986.

BIBLIOGRAPHY

Adomait, Martin and Brian Whiffin (1991), "In Situ Volatilization Technologies, R, D & D Scoping Study," In *Proceedings of 1st Annual Groundwater and Soil Remediation R, D & D Symposium*, Ottawa, Ontario.

Czupyrna, G. et al. (1989), *In Situ Immobilization of Heavy-Metal-Contaminated Soils*, Noyes Data Corporation, Park Ridge, NJ.

DOE (1988), "Recommendations for Holding Times of Environmental Samples," DE88 011831, prepared by M.P. Maskarinec, L.H. Johnson, and S.K. Holladay, Oak Ridge National Laboratory, Department of Energy, Oak Ridge, TN.

_____ (1990), "Environmental Technology Program Annual Report," DE91 009070, Lawrence Livermore National Laboratory, Department of Energy, Livermore, CA.

_____ (1990), "Final Report: Surface Radiation/Survey for the Phase I Remedial Investigation of the 300-FF-1 Operable Unit on the Hanford Site," EMO-1008, prepared by S.S. Teel and K.B. Olsen, Pacific Northwest Laboratory, Department of Energy, Richland, Washington.

_____ (1990), "Performance Evaluation of a Groundwater and Soil Gas Remedial Action," DE90 017659, prepared by Mary C. Hansen and Suzanne L. Hartnett, Argonne National Laboratory, Department of Energy, Argonne, IL.

———— (1990), "Soil-Gas Surveying at Low-Level Radioactive Waste Sites," EGG-M-89251, prepared by Alan B. Crockett, Kenneth S. Moor, and Lawrence C. Hull for Idaho National Engineering Laboratory, Idaho Falls, ID, 1990.

DOE and State of California (1987), "Measurement of Possible Cross-Contamination of Soil Samples During Soil Coring with the Split-Barrel or Bucket Auger Methods," State of California, Sacramento, California and Department of Energy, Argonne, IL.

GRI (1989), "Engineering-Scale Evaluation of Thermal Desorption Technology for Manufactured Gas Plant Site Soils," prepared by R. Helsel, E. Alperin, and A. Groen, IT Corporation, Knoxville, TN, for Gas Research Institute, Chicago, IL, and Illinois Hazardous Waste Research and Information Center, Savoy, IL.

King, R. Barry, Gilbert M. Long, and John K. Sheldon (1992), *Practical Environmental Bioremediation*, Lewis Publishers, Boca Raton, FL.

Kostecki, Paul, and Edward J. Calabrese (1991), *Hydrocarbon Contaminated Soils*, Vol. 1, "Analysis, Fate, Remediation, Health Effects, & Regulations," Lewis Publishers, Boca Raton, FL.

NIOSH (1984), *NIOSH Manual of Analytical Methods*, prepared by M.P. Maskarinec, L.H. Johnson, and S.K. Holladay for National Institute for Occupational Safety and Health, U.S. Department of Health and Human Services, Washington, D.C.

NTIS (1992), "Superfund: Technology and Cleanup: January 1980–November 1991," PB92-801505, National Technical Information Service, Department of Commerce, Washington, D.C.

Paul, E.A. and F.E. Clark (1988), *Soil Microbiology and Biochemistry*, Academic Press, Toronto.

Pedersen, T.A., and J.T. Curtis (1991), *Soil Vapor Extraction Technology*, Noyes Data Corporation, Park Ridge, NJ.

Petroleum Association for the Conservation of the Canadian Environment (1985), "The Persistence of Polynuclear Aromatic Hydrocarbons in Soil," PACE Report No. 85-2, prepared by T.L. Bulman et al., for Petroleum Association for Conservation of the Canadian Environment, Ottawa, Ontario.

Russell, David L. (1992), *Remediation Manual for Petroleum-Contaminated Sites*, Technomic Publishing Co., Lancaster, PA.

U.S. EPA (1979), "Methods for Chemical Analysis of Water and Wastes," EPA/600/4-79/020, U.S. Environmental Protection Agency, Washington, D.C.

———— (1979), "Survey of Solidification/Stabilization Technology for Hazardous Industrial Waste," EPA/600/2-79/056, U.S. Environmental Protection Agency, Washington, D.C.

———— (1983), Characterization of Hazardous Waste Sites—A Methods Manual, Vol. 2, "Available Sampling Methods," EPA/600/4-83/040, U.S. Environmental Protection Agency, Washington, D.C.

———— (1982), "Test Method 624 (Purgeables): Methods for Organic Chemical Analysis of Municipal and Industrial Wastes," EPA/600/4-82/057, Environmental Support Laboratory, U.S. Environmental Protection Agency, Cincinnati, OH.

———— (1984), "Soil Sampling Quality Assurance User's Guide, EPA/600/4-84/043, U.S. Environmental Protection Agency, Washington, D.C.

———— (1985), "Remedial Action at Waste Disposal Sites (Revised)," EPA/625/6-85/006, Hazardous Waste Engineering Research Laboratory, Cincinnati, OH, and

Office of Emergency and Remedial Response, U.S. Environmental Protection Agency, Washington, D.C.

_____ (1985), "Treatment of Contaminated Soils with Aqueous Surfactants," EPA/600/S2-85, prepared by W.D. Ellis, J.R. Payne, and G.D. McNabb for U.S. Environmental Protection Agency, Washington, D.C.

_____ (1986), "Systems to Accelerate In Situ Stabilization of Waste Deposits," EPA/540/2-86/002, prepared by M. Amdurer et al., for the Hazardous Waste Research Laboratory, U.S. Environmental Protection Agency, Cincinnati, OH.

_____ (1988), "Field Studies on In Situ Soil Washing," EPA/600/S2-87/110, Hazardous Waste Engineering Research Laboratory, U.S. Environmental Protection Agency, Cincinnati, OH.

_____ (1988), "Guidance for Conducting Remedial Investigation and Feasibility Studies Under CERCLA," EPA/540/G-89/004, Office of Emergency and Remedial Response, U.S. Environmental Protection Agency, Washington, D.C.

_____ (1988), "Results of Treatment Evaluations of Contaminated Soils," EPA/600/D-88/131, prepared by PEI Associates, Inc., for U.S. Environmental Protection Agency, Cincinnati, OH.

_____ (1988), "Treatment Potential for 56 EPA Listed Hazardous Chemicals in Soil," EPA 600/6-88/001, Robert S. Kerr Environmental Research Laboratory, U.S. Environmental Protection Agency, Ada, OK.

_____ (1989), "Bioremediation of Contaminated Surface Soils," EPA/600/9-89/073, prepared by J.L. Sims, R.C. Sims, and J.E. Matthews, Robert S. Kerr Environmental Research Laboratory, Ada, OK, U.S. Environmental Protection Agency and Dynamac Corporation, Ada, OK, and Utah State University, Logan, UT.

_____ (1989), "Cleaning Excavated Soil Using Extraction Agents: State-of-the-Art Review," EPA/600/2-89/034, U.S. Environmental Protection Agency, Washington, D.C.

_____ (1989), "Risk of Unsaturated/Saturated Transport and Transformation of Chemical Concentrations (RUSTIC)," Vol. 1, EPA 600/3-89/048a, prepared by J.D. Dean et al., for Environmental Research Laboratory, U.S. Environmental Protection Agency, Athens, GA.

_____ (1989), "Soil Sampling Quality Assurance User's Guide," 2nd ed., EPA 600/8-89/046, Environmental Monitoring Systems Laboratory, U.S. Environmental Protection Agency, Las Vegas, NV.

_____ (1989), "Stabilization/Solidification of CERCLA and RCRA Wastes: Physical Tests, Chemical Testing Procedures, Technology Screening, and Field Activities," prepared by PEI Associates, Inc., and Earth Technology Corporation for Center for Environmental Research Information, U.S. Environmental Protection Agency, Cincinnati, OH.

_____ (1989), "Technology Evaluation Report: SITE Program Demonstration Test, International Waste Technologies, In Situ Stabilization/Solidification, Hialeah, Florida," Vol. 1, EPA/540/5-89/004a, Risk Reduction Engineering Laboratory, U.S. Environmental Protection Agency, Cincinnati, OH.

_____ (1990), "Approach to Bioremediation of Contaminated Soil," EPA/600/J-90/203, prepared by Judith L. Sims, Ronald C. Sims, and John E. Matthews for Robert S. Kerr Environmental Research Laboratory, U.S. Environmental Protection Agency, Ada, OK.

———— (1990), "Evaluation and Testing of a Protocol to Determine the Aerobic Degradation Potential of Hazardous Waste Constituents in Soil," EPA/600/D-91/211, Risk Reduction Engineering Lab, U.S. Environmental Protection Agency, Cincinnati, OH and Agricultural Research Center, Beltsville, MD.

———— (1990), "Hazardous Waste & Hazardous Materials," Vol. 26, U.S. Environmental Protection Agency, Washington, D.C.

———— (1990), "Soil Washing: Engineering Bulletin," EPA/540/2-90/017, U.S. Environmental Protection Agency, Washington, D.C.

———— (1991), "Comparison of In Situ Vitrification and Rotary Kiln Incineration for Soils Treatment," EPA/600/J-91/255, Risk Reduction Engineering Laboratory, U.S. Environmental Protection Agency, Cincinnati, OH.

———— (1992), "Innovative Treatment Technologies: Semi-Annual Status Report," EPA/540/2-91/001 and OSWER-9380.0-19, Office of Solid Waste and Emergency Response, Technology Innovation Office, U.S. Environmental Protection Agency, Washington, D.C.

14

ASSESSING AIR QUALITY

Air emissions are regulated by EPA through the Clean Air Act (CAA) and the Resource Conservation and Recovery Act (RCRA) in 40 CFR 0–99 and 40 CFR Part 264 Subparts AA and BB, respectively. In general, these regulations were developed to reduce air emissions generated by industrial point sources and fugitive sources, incinerators, power plants, mobile emission sources, hazardous waste facilities, and other sources. The regulations also address areas that have extremely poor air quality, such as large cities like Los Angeles and Houston.

In general, impacts to air quality are assessed through two methods—contaminant dispersion modeling and emissions monitoring. This chapter discusses EPA's Industrial Source Complex model and other EPA-sponsored and non-EPA models used by industry. Additionally, monitoring of industrial point sources, fugitive emissions, and ambient air is addressed. National air quality and emissions trends, documented by EPA, also are included.

MODELS FOR ASSESSING IMPACTS TO AIR QUALITY FROM INDUSTRIAL SOURCES

EPA has published a model for assessing impacts to air quality from industrial sources. This model, the Industrial Source Complex (ISC2), dispersion model consists of two models, a short-term model and a long-term model. The model takes into account numerous factors not included in other, more simplistic, models. This model and other approved models are used during the permit process for assessing contaminant concentrations at the facility property line and other selected receptor points.

A complete technical description of the model is published by EPA in "Industrial Source Complex Dispersion Model User's Guide," Vols. 1 and 2 (see references). Key features of the model include:

- Polar or cartesian coordinate systems for receptors
- Rural or urban options, with a choice of three urban options
- Plume rise due to momentum and buoyancy as a function of downwind distance for stack emissions
- Evaluation of building wake effects with an option of using one of two downwash calculation methods
- Procedures for evaluation of stack-tip downwash
- Consideration of the effects of gravitational settling and dry deposition on ambient particulate concentrations and capability to calculate dry deposition
- Consideration of wind profile, including a procedure for calm-wind processing
- Concentration estimates for periods of 1-hour to annual averaging periods
- Adjustment procedures for elevated terrain
- Consideration of time-dependent exponential decay of pollutants
- Consideration of buoyancy-induced dispersion
- A default option to use EPA-established options and parameters
- Capability to treat height of receptor aboveground ("flagpole" receptors)

EPA-sponsors an electronic bulletin board service through the Office of Air Quality Planning and Standards that makes ISC2 and other EPA-approved models available to the public. Additionally, user's manuals, programs for emissions estimating, regulatory guidance memos, and other information is available from this electronic source. Selected air emissions models are presented in Table 14.1. Typical input for these and other models is presented in Table 14.2.

Specialized models that predict effects of specific source emissions such as gas and liquid jet releases and dense gas releases at grade are documented (Hanna and Drivas 1987). Incidents such as Bophal have emphasized the need for facilities to model potential effects from accidental releases of toxic gases and other chemicals in order to perform process risk assessments. In addition, Local Emergency Planning Committees (LEPCs) must plan for chemical emergencies that might occur in the local area, and modeling of accidental releases may help in the planning process. The topics of hazard assessment and emergency planning are explored further in Chapter 17.

TABLE 14.1 Selected EPA-Approved Models

Model	Description/Comments
Climatological dispersion model (CDM)	Gaussian plume dispersion model; long-term model uses average emission rates from point and area sources and a joint frequency distribution of wind direction, wind speed, and atmospheric stability; computations can be made for up to 200 point receptor sources and up to 2500 area sources at an unlimited number of receptor locations
Toxic screening models (TSCREEN)	Contains three models including the PUFF model, relief valve discharge (RVD) model, and SCREEN; PUFF is a gaussian model with a wide range of applications including modeling accidental releases; RVD can model point source releases under pressure including continuous and instantaneous releases; SCREEN is a gaussian model used for screening point sources
Urban airshed model (UAM)	Very sophisticated model used for assessing ozone precursors in the complex terrain of urban airsheds; used by states in developing state implementation plans
Plume visibility model (PLUVUE)	Plume-based model; predicts the transport, atmospheric diffusion, chemical conversion, optical effects, and surface deposition of point source emissions
Meteorological processor (RAMMET)	Utility program that processes hourly surface meteorological observations; normally used to process a calendar year of hourly data; compiles data into format for use with other models
VALLEY	Dispersion model for use where receptors are well above surrounding areas (i.e., in valleys); complexity and applications of the model are similar to ISC2

TABLE 14.2   Typical Input Required for Air Emissions Models

Meteorological Data	*Source Data*
• Wind velocity	• Physical characteristics
• Temperature	• Chemical characteristics
• Relative humidity	• Geometry of source
• Turbulence	• Release rates
• Net radiation flux	
Site Information	*Receptor Information*
• Topography	• Receptor locations
• Equipment operating parameters	• Distance between receptors
• Property boundaries	

AIR MONITORING

Industrial Source Monitoring

Unlike source emissions and transport modeling, which predicts concentrations of contaminants at facility boundaries or other receptors, emissions monitoring of industrial sources will assess actual emissions during facility operations. There are basically two types of emissions from an industrial facility—point source emissions and fugitive emissions. For many years, the main focus of air pollution management was on point source emissions, and much attention was given to monitoring these sources. However, when the toxic release inventory (TRI) reports submitted by facilities in the late 1980s showed fugitive emissions to be a substantial portion of releases to the atmosphere, these emissions came under focus.

Point Source Emissions Monitoring.   EPA has specified approved monitoring methods for point sources in 40 CFR Part 60 Appendix A. Several of these methods are described in Table 14.3. Examples of two sampling trains, one for sampling moisture content in a stack and one for sampling particulates, are shown in Figure 14.1 and 14.2.

A series of samples may be taken at the inlet and outlet of abatement equipment to ascertain equipment efficiencies as well as contaminant emissions during specific operational modes such as startup, full production, and emergency shutdown. Additionally, unabated stack emissions may require monitoring. The major drawback to stack sampling using EPA-approved methods is that the data gathered is from a noncontinuous, one-time event. Since the sampling is noncontinuous, operational changes or system efficiency problems may not be detected until the next stack sampling event. Because of sampling expense, it is not uncommon to sample major stacks only annually or biennially, or as specified in the facility's permit.

On-line monitors often are used to bridge data gaps between stack sampling events. These monitors give real-time emissions data and can be installed with an alarm system, which will notify facility personnel of higher than normal emissions. Some on-line monitors, such as flame ionization detectors (FIDs), analyze an air stream for a total concentration—in this case total hydrocarbons. Other monitors such as oxygen or carbon monoxide monitors can analyze for a particular chemical. Many on-line analyzers described in Chapter 5 can be used for continuous monitoring of stacks.

Fugitive Emissions Monitoring.   Fugitive emissions are more difficult to monitor since they are not captured and released through a stack or vent, but instead are dispersed into the air from small sources spread throughout a manufacturing facility. Industrial sources for fugitive emissions include leaking valves and connections, pumps, closed but leaking vents, door seals, individual cleaning operations that use small bottles of volatile solvents,

TABLE 14.3   Selected EPA-Approved Monitoring Methods

Method	Description/Applications
Method 2	Method uses a Type S pitot tube to determine stack velocity and volumetric flow rate
Method 4	Method uses sampling train with probe heater, consenser, and vacuum system to determine moisture content of stack gases
Method 5	Method uses sampling train with a glass fiber filter and impingers to collect particulates for analysis
Methods 6C and 7E	Methods allow for use of a gas analyzer to measure concentrations of sulfur dioxide and nitrogen oxides emissions, respectively, from stationary sources
Method 12	Method allows for use of inorganic lead sample train to collect samples for analysis by atomic absorption spectrophotometer
Method 17	Method allows for in-stack filtration (collection) of particulates
Method 21	Method allows for determination of volatile organic compound leaks
Method 25A	Method allows for determination of total gaseous organic concentration using a flame ionization analyzer

Source: Information from 40 CFR Part 60 Appendix A.

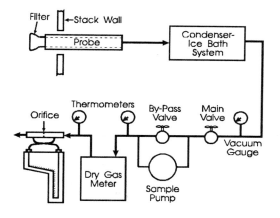

Source: Adapted from 40 CFR Part 60 Appendix A.

FIGURE 14.1   Method 4 Moisture Sampling Train.

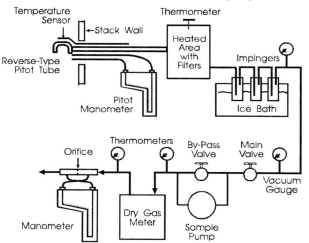

Source: Adapted from 40 CFR Part 60 Appendix A.

FIGURE 14.2   Method 5 Particulate Sampling Train.

poorly covered or uncovered process equipment, and poorly enclosed or inadequately vented spraying operations. Sources for fugitive emissions such as valves and connections can be monitored with portable instruments such as an organic vapor analyzer (OVA), a portable FID, or a photoionization detector. EPA Method 21 provides an accepted method for determination of volatile organic compound leaks—particularly from valves, flanges, other connections, and pumps. Fugitive emissions from industrial operations often can be quantified through workplace sampling, using a portable infrared spectrometer (**IR**) or diffusional monitors.

Ambient Air Monitoring

In addition to industrial emission sources, air quality can be affected by other sources such as mobile combustion sources, power-generating facilities, municipal and other incinerators, and construction activities. Thus, ambient air monitoring is performed in order to obtain data about the quality of the air in an entire region. Key parameters that are used to determine the general state of ambient air quality include carbon dioxide and nitrous oxide (greenhouse gases), ozone, carbon monoxide, nitrogen oxides, sulfur compounds, non-methane hydrocarbons, particulate matter including total suspended particulate matter and particulate matter less than 10 micrometers (µm) and less than 2.5 µm in aerodynamic diameter (PM_{10} and $PM_{2.5}$, respectively), and detectable concentrations of toxic pollutants including organics and metals. Additionally, in some regions acid rain and the deposition of acidic compounds on soils are key indicators of pollution. Ambient air quality trends are discussed in the next section.

Methods of monitoring for the ambient air quality parameters are documented by EPA and other sources such as ASTM and NIOSH (see references). Selected examples of analytical methods used for quantifying compounds are presented in Table 14.4.

Many innovative ambient air sampling and analytical methods are researched by EPA's Atmospheric Research and Exposure Assessment Laboratory. In particular, high-volume sampling, superior extraction methods such as superfluid extraction, and equipment that yields more accurate and lower quantitation limits are being developed and compared to more conventional methods and equipment.

NATIONAL AIR QUALITY AND EMISSIONS TRENDS

Air quality trends in the United States over time have been documented by EPA in "National Air Quality and Emissions Trends Report, 1990" (see references). Figure 14.3 shows a comparison of nationwide emissions for the years 1970 and 1990 for six air quality parameters—total particulates, sulfur oxides, carbon monoxide, nitrogen oxides, volatile organic compounds, and lead. As can be seen from the figure, the long-term trend in emissions reduction has been significant for most of these constituents. Lead showed the biggest reduction of 97%, with total particulates also showing a significant decrease of 59%. Carbon monoxide, sulfur oxides, and volatile organic compounds were reduced by 41%, 25%, and 31%, respectively. Nitrogen oxides, however, increased by 6%. Source categories emitting the largest amounts of these key pollutants in 1990 are presented in Table 14.5.

Ozone, a photochemical oxidant, is another key air quality parameter and is the main component in smog. Ozone is not emitted directly into the atmosphere, but is formed through complex chemical reactions. Major components which contribute to the formation of ozone include emissions of volatile organic compounds, nitrogen oxides, and sunlight. For air quality trending, the second highest daily maximum 1-hour concentration at 471

TABLE 14.4   Selected Examples of Analytical Methods Used for Monitoring Ambient Air Quality

Analytical Method	Application
Gas chromatograph/mass spectrometer (GC/MS)	Used for analyzing organic compounds collected in a tedlar bag, charcoal filter tube, or air canister
Preconcentration direct flame ionization detector (PDFID)	Used for analyzing nonmethane hydrocarbons
Ultraviolet-photometric ozone analyzer	Used for continuous analysis of ozone
Nondispersive infrared spectrometry	Used for continous analysis of carbon monoxide
Visible absorption spectrometry (VAS)	Used for analyzing metals such as hexavalent chromium that are collected on a PVC membrane or equivalent filter
Inductively coupled plasma-atomic emission spectrometry	Used for analyzing metals that are collected on a mixed cellulose ester filter or equivalent
Constant volume sampling (CVS) analyzer	Used to analyze carbon monoxide, carbon dioxide, and nitrogen oxides, generally from automobile emissions
Superfluid extraction/gas chromatograph	Used to analyze volatile and semi-volatile compounds collected on a sorbent bed
Colorimetric analysis	Used to analyze mercaptans collected with a midget bubbler with coarse porosity frit
Gravitation measurements	Used to analyze for dusts and other heavy particulates that are collected on filters
Optical sizing	Used for quantifying small particulates that are collected on filters

Sources: Information from ASTM (1992), NIOSH (1984), and EPA/600/4-89/017 (1989).

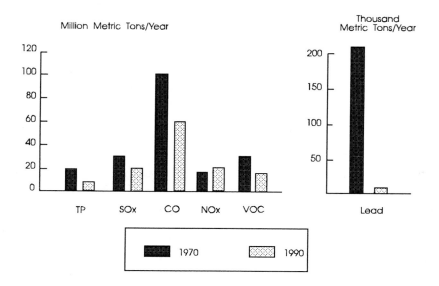

Source: Adapted from EPA-450/4-91-023 (1991).

FIGURE 14.3   A Comparison of Air Quality Parameters in the U.S. in 1970 and 1990.

monitoring sites has been used by EPA. A comparison between 1981 and 1990 showed that there was a decrease of approximately 10% in average ozone concentration levels. Data from 1983 and 1988, however, showed considerably higher concentration averages than 1990, and for most years during 1981 through 1990, the average concentrations were above the National Ambient Air Quality Standards (NAAQS).

Although concentration levels of most air quality parameters have declined over the past 20 years, many metropolitan areas exceed NAAQS for one or more parameters on a number of days each year. The Clean Air Act Amendments will address this problem by placing requirements on areas that cannot attain NAAQS, with the strictest requirements being placed on the areas with the worst air quality.

CONCLUSION

Air quality assessments can be performed by several means. These include pollution modeling, stack and fugitives sampling, and ambient air quality analysis. A combination of these methods is needed to assess fully the impact of point sources, mobile sources, and other sources on air quality. In the coming decade, as additional air quality regulations are promulgated, emphasis will be placed on reducing emissions in areas with the poorest air quality, such as large cities. National Ambient Air Quality Standards and emissions trends

TABLE 14.5   Source Categories Emitting the Largest Amounts of Key Air Quality Parameters in 1990

Air Quality Parameter	Source Categories Emitting Largest Amounts
Total particulate emissions—not including fugitives (7.5 million metric tons)	Industrial processes ($\sim$37%), fuel combustion ($\sim$23%), transportation ($\sim$20%), solid waste ($\sim$4%), miscellaneous ($\sim$16%)
Particulate matter (PM_{10}) emissions—not including fugitives (6.4 million metric tons)	Industrial processes ($\sim$42%), transportation ($\sim$23%), fuel combustion ($\sim$17%), solid waste ($\sim$3%), miscellaneous ($\sim$14%)
Particulate matter (PM_{10}) fugitive emissions (40.8 million metric tons)	Unpaved roads ($\sim$35%), construction ($\sim$22%), paved roads ($\sim$18%), agricultural tilling ($\sim$15%), wind erosion ($\sim$9%), mining and quarrying ($<$1%)
Sulfur oxides (21.2 million metric tons)	Fuel combustion ($\sim$81%), industrial processes ($\sim$15%), transportation ($\sim$4%), all other sources ($<$1%)
Carbon monoxide (60.1 million metric tons)	Transportation ($\sim$63%), fuel combustion ($\sim$12%), industrial processes ($\sim$8%), solid waste ($\sim$3%), miscellaneous ($\sim$14%)
Nitrogen oxides (19.6 million metric tons)	Fuel combustion ($\sim$57%), transportation ($\sim$38%), industrial processes ($\sim$3%), solid waste ($<$1%), miscellaneous ($\sim$2%)
Volatile organic compounds (18.7 million metric tons)	Industrial processes ($\sim$43%), transportation ($\sim$34%), fuel combustion ($\sim$5%), solid waste ($\sim$3%), miscellaneous ($\sim$14%)
Lead (7.1 thousand metric tons)	Transportation ($\sim$31%), industrial processes ($\sim$31%), solid waste ($\sim$31%), fuel combustion ($\sim$7%)

Note: The sum of the percentages may not equal 100% because of rounding.
Source: EPA (1991).

will remain a focus, as will emissions abatement and control, particularly for the most toxic chemicals. For point sources, such as industrial processes, chemical emissions reduction strategies will be integral to most air management programs. The topics of chemical reduction and waste minimization are reviewed in depth in the next chapter.

REFERENCES

American Society of Testing and Materials, *Annual Book of ASTM Standards*, Vol. 11.03, "Atmospheric Analysis; Occupational Health and Safety," Philadelphia, 1992.

Hanna, Steven R., and Peter J. Drivas, *Guidelines for Use of Vapor Cloud Dispersion Models*, Center for Chemical Process Safety, American Institute of Chemical Engineers, New York, 1987.

National Institute for Occupational Safety and Health, *NIOSH Manual of Analytical Methods*, 3rd ed., prepared by M. Millson and R. Delon Hull, Cincinnati, OH, 1984.

U.S. Environmental Protection Agency, "Advanced Methodologies for Sampling and Analysis of Toxic Organic Chemicals in Ambient Outdoor, Indoor, and Personal Respiratory Air," EPA/600/J-89/315, Atmospheric Research and Exposure Assessment Laboratory, Research Triangle Park, NC, 1989.

U.S. Environmental Protection Agency, "Compendium of Methods for the Determination of Toxic Organic Compounds in Ambient Air," EPA/600/4-89/017, Research Triangle Park, NC, 1989.

U.S. Environmental Protection Agency, "Industrial Source Complex (ISC2) Dispersion Model User's Guide," Vols. 1 and 2, EPA/450/4-92/008A and EPA/450/4-92/002B, U.S. Environmental Protection Agency, Washington, D.C., 1987.

U.S. Environmental Protection Agency, "National Air Quality and Emissions Trends Report, 1990," EPA 450/4-91-023, Office of Air Quality Planning and Standards, Research Triangle Park, NC, 1991.

U.S. Environmental Protection Agency, "UNAMAP—User's Network for the Applied Modeling of Air Pollution," EPA/600/D-87/330, prepared by Thomas E. Pierce and D. Bruce Turner, Atmospheric Sciences Research Laboratory, Office of Research and Development, Research Triangle Park, NC, 1987.

BIBLIOGRAPHY

Bernarde, Melvin (1992), *Global Warning... Global Warming*, John Wiley & Sons, New York.

Bubenick, David V., ed. (1984), *Acid Rain Information Book*, Noyes Data Corporation, Park Ridge, NJ.

Ehrenfeld, John R., et al. (1986), *Controlling Volatile Emissions at Hazardous Waste Sites*, Noyes Data Corporation, Park Ridge, NJ.

Godish, Thad (1990), *Air Quality*, 2nd ed., Lewis Publishers, Boca Raton, FL.

Lodge, James P., ed. (1989), *Methods of Air Sampling and Analysis*, 3rd ed., Lewis Publishers, Boca Raton, FL.

National Research Council (1991), *Rethinking the Ozone Problem in Urban and Regional Air Pollution*, National Academy Press, Washington, D.C.

Stensvaag, John-Mark (1991), *The Clean Air Act of 1990*, John Wiley & Sons, New York.

U.S. EPA (1984), "Compendium of Methods for the Determination of Toxic Organic Compounds in Ambient Air," EPA 600/4-84-041, U.S. Environmental Protection Agency, Research Triangle Park, NC.

_____ (1986), "Supplement to the Compendium of Methods for the Determination of Toxic Organic Compounds in Ambient Air," EPA 600/4-87-006, prepared by R.M. Riggin et al., Quality Assurance Division, Environmental Monitoring Systems Laboratory, U.S. Environmental Protection Agency, Research Triangle Park, NC.

_____ (1988), "Second Supplement to Compendium of Methods for the Determination of Toxic Organic Compounds in Ambient Air," prepared by W.T. Winbery, N.T. Murphy, and R.M. Riggin, Atmospheric Research and Exposure Assessment Laboratory, U.S. Environmental Protection Agency, Research Triangle Park, NC.

_____ (1991), "The Clean Air Act Section 183(d) Guidance on Cost-Effectiveness," EPA-450/2-91-008, Office of Air Quality Planning and Standards, Office of Air and Radiation, U.S. Environmental Protection Agency, Research Triangle Park, NC.

_____ (1991), "Evaluation of High Volume Particle Sampling and Sample Handling Protocols for Ambient Urban Air Mutagenicity Determinations," EPA/600/D-91/128, Health Effects Research Laboratory, U.S. Environmental Protection Agency, Research Triangle Park, NC.

Wuebbles, Donald J. (1991), _Primer on Greenhouse Gases_, Lewis Publishers, Boca Raton, FL.

15

CHEMICAL REDUCTION AND WASTE MINIMIZATION STRATEGIES

In the late 1980s and early 1990s, pollution prevention became a national initiative. Voluntary programs aimed at reducing chemical usage, waste, air emissions, and other releases to the environment were jointly conceived between EPA and industry leaders. This chapter details chemical reduction strategies and waste minimization techniques that can be applied to large and small waste-generating facilities. Included is information about source reduction, recycling, neutralization/detoxification, and other methods of waste minimization. Additionally, state and federal programs and incentives for pollution prevention are described.

CHEMICAL REDUCTION STRATEGIES

Process management is an appropriate place to start any chemical reduction program. Typically, process engineers are focused on product volume and yields and have built conservative factors into chemical activities that can be trimmed away for initial chemical reductions. The potential for recycling and reclamation, either on-site or off-site, is the next logical area for achieving reductions. Finally, although the most difficult to achieve, nonchemical substitutions in a process can yield the most dramatic results.

Innovative Chemical Reductions Using Statistical Process Control and Automated Equipment

With recent focus in industry on total quality management (TQM), many manufacturing and industrial engineers are beginning to focus on all aspects of the product, including environmental aspects such as chemical control and chemical reduction programs. Methods of analysis, such as statistical process control (SPC), which are commonly used in TQM programs to assess the quality of the product, can be used in an innovative way to control the chemicals in the process. In particular, SPC can be used to control activities such as chemical additions and bath drops in some processes such as cleaning processes, plating processes, and others (Woodside, Slapik, and Prusak 1991).

It is not uncommon for chemical adds and bath drops to be based on part count or time. Preliminary design experiments or conservative engineering calculations typically are used to determine the proper part count or time factors for chemical activities. Thus, using these conservative factors for chemical adds or bath drops can result in excessive chemical usage and waste generation. By performing additional experiments and analyzing yield data, the process can be enhanced. Part count and time factors can be optimized through the SPC process, with statistical confidence levels of ± 3 standard deviations usually considered adequate. By using real-time process data, the life of a process chemical bath can be extended by 10% to 25%, thus significantly reducing chemical usage and waste generation immediately.

Additionally, processes that utilize manual analysis and a preset list of chemical specifications can be enhanced using an automated system and SPC. Although equipment purchases are necessary, the automated system quickly can pay for itself through savings in chemical usage. The automated system includes on-line chemical analysis equipment such as a conductivity cell or an auto-titrator. The chemical analysis equipment is tied to a programmable logic controller and a well-metered feed system. Preset manufacturing specifications normally allow for a wide range of operation for a particular process. With the automated system, the same process can be controlled at much tighter tolerances. SPC analysis of chemical data and yields will allow a process optimum to be established using real-time data. Over time, these enhanced process controls can allow for substantial chemical usage reductions.

Recycling or Reclamation through Closed-Loop Systems

Recycling or reclamation typically applies to solvents, metals and metal solutions, and contaminated water. Closed-loop recycling is the most efficient method of recycling. The waste never leaves the process loop; hence, no drumming or handling of the material is necessitated. Additionally, this form of recycling is not currently regulated by EPA rules as defined in 40 CFR §261.6. Applications for closed-loop and on-site recycling can pertain to solvents, process chemicals, and waste streams.

Solvents from Cleaners and Defluxing Operations.   Solvents from cleaning processes can be hard-piped to a distillation unit, recycled, and pumped back to the process. This type of distillation can allow a process material to be reused an infinite number of times, with an average material make-up of 10% to 15% per month for continuous processes. The distilled bottoms are collected in the distillation unit. Unlike the recycled solvent, when the bottoms are removed, they are regulated as a listed hazardous waste from a nonspecific source under RCRA.

Solvents from some operations such as electronic assembly defluxing operations may be pure enough for usage in maintenance degreasing operations or other operations. These solvents can be transported via internal pipelines or in drums. This allows for reuse of the material without reclamation.

Process Chemicals.   If process chemicals have been mixed with water in the manufacturing process, these chemicals can be filtered through a filtering device, such as a carbon canister, molecular sieve, or ion exchange column and recycled back to the process. Process chemicals also can be replenished or revitalized by reacidification, reconcentrating, or dosing with inhibitors or metal salts.

Waste as a Feedstock.   Another recycling option is the use of a waste as a feedstock or process chemical in another operation on-site. An example of this type of application is the use of used ultra-pure rinse water in other operations on-site where high purity water is not required. Reuse will allow for savings in treatment and/or disposal costs.

A second application might include reuse of spent acids and bases. These materials may be used in a wastewater treatment plant for initial neutralization. Depending on the volume generated, the waste stream might warrant being hard-piped to an equalization tank or neutralization unit. For low-flow waste streams, the material can be drummed and accumulated before being used.

Other Recycling or Reclamation Methods

Other recycling or reclamation options include distillation, reclamation, and use of sludge as fertilizer. Additionally, the concept of designing environmentally conscious products (ECP)) is gaining momentum.

Distillation.   Spent solvents can be distilled into usable or salable products by a commercial vendor. Solvents to be reclaimed in this manner may need to be manifested. Reclamation facilities must obtain a RCRA identification number to operate and must meet certain requirements in order to be exempt from permitting.

Metals and Metal Solutions Reclamation.   Scrap metals can be smelted into usable products. Likewise, processes such as chemical precipitation or electrowinning can allow for reclamation of metals or metal salts from spent metal-containing solutions. Metals reclamation facilities are regulated in the same manner as commercial distillation facilities.

Sludge to Fertilizer. Over 130 cities in the United States have operational programs for composting municipal wastewater treatment sludges for use in land application. Over 100 additional municipal programs are in the design or pilot stage for this activity. The composting technology includes aerated static piles, in-vessel operations, and windrows. Bulking agents include wood chips, sawdust, leaves, grass clippings, and brush (Goldstein 1990).

This type of recycling program has been implemented successfully by the City of Austin, Texas, which processes 100% of its sludge into biosolids. These biosolids are beneficially reused on agricultural land application or are composted into "Dillo Dirt," which is distributed and marketed back to the general public as a soil conditioner. The City of Austin also offers curbside pickup of leaves, grass, and tree trimmings, and by the fall of 1993 will reuse 100% of this waste as bulking agents in its windrow process.

Product Recycling and Environmentally Conscious Design. In an effort to move from a "cradle to grave" waste management philosophy to a "cradle to cradle" philosophy, many industries are now designing products with the environment as a key factor. This includes the ability to recycle product components into other products or materials once the product has reached its end of life, as well as the selection of environmentally conscious materials during manufacture. This concept of "design for the environment" likely will be a major thrust during the 1990s.

Material Substitutions

A General Approach. Research, development, and manufacturing engineers all must play a part in order to change a manufacturing process from a high-use and toxic chemical process to one that uses aqueous materials or alternate chemicals that are less toxic. Normally, chemical reductions by material substitutions are process specific, and many require process equipment changes. The new equipment may prove to be quite costly, and return on capital investment for the changes may take several years. There are immediate positive effects from making the changes, however. These include a positive public image for the facility, and an aggressive pollution prevention program that will satisfy the regulatory agencies.

Research and development engineers should be given several key lists from those defined in Chapter 2 containing names of chemicals to avoid when creating new products or processes. Additionally, research and development engineers should be made aware that EPA has targeted 17 chemicals for immediate reduction in their "Clean Products Research Program" (EPA/600/D-91/029 1991). These chemicals and other chemicals of concern to EPA are listed in Table 15.1.

Typically, research and development engineers plan, develop, and implement processes with familiar and proven chemistries, and may not be aware of potential regulatory and environmental concerns associated with a particular chemical. By providing a listing of chemicals to avoid, toxic chemical reduc-

TABLE 15.1   Chemicals of Concern to EPA

Chemicals Targeted for Immediate Reduction

Chlorinated compounds	Metals
• Carbon tetrachloride	• Cadmium
• Chloroform	• Chromium
• Dichloromethane	• Lead
• Tetrachloroethylene	• Mercury
• Trichloroethane	• Nickel
• Trichloroethylene	
Other toxic compounds	Ketones
• Benzene	• Methyl ethyl ketone
• Cyanide	• Methyl isobutyl ketone
• Toluene	
• Xylene	

Other Chemicals of Concern to EPA

Other toxic chemicals	Pesticides
• Acrylonitrile	• Alachlor
• Arsenic	• Carbaryl
• Carbon disulfide	• Captan
• Ethylene oxide	• Chlorpyrifos
• Formaldehyde	• 2,4-Dichlorophenoxyacetic acid
• Styrene	
• Vinyl chloride	

Source: Information from EPA/600/D-91/029 (1991).

tions can be implemented at the front end, instead of the back end, of the process. Additionally, successes with alternative, nonhazardous chemicals and materials can be translated to existing processes as well.

Material Substitutions for Chlorofluorocarbons (CFCs) and other Ozone Depleting Chemicals (ODCs). The damaging environmental effect of CFCs and other ODCs on the ozone layer has made the search for alternative chemistries to these chemicals top priority around the world. Process cleaners using CFCs and other ODCs have been used in industry for decades, but will be phased out in the coming years, as new chemistries are developed. Some alternative cleaning agents that are being developed and implemented across industry are documented by EPA (EPA/450/3-89/030 1989). These alternatives include:

- Aqueous cleaners—Aqueous and emulsion cleaners are being utilized as substitutes for trichloroethylene and 1,1,1-trichloroethane vapor degreasers and freon cleaners. They are also effective as a substitute for ODC-based ultrasonic cleaning.

- Detergent cleaners—Biodegradable detergents are successful substitutes for CFCs in some navigational equipment cleaning operations.
- Terpene-based formulation—A terpene-based formulation has been used in some printed circuit board cleaning operations to replace ODCs and solvents.

Additionally, a "no clean" process that eliminates the need for several cleaning steps has been developed and is being implemented by the electronics and other industries.

HAZARDOUS WASTE MINIMIZATION

Establishing a Waste Minimization Program

Waste minimization techniques can be utilized as a means of pollution prevention once chemical source reduction and recycling possibilities are exhausted. To optimize a hazardous waste minimization program, several key elements must be addressed. First, data must be gathered on waste volumes and disposal costs. This gives a starting point for prioritizing waste minimization projects. Second, leverage as a function of environmental priority and feasibility should be defined so that resources are spent wisely. And, finally, liabilities need to be balanced against cost (Stuckey and Prusak 1990).

Data Gathering.   The goal of data gathering is to size the volume and cost of unique waste streams. Three years of data is usually sufficient to categorize major waste streams. Production data should be included so that waste generation relative to production can be assessed. The ideal data gathering mechanisms for assessing waste generation are manifests, billing records, and/or other waste tracking records. Wastes should be summarized by waste type (i.e., flammable, corrosive, toxic, etc.) and should include a breakdown of the processes generating the waste, as well as the final destination of the waste. Off-spec production and spills also should be identified.

Defining Leverage.   Establishing leverage is the process of determining which waste minimization efforts will allow the greatest results. Leverage, as a function of priority and feasibility can be determined as follows:

$$leverage = priority \times feasibility$$

where priority is a function of potential environmental impact and feasibility is the probability of minimizing a waste. Priority can be further defined as:

$$priority = environmental\ units \times fate\ factor$$

where environmental units are assigned to a waste stream as a function of its mass and its toxicity:

$$\text{environmental units} = \text{mass} \times \text{toxicity}$$

Toxicity is determined by health data such as a waste's LD_{50}, permissible exposure limit, leachability as determined by the toxicity characteristic leaching procedure (TCLP), or other data consistently applied to all of the wastes being ranked. Other factors can be included in the toxicity factor such as the waste's presence on EPA's CERCLA, SARA, or other list, and contribution to the green-house effect or ozone depletion.

The fate factor should take into account the waste's potential impact to health and the environment based on disposal method. Wastes that are recycled/reused or that are destroyed in an incinerator will be given a lower fate factor than wastes discharged to air or waterways. A waste hierarchy of fate factors includes:

- Source reduction
- Recycle or reclamation
- Incineration
- Treatment or detoxification
- Discharge to atmosphere or streams
- Land disposal

Feasibility can be viewed as the possible percentage reduction in a waste stream. In order to evaluate reduction potential, applicable engineering methods must be assessed. Finally, cost leverage can be evaluated using the same equation, with "disposal cost" being substituted for "priority."

Developing Waste Trends. Waste trends can be developed to assure correct waste minimization priorities are established. When developing trends, one-time or anomalous waste generation should be subtracted from existing data. Future production and product mix also should be factored into waste projections. Adjustments to current data should be made for:

- Excessive waste from process startup and debug
- Excessive waste from a one-time process upset
- Waste from a spill
- Waste from construction or demolition

Adjustments to future trends should be made for:

- Anticipated product demand and product/production life
- Known or anticipated process changes
- Known or anticipated regulation effects on disposal options and disposal cost

An example of waste trending for an industrial facility including waste history, specification of minimization projects, and anticipated trends, is shown in Figure 15.1.

Waste Minimization Techniques

Once source reduction and alternative chemistry options have been exhausted, waste minimization techniques can be utilized. These techniques include segregation, volume reduction, neutralization/detoxification, use of waste as a fuel, and use of waste as a product.

Segregation.   An often-used waste minimization technique is segregation. Segregation serves two purposes. First, it prevents the mixing of nonhazardous waste with hazardous waste, thus decreasing the overall hazardous waste stream. Second, it prevents the mixing of different wastes, which, if left unmixed, are recyclable. Candidates for segregation include halogenated and

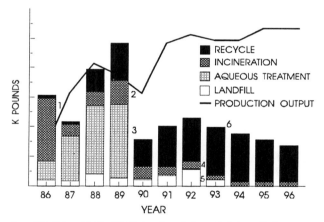

1. On-site carbon filtration of water containing dilute organic solvents reduced incineration volumes by approximately 80% from 1986 to 1987.

2. Off-site reclamation of oils and solvents reduced incineration volumes by an additional 50% from 1989 to 1990.

3. On-site neutralization and treatment of wastewater containing polymers reduced production-adjusted waste shipments by approximately 95% from 1989 to 1990.

4. Reclassification of hazardous waste as a product reduced incineration volumes by 50% from 1992 to 1994.

5. Off-site reclamation of metal-bearing sludge virtually eliminated landfill disposal between 1992 and 1994.

6. Chemical conservation and pollution control prevention strategies are expected to reduce recycling by 25% from 1993 to 1996.

FIGURE 15.1   Example of Waste Trending.

flammable solvents. These wastes should not be mixed together because the individual streams may be recyclable. Additionally, solvent-soaked rags, gloves, and filters should not not be mixed with other nonhazardous solids since mixing the two types of waste can render the entire waste stream hazardous.

Volume Reduction.   Although volume reduction does not minimize the amount of waste produced, it does limit the number of waste containers and/or space needed in a landfill for final disposal of the waste. Volume reductions usually result in substantial disposal cost reductions. Several techniques can be used to reduce volume including compaction, drying, and consolidation.

Compaction is implemented through crushers or compacting devices and can be used to reduce the volume of waste as much as 75%. Applicable wastes for this minimization include rags, filters, plastic and glass bottles, and empty containers. Drying reduces the water content of wastes, particularly sludges. This is applicable for situations where disposal vendors charge by total tonnage rather than dry content of a waste. Drying can reduce volumes by 20% to 50%.

Consolidation techniques allow for the mixing of similar wastes with low generation volumes. This is applicable to high-solids wastes, such as rags, filters, and gloves in situations where segregation does not offer any benefit. Also, consolidation may be utilized for some combustible liquids, if recycling is not an option. Likewise, consolidation is used for lab wastes.

Neutralization/Detoxification.   This process is best implemented on site for maximum cost savings. Corrosive wastes such as acids and bases are prime candidates for neutralization. Other wastes such as metal-bearing solutions and spent aqueous solutions can be treated and detoxified through an industrial wastewater treatment plant and discharged to a Publicly Owned Treatment Works (POTW) or to a stream. Industrial detoxification techniques include thermal destruction, biological treatment, and chemical treatment. These and other hazardous waste treatment techniques are described in detail in Chapter 10.

Waste-Derived Fuels.   Many used solvents or off-spec solvents with a high heating value can be used as a fuel blend in waste incinerators or other approved thermal destruction systems. The waste itself is not minimized, but the fuel value is reclaimed. Used oil also can be burned as fuel for energy recovery. This activity is regulated under RCRA in 40 CFR 266 Subpart E. New technology is available for processing municipal solid waste into combustible fuels that fire electric power plants. There are currently over 100 of these facilities across the nation (DOE 1988).

Waste as a Product.   In order to make use of a waste stream as a feedstock for another product off site, a production facility that can use the waste as a feedstock must be found. Several states and regions sponsor waste exchanges that match wastes to product needs and buyers in the area. Four waste exchanges that provide services to highly industrialized sectors of the nation include: Northeast Industrial Exchange in Syracuse, NY; Southeast Waste Exchange in Charlotte, NC; California Waste Exchange in Sacramento, CA; and Southern Waste Information Exchange in Tallahassee, FL.

Once an applicable production facility is located, the waste generator can develop a strategy for marketing a given waste stream. Bidding a predetermined volume of profiled waste guaranteed for a long period of time is one way to market nonvariable waste streams. These types of streams are typically advantageous to a vendor who is looking for a constant material supply. Other common recyclables such as batteries, used solder dross, off-spec chemicals, and scrap metal typically are bid by the shipment. In addition to pollution prevention aspects, there are added benefits to finding a production facility to process waste, such as potential cost savings and liability reduction as a result of avoiding disposal.

STATE AND FEDERAL POLLUTION PREVENTION PROGRAMS AND INCENTIVES

Technical Assistance Programs

EPA has established a formal pollution prevention program through the Office of Research and Development (EPA 1990). EPA's six basic pollution prevention program goals include:

- Stimulate the development and use of products that result in reduced pollution
- Stimulate the development and implementation of technologies and processes that result in reduced pollution
- Expand the reusability and recyclability of wastes and products and the demand for recycled materials
- Identify and promote the implementation of effective nontechnological approaches to pollution prevention
- Establish a program of research that will anticipate and address future environmental problems and pollution prevention opportunities
- Conduct a vigorous technology transfer and technical assistance program that facilitates pollution prevention strategies and technologies

These goals are carried out through numerous pollution prevention programs that mainly focus on research and technical assistance for industrial and other applications. Several of the key programs and a short description of each are presented in Table 15.2.

In addition to EPA's research and technical assistance programs, almost all states with heavy industrial activity provide the regulated community with some sort of technical assistance. Typically, this assistance is provided through guidance manuals, expert staff, and regulatory and technology transfer seminars.

In addition to conventional methods of technical assistance, some states and local communities are sponsoring additional programs. An example is the State of Texas' pollution prevention program called Clean Texas 2000, which was initiated by the governor in 1992. Clean Texas 2000 is a partnership of government, industry, cities, and environmental groups located in all parts of the state that are working together for pollution prevention. Efforts are focused on recycling, technology transfer, education, and cooperation of all toward a common goal of a cleaner environment in Texas by the year 2000.

TABLE 15.2   Selected Examples of EPA's Pollution Reduction and Clean Technologies Programs

Program	Program Objectives
Waste Reduction Innovative Technology Evaluation (WRITE)	• This program is composed of a set of pilot projects performed in cooperation with state and local governments.
	• Program goals include identifying and evaluating new or improved, economically favorable technologies for the prevention of pollution in industries presently experiencing waste problems.
	• New source reduction and recycling technologies are priorities.
	• The program will assist federal, state, and local governments, as well as small and midsized industries by providing performance and cost information on pollution prevention technologies.
Industrial Pollution Prevention Program (IPPP)	• This program focuses on mature pollution prevention technologies that have the potential for application industry-wide.
	• The technologies are evaluated via cooperative agreements between EPA and other non-profit organizations and institutions.
	• Projects in this program include substitutions for chlorinated and organic solvents, evaluation of equipment for recycling antifreeze and freon, and oil life extension for internal combustion engines.
Waste Reduction Assessments Program (WRAP)	• This program was designed to encourage industrial operations to utilize waste minimization assessments to identify sources of waste and implement changes to reduce waste generation.

TABLE 15.2 *Continued*

Program	Program Objectives
	• Targeted facilities are small- to medium-sized operations such as mini-photo labs, printing operations, and industrial vehicle maintenance shops.
Waste Reduction Evaluations at Federal Sites (WREAFS)	• This program focuses on waste minimization efforts at federal sites such as the Philadelphia Naval Shipyard, the Naval Undersea Warfare Engineering Station (Keyport, WA), and the Veterans Affairs Medical Center (Cincinnati).
	• Under this program, agencies such as DOE and DOD are focusing on pollution prevention in operations such as metal and other cleaning operations, solvent degreasing, and equipment overhaul.
	• All information obtained under this program will be transferred to the private sector.

Sources: Information from EPA (1990) and EPA/600/8-89/070 (1989).

Waste Reduction Incentives

In addition to offering technical assistance for pollution prevention and waste reduction, federal and state governments utilize other incentives to reduce waste (Iowa DNR 1991). These include waste fees, production and storage fees, waste reduction goals, and pollution prevention plans.

- Waste fees—Waste fees are levied on the waste generator, and usually include a per ton assessment on waste transported off-site, treated off-site, incinerated, or disposed of in landfill. Fees can range from as little as $5 per ton of waste to as much $100 (or more) per ton of waste. Additional annual fees for waste generators also are common. In general, fees do not apply to waste that is recycled. This incentive is used in some form in almost every state. In addition to providing incentive for pollution prevention, fee assessments raise revenue for states. The revenue usually is funnelled into the technical assistance and other waste programs.
- Production and storage fees—Production fees, or front-end fees, are fees that are levied on producers of hazardous substances. Additionally, fees are charged (by the federal government and some states) for year-end inventories of CFCs, in order to prevent stockpiling of a soon to be banned material.
- Waste reduction goals—Some states have adopted voluntary or involuntary waste reduction goals from a base year, usually 1987. Waste reduction

goals of 33% by a specified year, such as 1995, and of 50% a few years afterward is typical.

- Pollution prevention planning by industrial facilities—Regulations require permitted waste generators to have a pollution prevention or waste minimization plan that specifies opportunities for waste reduction and specific waste reduction goals. Additionally, some required reports include waste minimization information.

In addition to these incentives, the toxic release inventory report has provided the public with information about industrial releases to the environment. This, in turn, has brought public pressure upon the industrial sector to be aggressive in chemical source reduction and waste minimization.

Regulatory Barriers

Although EPA and industry are both interested in achieving pollution prevention to the greatest extent possible, some regulatory barriers have been identified that make pollution prevention options and other best management practices more difficult to implement (EPA/600/J-91/109 1991). These include:

- Regulation of secondary materials—There are many materials generated as waste from an industrial process, but which have intrinsic value and can be recycled, or used in an operation off-site. Once wastes are generated and removed from the process, they fall out of the pollution prevention umbrella and must be managed as a hazardous waste. This makes marketing of these materials more difficult.
- Derived-from waste rules—These rules state that any material derived from a listed hazardous waste is itself a hazardous waste. This makes beneficial reuse of many materials subject to tight RCRA control. Also, wastes such as inert ash from incineration of certain listed wastes must be managed as hazardous waste, even when there is no threat to human health or the environment. This rule may be revised in the near future.
- Delisting—The process for delisting a waste is long and tedious. Thus, many facilities will opt to manage certain nonhazardous wastes as hazardous, which increases demand on capacity-restrained disposal facilities such as landfills.
- Permitting—Requirements for permitting are burdensome for recycling facilities that need to store secondary materials before recovery. Present rules do not allow 90-day storage for recycling facilities and facilities are expected to reclaim all material as it arrives.

Some or all of these barriers may be mitigated by future EPA policies and/or regulations.

CONCLUSION

The need for pollution prevention efforts gained national attention when the first toxic release inventory (TRI) reports were made public in 1988. Amounts of pollutants documented as being released to the environment in the form of waste, wastewater discharges, and air emissions were dramatically higher than the public expected. Thus, public pressure to reduce these environmental releases led to the passage of the Pollution Prevention Act of 1990 and to voluntary efforts to reduce releases during the 1990s.

In order to accomplish stringent pollution reduction goals, source reduction efforts must be combined with recycling efforts and waste minimization techniques. Additionally, manufacturing design efforts that emphasize alternative, nontoxic chemistries and product reuse options are being pursued by many industries.

In addition to pollution prevention, environmental self-auditing has been a topic of interest in the 1990s. EPA and other agencies are encouraging facilities to perform facility assessments for environmental compliance. The next chapter addresses this topic and presents a strategy for developing and conducting a facility environmental assessment.

REFERENCES

Department of Energy, "Waste-to-Energy Compendium: Revised 1988 Edition," DOE/ CE/30844–H1 and DE89 004767, prepared by Science Applications International Corporation and Meridian Corporation for Office of Conservation and Renewable Energy, Washington, D.C., 1988.

Goldstein, Nora, and David Riggle, "Sludge Composting Maintains Momentum," *Biocycle*, December 1990.

Iowa Department of Natural Resources, "Hazardous Waste Reduction Strategies: A Discussion and Recommendations for the General Assembly," prepared by Robert W. Craggs, Waste Management Authority Division, Des Moines, IA, 1991.

Stuckey, Mark, and John J. Prusak, "The Basic Principles of Waste Minimization," *Institute of Environmental Sciences 1990 Proceedings*, May 1990.

U.S. Environmental Protection Agency, Alternative Control Technology Document: Halogenated Solvent Cleaners, EPA/450/3-89/030, prepared by Radian Corporation, Research Triangle Park, NC, 1989.

U.S. Environmental Protection Agency, "Encouraging Clean Technologies," EPA/ 600/D-90/179, prepared by Ivars J. Licis, U.S. EPA Pollution Prevention Program, Office of Research and Development, Risk Reduction Engineering Laboratory, Cincinnati, OH, 1990.

U.S. Environmental Protection Agency, "EPA Clean Products Research Program," EPA/600/D-91/029, prepared by M.A. Curran and A.R. Robertson, Risk Reduction Engineering Laboratory, Cincinnati, OH, 1991.

U.S. Environmental Protection Agency, "Guidance Document for the WRITE Pilot Program with State and Local Governments," EPA/600/8-89/070, prepared by M. Lynn Apel et al., Risk Reduction Engineering Laboratory, Cincinnati, OH, 1989.

U.S. Environmental Protection Agency, "Regulatory Barriers to Pollution Prevention: A Position Paper of the Implementation Council of the American Institute for Pollution Prevention," EPA/600/J-91/109, prepared by Aluminum Company of America for Risk Reduction Engineering Laboratory, Cincinnati, OH, 1991.

Woodside, Gayle, Michael A. Slapik, and John J. Prusak, "Innovative Statistical/Process Controls for Chemical and Waste Reductions," presented at American Chemical Society Conference on Pollution Prevention, Atlanta, GA, April 17, 1991. Represented at AIChE Spring Annual Conference, New Orleans, LA, March 30, 1992.

BIBLIOGRAPHY

Advisory Council for Research on Nature and Environment (1988), "Design and Waste Prevention," prepared by J. Eekels et al., Advisory Council for Research on Nature and Environment, Rijswijk, Netherlands.

Boley, Gary L. (1990), "Resource Recovery: Turning Waste into Watts," *Mechanical Engineering*, December 1990.

DOE (1991), "Engineering Technology Division Long-Range Plan," DE91 007886, Oak Ridge National Laboratory, Department of Energy, Oak Ridge, TN.

_____ (1991), "Targeting Industrial Waste Reduction," DE91 018221, Pacific Northwest Laboratory, Department of Energy, Richland, WA.

_____ (1990), "Waste Segregation Procedures and Benefits," DE91 000765, prepared by Jim D. Fish et al., Sandia National Laboratories, Department of Energy, Albuquerque, NM.

Donahue, Bernard A., et al. (1989), *Reclamation and Reprocessing of Spent Solvents*, Noyes Data Corporation, Park Ridge, NJ.

D'Ruiz, Carl D. (1991), *Aqueous Cleaning as an Alternative to CFC and Chlorinated Solvent-Based Cleaning*, Noyes Data Corporation, Park Ridge, NJ.

Higgins, Thomas E. (1989), *Hazardous Waste Minimization Handbook*, Lewis Publishers, Boca Raton, FL.

Illinois DOE (1991), " Alternatives for Measuring Hazardous Waste Reduction," prepared by Rachel Dickstein Baker, Richard W. Dunford, and John L. Warren for the Center for Economics Research and Illinois Department of Energy and Natural Resources, Research Triangle Park, NC.

Mueller Associates, Inc. (1989), *Waste Oil: Reclaiming Technology, Utilization and Disposal*, Noyes Data Corporation, Park Ridge, NJ.

PRC Environmental Management, Inc. (1989), *Hazardous Waste Reduction in the Metal Finishing Industry*, Noyes Data Corporation, Park Ridge, NJ.

Rogoff, Marc J. (1987), *How to Implement Waste-to-Energy Projects*, Noyes Data Corporation, Park Ridge, NJ.

U.S. EPA (1988), "The Waste Minimization Opportunity Assessment Manual," Office of Research and Development, Risk Reduction Engineering Laboratory, U.S. Environmental Protection Agency, Cincinnati, OH.

_____ (1989), "EPA's Approach to Pollution Prevention," EPA/600/J-89/004, Risk Reduction Engineering Laboratory, U.S. Environmental Protection Agency, Cincinnati, OH.

_____ (1989), "EPA's Research and Development Program for Waste Minimization," EPA/600/D-89/207, Risk Reduction Engineering Laboratory, U.S. Environmental Protection Agency, Cincinnati, OH.

_____ (1990), "EPA's Pollution Prevention R&D Approaches and Insights into the Chemical Process Industry," EPA/600/D-90/142, prepared by Paul M. Randall, Risk Reduction Engineering Laboratory, Cincinnati, OH.

_____ (1990), *Solvent Waste Reduction*, Noyes Data Corporation, Park Ridge, NJ.

_____ (1991), "Industrial Pollution Prevention Opportunities for the 1990's," EPA/600/8-91/052, prepared by Ivars J. Licis, Office of Research and Development, Risk Reduction Engineering Laboratory, U.S. Environmental Protection Agency, Cincinnati, OH.

16

FACILITY ENVIRONMENTAL ASSESSMENTS

During the past several years, managers of industrial facilities have found it advantageous to assess facility operations in terms of regulatory compliance and good management practices. A comprehensive environmental assessment program can aid in detecting potential regulatory problems early, before they become noncompliance issues. In addition, a self-auditing program may be beneficial during regulatory prosecution, as stated in the introduction to a guidance document prepared by the U.S. Department of Justice, dated July 1, 1991:

> It is the policy of the Department of Justice to encourage self-auditing, self-policing and voluntary disclosure of environmental violations by the regulated community by indicating that these activities are viewed as mitigating factors in the Department's exercise of criminal environmental enforcement discretion (Environmental Law Reporter 1991).

Once developed, the assessment program also can be used by the waste-generating facility to evaluate vendored operations for regulatory compliance. Target vendors might include off-site waste treatment and disposal facilities, waste brokers, and recycling vendors. This chapter outlines a methodology for developing an organized approach to hazardous materials and hazardous waste compliance assessments.

REVIEW OF PERMITS AND OPERATIONS

The first step to beginning an environmental assessment is to perform a permit review and an operations field check. If the permits are several years old and if the facility is one that has been upgraded recently with new chemical processes, there may be discrepancies in the permits that need to be addressed. Some key items that should be included during a permits review are presented in Table 16.1. Other items to be reviewed that may not be related directly to a permit or facility changes, but that affect the overall regulatory compliance or potential environmental liability of a facility are presented in Table 16.2.

Depending on the size of the facility, the permit and operations review could take several days. All discrepancies or inadequacies found should be addressed with management for corrective action.

TABLE 16.1   Items to Review During a Permits Assessment

Equipment Review Items

- Process equipment and associated exhausts
- Hazardous waste tanks, and certifications by a registered professional engineer
- Chemical ventilation systems in all buildings and process centers
- Equipment preventative maintenance schedules

Process and Operations Review Items

- Hazardous materials storage, treatment, and disposal operations
- Chemical usage, including individual chemicals and quantities
- Chemical dispensing or waste consolidation activities that are performed under local exhaust, including chemicals and quantities
- Identification of waste streams and wastewater treatment efficiencies, including end-of-pipe discharges
- Identification of lab, bench scale, or pilot operations
- Occurrence (or recurrence) of compliance problems and corrective actions

Review of Changes to Process or Equipment

- Conversions of nonhazardous waste tanks to hazardous waste tanks and certifications by a registered professional engineer
- Closure of storage and treatment tanks or surface impoundments
- Conversion of a warehouse or storeroom space to a hazardous materials or hazardous waste storage area
- Relocation of process centers
- Increases in production rates resulting in greater chemical usage
- Changes in work day in terms of adding overtime hours or additional shifts
- Changes in on-site vendor operations

TABLE 16.2   Other Items to Review During a Facilities Assessment

Items Related to Safety and Security

- Adequacy of fire suppression systems, in terms of changes in chemicals, quantities, and location of operations
- Security of the site, individual buildings, hazardous materials and waste storage areas, and docks
- Designation of adequate evacuation routes to be used in case of an emergency

Items Related to Spill Containment

- Adequacy of docks and load/unload spill containment to contain catastrophic failures
- Adequacy of tank system spill containment, including those containing product materials and waste, to contain catastrophic failures.
- Adequacy of spill containment in container storage areas

Items Related to Run-off Management

- Identification and proper control of storm drains near docks, load/unload stations, and tank systems and identification of other nearby migration paths to surface or groundwater
- Run-off controls or diversions to contain a release within the property boundaries
- Identification of land use adjacent to the facility and within a 2 to 5 mile radius

Other Items

- Control of fill material brought on-site for construction or other purposes and certification that the material is not contaminated
- Control of contractor chemical and waste activities and other controls to protect the environment

DOCUMENTATION REVIEW

Along with permits and general operational compliance with regulatory standards, proper documentation is another important area to be assessed during a facility assessment or self-audit. All documents and plans should be up-to-date and should reflect current operating rates and practices. Some of the documents that might be required of an industrial facility are listed in Table 16.3. As can be seen from the table, documentation requirements can be extensive. In addition to assessing the completeness of the documentation, the organization of the documents for easy retrieval also can be evaluated.

FIELD REVIEW FOR REGULATORY COMPLIANCE AND GOOD MANAGEMENT PRACTICES

No facility environmental assessment is complete until a field review is conducted. This review, or physical inspection, is necessary to ensure that com-

TABLE 16.3 Documentation that Could be Required to Meet Current Regulations

Regulation	Documents that Could be Required
RCRA	RCRA Part A, RCRA Part B, waste analysis plan, waste analysis records and waste determinations, inspection plan, emergency and contingency plan, groundwater monitoring plan, closure plan, post-closure plan, training plan, financial bond or certification, tank assessments and certifications, closure documentation, post-closure documentation, groundwater monitoring data and reports, training records, a list of job titles that include hazardous waste operations and names of employees occupying those job positions, manifests and land disposal restriction notifications from the past three years, waste inventory, waste container and tank inspection logs, biennial or annual reports to EPA or equivalent, waste minimization reports, the facility operating record, fugitives monitoring records, records of noncompliance release reporting, documentation of incidents where the contingency plan was invoked, registration of petroleum underground storage tanks
OSHA	Hazard communication plan and training records, lead training records, other training records for handling chemicals such as asbestos, industrial hygiene plan for chemical laboratories, personal monitoring records, respirator training and fit test records, respirator inspection logs, medical surveillance records, laser training records, hazardous waste operations and emergency response training records and certifications, documentation of tasks requiring personal protection, documentation of proper removal of asbestos, documentation of location of ionizing radiation producing equipment; documentation of process hazard reviews
SARA	Notice to SERC and LEPC of chemicals subject to section 312 reporting including a list of chemicals or MSDS sheets, notice of applicability, tier I and/or tier II reports, toxic release inventory reports, records of noncompliance release reporting
TSCA	Premanufacturing notification submittals, documentation of allegations of adverse effects of chemicals, health and safety data reports pertaining to new chemicals, production records of TSCA listed chemicals, PCB storage and disposal logs
Clean Water Act	Wastewater discharge monitoring data and reports, spill prevention control and countermeasure plan, pretreatment compliance reports, local POTW limits compliance reporting or certification, notification to local POTW of hazardous waste discharges to sewer system, stormwater monitoring data and reports, stormwater pollution prevention plans, documentation of best management practices, whole effluent toxicity test data and reports, records of noncompliance reporting

continued

309

TABLE 16.3 *Continued*

Regulation	Documents that Could be Required
Clean Air Act	Construction and operating permits, emissions monitoring data and reports, new source performance (NSPS) test results and other NSPS data and records, hours of operation, records of control system efficiency, chemical consumption tracking logs, production tracking logs, fuel oil or natural gas consumption tracking logs, fugitives monitoring records, emissions inventory
Nuclear Regulatory Commission	Training records, quality control plan, records of defects or noncompliance with quality control plan, waste analysis plan and records, manifests and disposal records, groundwater monitoring records, compliance with other facility permit requirements

pliance programs and good management procedures are being followed in manufacturing areas, chemical and waste control areas, and in other parts of the facility. This part of the assessment is perhaps best handled with a standard checklist that can be duplicated and filled out for every major process center, laboratory, and chemical and waste support area. Table 16.4 details some major categories and items that might be included on a checklist.

The checklist can be customized to the facility activities and can be used as a periodic "self-test" by the area manager. An example of a waste management checklist for waste accumulation areas is presented in Table 16.5. Additional checklists would be necessary for chemical-using departments and departments that require special safety measures such as those using lasers.

TABLE 16.4 **Categories and Items to Include on a Field Inspection Checklist**

Category	Inspection Items
Labeling	Chemical and waste containers are labeled for content and hazards; hazardous waste containers are dated clearly as to when the waste container was filled; tanks and containers containing hazardous waste are marked *Hazardous Waste*; tanks are labeled for quantity, content, and hazard; gas cylinders are labeled for content and hazard and have *In Use* or other appropriate tag; radioactive waste is labeled with appropriate decals; biomedical waste is labeled with appropriate decals; pipes are labeled for content; asbestos-containing materials are labeled to indicate such; tank trucks and other on-site carriers are placarded properly; chemical and waste containers are labeled with DOT labels prior to shipment
Chemical information	A written hazard communication plan is available to all employees; a written hazardous materials response plan

TABLE 16.4 *Continued*

Category	Inspection Items
	is available to all employees; MSDSs or the equivalent are available in the work area for all chemicals used in the process center, chemical support area, or lab; documentation shows employees have been trained on all hazards in the area
Signs	Exit signs are placed properly and lit at all times; emergency evacuation routes are posted with location of fire extinguishers and emergency response equipment; emergency contact numbers are posted on all phones; personal protective equipment signs are posted on doors of process centers requiring such protection; *No Eating, Drinking, and Smoking* signs are posted in lead-using and other chemical areas; *Danger, Keep Out* signs are posted at permitted hazardous waste operations sites; *No Smoking* signs are posted in flammable areas; *Keep Out, Asbestos Containing Materials* signs are posted during asbestos removal projects; PCB warning signs are posted where PCB-containing equipment is used or PCB waste is stored
Ionizing radiation sources	On-site inventory of ionizing radiation sources matches inventory of Laser Safety Officer
Personal protection	Requirements for personal protection are documented at process center; inventory of personal protective equipment in the area meets criteria documented in personal protection plan; personal protective equipment is being used, cleaned, and inspected, as specified
Local exhaust/ ventilation	Alarm set points are calibrated on schedule; make-up air vents are open; system balancing has been performed on schedule
Storage and containment	Chemical and waste containers are segregated properly and stored in cabinets or in containment pans; chemical tanks are segregated properly with vaults, berms, or other mechanisms to ensure incompatibles cannot mix if a failure occurs; containments are clean, dry, and in good physical condition; required inspection logs are available to indicate inspection of waste tanks, containers, and associated containments; inspection logs are available indicating inspection for a visible sheen before releasing rainwater from fuel oil or other containments
Fire protection	All fire suppression systems are tested on schedule; all portable fire extinguishers are checked regularly to ensure availability

TABLE 16.5   Waste Management Checklist for Waste Accumulation Areas

Department Managers: This questionnaire must be filled out monthly. If you have questions, call Environmental Engineering on extension 6924. Completed checklists should be reviewed with employees and kept in the department's chemical safety notebook.

1. Is the accumulation date written on all hazardous waste containers? _____ yes _____ no   (If no, do so immediately, using the date when container was first used.)

2. Is the label "HAZARDOUS WASTE" attached to all containers that receive hazardous waste? _____ yes _____ no   (If no, attach immediately. If labels are not available in the area, call chemical distribution on extension 6870.)

3. Is the label "NONHAZARDOUS WASTE" attached to all containers that receive normal trash and waste not contaminated with hazardous materials? _____ yes _____ no   (If no, attach immediately. If labels are not available in the area, call chemical distribution on extension 6870.)

4. Are the hazards of the materials in the hazardous waste containers identified (i.e. rags contaminated with solvent, discarded lab chemicals, gloves contaminated with isopropyl alcohol, etc.) _____ yes _____ no   (If no, do so immediately. If you are not sure of the hazard, call the waste management engineer on extension 6954.)

5. Has the waste been in the area more than 90 days? _____ yes _____ no   (If yes, call waste pickup immediatley on extension 6872.)

6. Has the waste been in the area more than 60 days? _____ yes _____ no   (If yes, arrange for pickup before next month's waste management audit or before 90 days.)

7. Are weekly inspectious performed on waste containers to assess condition of containers, spill containment, labeling, proper segregation, accumulation time, and other appropriate items? _____ yes _____ no   (If no, call the waste management engineer on extension 6954 immediately to report deviations. Forms for the inspections can be viewed on-line under "Chemical Library" and can be printed directly or ordered from the waste management engineer.)

8. Are waste handling procedures available in the area and have all employees been trained on these procedures? _____ yes _____ no   (If no, correct immediately.)

9. Are procedures being followed to minimize the amount of waste generated in the area? _____ yes _____ no   (If no, institute procedures immediately. If there are no procedures in the area, call the waste management engineer on extension 6954.)

TABLE 16.5 *Continued*

Comments: _____

_____

_____

_____

REVIEW OF CONTRACTOR ACTIVITIES

Often overlooked, but very important in terms of overall hazardous materials and hazardous waste regulatory compliance, is control of contractor activities. Although many contractors are aware of basic regulatory requirements, they may not be aware of the specifics of the facility's permits or of the facility's general procedures for chemical and waste handling. In particular, contractors who bring chemicals on site and who generate waste should be candidates for periodic field inspections, just like other chemical using/waste generating areas of the facility.

Just as a checklist is useful for process centers, labs, and chemical and waste support areas, a checklist can be useful in identifying items to assess during a contractor review. These may differ slightly from the facility checklist, but the basic compliance requirements will be similar. Examples of categories and items that might be included in a contractor inspection checlist are listed in Table 16.6.

ADDITIONAL REVIEW ITEMS

For a complete facility assessment, there are several other items that could be reviewed during an environmental assessment. These include:

- Facility plans for chemical reductions and/or chemical substitutions with less toxic chemicals
- Waste minimization goals and schedules
- Facility plans for fugitives management
- Process hazard reviews and other risk assessments
- Incident investigation process and documentation
- Preventative maintenance plans and schedules
- Personal monitoring plan
- Stack sampling plan
- Spill or release documentation and reduction plan
- Facility assessment documentation, including corrective actions and schedules for subsequent assessments

Goals or schedules can become outdated quickly , so these items should be reviewed at least annually. Discrepancies to plans should be investigated and resolved as soon as possible.

TABLE 16.6   Categories and Items to Include in a Contractor Inspection Checklist

Category	Inspection Items
Training	Documentation is available that demonstrates the contractor has given employees appropriate training, including hazard communication training and other job-specific training such as hazardous waste operation training, personal protective equipment training, and asbestos removal training
Labeling	Contractor chemical and waste containers are labeled with contractor name (in case of leakage or other problems), content, and hazard; contractor gas cylinders are labeled with contractor name, content, and hazard and have *In Use* or another appropriate tag
Chemical information	Contractor chemicals are approved before use on site and the contractor has MSDSs or the equivalent available for review by the facility's contractor coordinator or other persons working in the area; contractor employees have access to MSDS or equivalent for chemicals in the workplace
Signs	The contractor has the work area roped off and signs posted with the contractor's name and contact number; *No Smoking, Hard Hat Area,* and other safety signs are posted outside the roped off area; any other signs required for specialty jobs, such as asbestos removal, are posted
Permits	Contractor has obtained necessary control permits such as confined space entry permits, permits for powdered-actuated tools, and welding or other hot work permits
Ionizing radiation sources	Contractors have received approval and have documented training before bringing on site and using ionizing radiation equipment; contractors are approved and have demonstrated training before operating ionizing radiation equipment
Personal protection	Required personal protection is identified and worn by the contractor; appropriate workplace monitoring is performed
Local exhaust/ ventilation	Contractors are approved to use existing local exhaust systems for dispensing operations; contractors are approved to set up a portable local exhaust system
Storage and containment	Chemicals and wastes are segregated properly in work area; adequate spill containment is used for temporary storage of drums and other containers
Fire protection	Fire extinguishers are available and adequate for suppressing fires in the work area

CONCLUSION

A facility assessment program is used as a tool for measuring the compliance posture of a facility. Once the program is developed, it can be applied sitewide on an on-going basis to assess proper management of administrative and permit requirements pertaining to chemical and waste management. Any deviations from requirements can be corrected in a timely fashion—before an agency inspector performs an enforcement audit.

An extension of the facility assessment program in the form of a process hazard assessment is now a requirement for certain facilities a as a result of the promulgation of OSHA's regulations pertaining to process safety management of highly hazardous chemicals. Additionally, emergency planning at the community level is a regulatory requirement. These topics are discussed in the next chapter. Included is information pertaining to accepted methods for process hazard assessment, elements of an emergency plan, and release mitigation techniques.

REFERENCES

Environmental Law Reporter, "Factors in Decisions on Criminal Prosecutions for Environmental Violations in the Context of Significant Voluntary Compliance or Disclosure Efforts by the Violater," *Environmental Law Reporter,* (21 ELR 35400), November 1991.

BIBLIOGRAPHY

Government Institutes (1989), Environmental Audits, 6th ed., Government Institutes, Rockville, MD.

Government Institutes (1985), *Good Laboratory Practice Compliance Inspection Manual,* Government Institutes, Rockville, MD.

Government Institutes (1991), *OSHA Field Operations Manual,* 4th ed., Government Institutes, Rockville, MD.

Greeno, J.L., G.S. Hedstrom, and M.A. DiBerto (1988). *The Environmental, Health, and Safety Auditor's Handbook,* Arthur D. Little, Inc., Cambridge, MA.

Ortolano, Leonard (1984), Environmental Planning and Decision Making, John Wiley & Sons, New York.

Tarantino, John A. (1992), *Environmental Liability Transaction Guide: Forms and Checklists,* John Wily & Sons, New York.

U.S. Army COE (1990), "Environmental Review Guide for Operations," Construction Engineering Research Laboratory, U.S. Army Corps of Engineers, Champaign, IL.

WEF (1989), "Environmental Audits—Internal Due Diligence," Water Environment Federation (Formerly Water Pollution Control Federation), Washington, D.C.

SECTION V

HAZARD ASSESSMENT AND EMERGENCY RESPONSE

17

HAZARD ASSESSMENT AND EMERGENCY RESPONSE PLANNING

EPA and OSHA have focused recently on hazard assessment and emergency response planning, and regulations pertaining to this part of hazardous materials and hazardous waste management were promulgated in the late 1980s and early 1990s. This chapter provides the regulatory background for this topic and describes the elements involved in successful hazard assessment and emergency response planning. Additionally, release mitigation techniques for industrial facilities and for transporters of hazardous materials are discussed.

REGULATORY BACKGROUND

Accidental releases of hazardous constituents and the prevention of these releases have been a focus in American and international societies for several decades. Incidents like those that have occurred in Chernobyl, Seveso, and Bophal have made the public aware that there are risks associated with activities that use hazardous materials. As a result, legislation aimed at decreasing or eliminating hazard potential and developing or improving emergency response planning has been developed in the United States, as well as internationally through the European Economic Community and other organizations.

EPA regulations that address the issue of hazard assessment and emergency response planning were developed as part of the Emergency Planning and Community Right-to-Know Act (EPCRA) incorporated into Title III of the Superfund Amendments and Reauthorization Act of 1986 (known as SARA Title III). These regulations required each community to set up a formal Local Emergency Planning Committee (LEPC), which has the mission of assessing hazards in the area and developing an emergency response plan. The LEPC includes representation from locally-elected officials, the fire department, police, the civil defense unit, hospitals, health and first aid groups, local environmental entities, broadcast and print media, industry, and citizens from the community. This extensive representation allows input into the planning process from all affected parties in the local area. Additionally, a State Emergency Response Commission (SERC) is required in every state to aid in emergency response coordination.

OSHA passed a rule that addresses process safety management of highly hazardous chemicals. This rule identifies highly hazardous chemicals and respective threshold quantities for regulation of these materials. If a facility stores, uses, handles, or manufactures the listed chemicals above the threshold values, the facility must provide safety information that enables the employer and employees to identify and understand the hazards posed by processes involving these chemicals. Additionally, a process hazard analysis is required for each process involving highly hazardous chemicals.

LOCAL EMERGENCY RESPONSE PLANNING

The EPCRA or SARA Title III regulations required all LEPCs to develop a plan for emergency response to incidents that could occur in a designated local district such as a county or township. The deadline for the plans to be submitted to the EPA and the SERC was October 1988. Updates to the original plan are required if there are changes in local hazards or emergency response procedures.

Developing a Plan

Several guidance documents are available on the subject of local emergency planning including:

- *Hazardous Materials Emergency Planning Guide*, published by the National Response Team (NRT) of the National Oil and Hazardous Substances Contingency Plan
- *Criteria for Review of Hazardous Materials Emergency Plans*, also published by the NRT
- *Guide for Development of State and Local Emergency Operations Plans*, published by the Federal Emergency Management Agency

Developing a thorough, workable emergency response plan can be a large task, depending on the number of facilities and amounts of chemicals stored in the local area. Elements to be included in a local emergency response plan are summarized in Table 17.1.

TABLE 17.1   Elements to Include in a Local Emergency Response Plan

Planning Factors

- Identification and description of all facilities in the district that possess extremely hazardous substances

- Identification of other facilities that may contribute to risk in the district or may be subject to risks as a result of being within close proximity of facilities with extremely hazardous substances

- Documentation of methods for determining that a release of extremely hazardous substances has occurred, and the area of population likely to be affected by a release

- Other information such as findings from the hazard analysis, geographical features, demographical features, and other planning information

Concept of Operations

- Designation of community emergency coordinator and facility emergency coordinators, who will make determinations necessary to implement the plan

Emergency Notification Procedures

- A description of procedures for providing reliable, effective and timely notification by the facility coordinators and community emergency coordinator to persons designated in the plan and to the affected public that a release has occurred

- Other information such as emergency hotline numbers and lists of names and numbers of organizations and agencies that are to be notified in the event of a release; optional description of methods to be used by facility emergency coordinators to notify community and state emergency coordinators of a release

Direction and Control

- Descriptions of methods and procedures to be followed by facility owners and operators and local emergency and medical personnel to respond to a release of extremely hazardous substances

- Information identifying organizations and persons who provide direction and control during the incident, including chain of command

Resource Management

- A description of emergency equipment and facilities in the community, and identification of persons responsible for such equipment and facilities

- List of all personnel resources available for emergency response

continued

TABLE 17.1 *Continued*

Health and Medical

- Descriptions of methods and procedures to be followed by facility owners and operators and local emergency and medical personnel to respond to a release of extremely hazardous substances
- Information on major types of emergency medical services in the district and neighboring districts, including emergency medical services, first aid, triage, ambulance service, and emergency medical care

Personal Protection of Citizens

- Descriptions of methods in place in the community and at each affected facility for determining areas likely to be affected by a release
- Description of methods for indoor protection of the public

Personal Protective Measures/Evacuation Procedures

- A description of evacuation plans, including those for precautionary evacuations and alternative traffic routes
- Information on precautionary evacuations of special populations and information on mass care facilities that provide food, shelter, and medical care to relocated populations

Procedures for Testing and Updating the Plan

- Descriptions of methods and schedules for exercising the emergency response plan and procedures for evaluating and updating the plan

Training

- Descriptions of the training programs, including schedules for training of local emergency response and medical personnel

Other Information Regarded as Essential by NRT

- General information, including a description of essential information to be recorded in an actual incident and signatures of LEPC chairperson and other officials/industry representatives endorsing the plan
- Instructions for plan use and record of amendments, including listings of organizations and persons receiving the plan or plan amendments, and other data about the dissemination of the plan
- Description of communication methods among responders
- Identification of warning systems and emergency public notification
- Description of methods used for public information and community relations, prior to any emergency
- Descriptions of procedures for responders to enter and leave the incident area, including safety precautions, medical monitoring, sampling procedures, and designation of personal protective equipment

TABLE 17.1 *Continued*

- Identification of major tasks to be performed by fire fighters, including a listing of fire response and HAZMAT personnel

- A description of the command structure for multi-agency/multi-jurisdictional incident management systems

- Descriptions of major law enforcement tasks related to responding to releases, including security-related tasks

- Descriptions of methods to assess areas likely to be affected by an on-going release

- A description of agencies responsible for providing emergency human services

- A description of the chain of command for public works actions and a listing of major tasks

- A description of major containment and mitigation activities for major types of HAZMAT incidents

- Descriptions of major methods for cleanup

- A listing of reports required following and incident and methods of evaluating response activities

- Other information outlined by the National Response Team as appropriate for inclusion in an emergency plan

Note: Information includes criteria with a SARA Title III reference and elements regarded as essential by NRT.

Source: Information from NRT (1988).

Assessing Local Hazards

An integral part of developing a local emergency response plan is a hazard analysis. A hazard analysis is a systematic method of identifying hazards that could affect the community and types of emergencies that could occur as a result of these hazards. Basic elements included in the analysis as identified in EPA's emergency planning guide, "Technical Guide for Hazard Analysis" (see references), are: hazards identification, vulnerability analysis, and risk analysis. Utilizing all three elements in a hazard analysis allows for evaluation of:

- The potential for a situation to cause injury to life or damage to the environment
- The susceptibility of life or property to such injury or damage
- The probability that such injury or damage will occur.

There are many complexities associated with an inclusive analysis, particularly for large industrialized areas; thus, this type of analysis may be too costly or impractical to perform. For these cases, an analysis of major hazards may be sufficient.

Hazard Identification.   This part of the hazard analysis includes information about facilities in the area and transportation situations that have the potential for causing injury to the public or damage to the environment. To develop this information, hazardous materials that are (or could be) maintained in the area should be considered. Facilities that store or use hazardous materials might include:

- Industrial facilities
- Storage facilities/warehouses
- Public works facilities such as water and wastewater treatment plants
- Hospitals, education, and governmental facilities
- Trucking, rail, air and other transportation terminals
- Waste disposal and treatment facilities
- Nuclear facilities

Additionally, transportation corridors and types of hazardous chemicals likely to be transported through the area via highways, railways, and waterways should be included in the hazard identification.

Vulnerability Analysis.   This part of the analysis identifies local area vulnerabilities that are susceptible to damage if a release occurs. Normally, worst case conditions are assumed when making this assessment. Aspects to consider include:

- Location of release or spill
- Potential size of release or spill
- Wind direction and speed
- Population zones and locations of schools, nursing homes, and other institutions
- Location of essential life support areas within the affected zone, such as power plants, major transportation corridors, and water supplies
- Environmentally sensitive areas within the zone, such as wildlife refuges and endangered species habitat

For releases to air, the use of dispersion models can aid in predicting the zone of impact and contaminant concentrations. For releases to the ground, geographical maps can aid in predicting the zone of impact if the release reaches rivers or streams. Potential groundwater contamination also should be reviewed.

Risk Analysis.   During this part of the analysis, the probability of a major release occurring, along with the type of harm to people or damage to property expected from a release, is evaluated. In order to appropriately understand risks, the facility of concern should be contacted to assess the following:

- Safety features of equipment, including leak detection systems, automatic shut off features, fail safe systems, and alarm capabilities for early warning
- Training programs for safe handling techniques
- Spill prevention and countermeasure plans
- Facility emergency plans
- Completion of process hazard analyses for highly toxic chemical processes and/or chemical bulk storage areas

For planning on a community-wide scale, the hazard analysis can provide an overall understanding of hazards in the area. Populations affected, transportation corridors affected, incident probability, and incident severity all must be evaluated in order to help a community plan its emergency preparedness and response activities.

PROCESS HAZARD ANALYSIS

Methodologies for performing a process hazard analysis are well documented and have been used throughout industry for many years. An excellent review of methods for performing a process hazard assessment is presented in AIChE's *Guidelines for Hazard Evaluation Procedure* (see references). OSHA's rule pertaining to process safety management of highly hazardous chemicals, detailed under 29 CFR §1910.119, has required the use of one or more of the following methodologies for reviewing processes involving highly hazardous chemicals over threshold amounts:

- What-if—This methodology is used for relatively uncomplicated processes. "What if . . ." questions are used to evaluate the effects of component failures at each process or material handling step.
- Checklist—A checklist approach is used for more complex processes. During this analysis, an organized checklist is developed, and certain aspects of the process are assigned to committee members having the greatest experience or skill in evaluating those aspects. Operator practices and job knowledge are audited in the field, and suitability of equipment and materials of construction are studied. Additionally, process chemistry and control systems are reviewed, and operating and maintenance records are audited. A checklist evaluation typically precedes the use of the more sophisticated processes.
- What-if/checklist—This combination methodology is a comprehensive approach that combines the creative thinking of a selected team of specialists with the methodical focus of a prepared checklist. The review team methodically examines the operation from receipt of raw materials to delivery of the finished product to the customer's site. Information acquired from this method can be used in training of operating personnel.

- Hazard and operability study (HAZOP)—HAZOP is a formal method of systematically investigating each element of a system for all potential deviations from design conditions. Parameters such as flow, temperature, pressure, and time are reviewed in conjunction with material strength, locations of connections, and other design aspects. Piping and instrument designs (or plant model) are analyzed critically during this process for potential problems that could arise in each vessel or pipeline in the process. Potential causes for failure are evaluated, consequences of failure are reviewed, and safeguards against failure are assessed for adequacy.
- Failure mode and effect analysis—This technique is a methodical study of component failure. The analysis includes all system diagrams and reviews all components in the system that could fail such as pumps, instrument transmitters, seals, temperature controllers, and other components. Consequences of failure, hazard class, probability of failure, and detection methods are evaluated, and multiple concurrent failures also are assessed.
- Fault tree analysis—This methodology is a qualitative or quantitative model of all undesirable outcomes from a specific event. It includes catastrophic occurrences such as explosion, toxic gas release, and rupture. A graphic representation of possible sequences results in a diagram that looks like a tree with branches. Assessment of probability using failure rate data is used to calculate probability of occurrence.

Table 17.2 presents examples of types of processes that might be suitable for each of the above-mentioned methodologies.

There are other recognized hazard analysis methods that are not detailed specifically in the OSHA regulation, but that are used for hazard assessments (EPA/600/8-87/028 1987 and EPA 1991). Most can be used to augment the six methodologies described above. These include:

- Preliminary hazard analysis—This methodology is used for hazard identification during the preliminary phase of plant development. It is especially useful for new processes where there is limited past experience.
- Human error analysis—This analysis is a systematic evaluation of the factors that influence the performance of persons involved in operating and maintaining the system. Human errors that are likely to occur and that could cause an incident are identified.
- Event tree analysis—This analysis considers operator or safety system response to the initiation of an event. Incident consequences from varied responses (or failures in response) are assessed.

**TABLE 17.2   Examples of Applications Suitable for Selected Process Hazard
Analysis Methodologies**

Methodology	Applications
What-if	Applicable to all parts of a facility such as bulk chemical systems, storage areas, and processing or manufacturing areas; useful during process development, pre-startup, and operation.
Checklist	Applicable to equipment, materials, and procedures in a chemical process facility; useful during design, construction, startup, operation, and shutdown
What-if/checklist	Applicable to all parts of the facility, including equipment and operations of a process; useful during design, construction, and operation
Hazard and operability study	Applicable to systems and equipment with a high degree of operating variability; useful during late design and operation phases
Failure mode and effect analysis	Applicable to individual system components; useful during design, construction, and operation
Fault tree analysis	Applicable to equipment and system design features and operational procedures that are interrelated; useful during design and operation phases

Source: Information from EPA/600/8-87/028 (1987).

- Cause-consequence analysis—This methodology uses a blend of fault tree and event tree analysis to diagram the interrelationships between accident outcomes and their basic causes.
- Dow and mond indices—These indices provide an empirical method for ranking the risks in a chemical process plant. Penalties are assigned to process materials and conditions that can cause accidents, while credits are assigned to plant safety procedures that can mitigate the effects of an accident. These indices are useful in the early stages of plant design.
- Probabilistic risk assessment—This assessment measures the overall risk through numerical evaluation of both accidental consequences and probabilities. This method generally is used to compare risks when alternative designs exist or to assess risk reduction strategies.
- Safety review—This review is used as a comprehensive facility inspection to identify facility conditions or procedures that might allow an accident or incident.

Other preliminary or screening analyses that can be used during process hazard analysis include review of chemical reactions caused by mixing of materials, evaluation of momentum forces such as pressure fluctuations or water hammers, and evaluation of physical characteristics such as liquid levels, temperature, and mass flow rates.

RELEASE MITIGATION TECHNIQUES

Mitigation Techniques for In-Plant Incidents

NFPA has summarized an EPA survey of nearly 7,000 incidents showed that almost 75% were in-plant (fixed facility) incidents (NFPA 1989). Thus, it is appropriate during a hazard assessment to focus on methods that mitigate potential incidents at fixed facilities. Process hazard analysis is one of the best methods of pinpointing factors that can cause (or avert) a disastrous incident or accident. Other release mitigation techniques can be used, and include items such as pre-release controls, pre-release protection equipment, management activities, and safety systems and procedures. Table 17.3 provides information about these release mitigation techniques.

Other factors contributing to safe storage and handling of chemicals and release mitigation are documented by AIChE in *Guidelines for Safe Storage and Handling of High Toxic Hazard Materials* and *Vapor Release Mitigation* (see references). Items to consider include:

- Facility siting and sizing—Considerations include buffer zones to population areas, plant layout to minimize zone of impact, drainage controls, and sizing of systems to minimize zone of impact and to reduce inventories.
- Facility security—Considerations include routine patrols of plant security personnel, fencing, good lighting, controlled access at the main gate, locks on drain valves and pump starters, and television monitoring.
- Chemical selection and storage practices— Considerations include use of low-toxicity chemicals when possible, inventory control, adequate segregation of incompatibles, use of low-pressure vessels when possible, and formalized load/unload practices, including equipment inspection before use and containment of all load/unload hoses, pumps, and vessels.
- Emergency preparedness—Considerations include adequacy of local emergency groups to assist during an incident, adequacy of emergency systems on site such as fire suppression systems and isolation capability, the need for contracts with emergency cleanup vendors, and adequate field training for in-plant emergency response teams.

TABLE 17.3 Examples of Release Mitigation Techniques

Category	Specific Mitigation Measures
Pre-release controls	Preventative maintenance of equipment; regular equipment inspections; regular equipment testing; comprehensive safety audit; pre-installation assessment of equipment designs; process controls for operations monitoring; alarm capability for early warning of problems or out-of-spec conditions; regular upgrading of equipment; investigation of all spills and releases and assessment of similar equipment for the same problems; development of standard operating procedures; pre-startup equipment checks; release prevention equipment
Pre-release protection equipment	Containment for tanks, process equipment, overhead piping; neutralization capabilities; flares or incinerators; adsorbers; scrubbers; spray curtains; emergency equipment including spill cleanup supplies, fire fighting equipment, air monitors, and other necessary equipment
Management activities	Employee safety training, hazard communication training, emergency response training, certification of operators on equipment or system; membership in community emergency planning groups; development of a release control program; development of an accident/incident investigation program; participation in research/conferences/development of a safety loss prevention program; formalized corrective action process for deviation from rules; formalized notification procedures for accidental releases; initiation of a program to improve system designs
Safety systems and procedures	Backup systems; redundant systems; valve lock-out program; automatic shut offs; bypass and surge systems; manual overrides; interlocks; alarms; formalized safety procedures; formalized testing program for all safety equipment; formalized color coding/labeling program

Source: Information from EPA (1991).

Release Mitigation Techniques during Transport of Hazardous Materials

The potential for a release of hazardous material during transport is a concern of many cities and communities, since railways, waterways, and federal or state highways are not controlled locally. NFPA's summary of an EPA survey showed that approximately 25% of the incidents are in-transit incidents, with incidents involving trucks being the most common, followed by railcars, water vessels, and pipelines (NFPA 1989).

Mitigating techniques for the transport of hazardous chemicals and wastes include proper maintenance of trucks and rail cars, proper maintenance of highways and railroad tracks, and adequate training for the transport operators. Proper shipping papers, which define the chemicals in transit, are a necessity in the event of a release. These include bills of lading for highway transportation, waybills for rail shipments, and dangerous cargo manifests for shipment by water. Additionally, manifests are required when shipping hazardous waste, PCBs, and low-level radioactive waste. The Chemical Manufacturer's Association operates a chemical information service—CHEMTREC—which was established during the 1970s. This service provides a toll-free hotline to persons needing spill response and other emergency information.

Mitigating techniques have been established for the transport of radioactive wastes in terms of waste packaging requirements. Type A and type B packages for transporting intermediate and large quantities of radioactive materials, respectively, are specified by international radioactive transportation regulations (DOE 1990), as well as by DOE in 10 CFR Part 71. These packages must pass tests in terms of free drop, compression, penetration, water immersion, and extended heating to ensure that they will not be damaged in the event of a transportation accident.

THE COMMUNITY'S ROLE IN EMERGENCY PLANNING AND RESPONSE

Under the SARA Title III legislation, individuals from the community have a right to know what hazards exist in the local area. Additionally, local citizens are represented on the LEPC, which has the responsibility for local emergency planning. In some heavily industrialized areas, public participation is invited further through periodic facility tours, through industry mailouts detailing current products or technologies, and through industry participation in neighborhood meetings.

In order to create an effective emergency plan, the public must be made aware of emergency signals, evacuation routes, and other emergency actions necessary for individual safety. Additionally, the public should be familiar with area rail and truck transportation corridors that allow the transport of hazardous materials. Many cities throughout the nation have designated routes for hazardous materials transport that avoid highly-populated areas or constantly congested arteries.

The role of the community in emergency planning and response is still evolving. Certainly, the public has an ever-increasing awareness of existing hazards in the community. As the awareness level grows, it is hoped that risk potential can be communicated effectively, so that the public can have an educated understanding of potential exposures.

CONCLUSION

Emergency planning has been a focus of industry for many decades, and has included site evacuation plans, emergency response plans, and facility reviews by emergency response groups such as the fire department or other emergency response groups. In the late 1980s, the community was brought into the emergency planning process through the enactment of SARA Title III. This law required the formation of LEPCs. These committees include citizens, media, and other representatives that traditionally were not considered part of the planning process. Other more traditional emergency representatives such as fire department, medical entities, and industry representatives are also part of the committee. The committee's responsibility is to assess risk to the community from local facilities and to provide a community emergency response plan. Should there be a chemical incident or emergency, trained personnel within the community would respond. Incident response and management is discussed in the next chapter.

REFERENCES

American Institute of Chemical Engineers, *Guidelines for Hazard Evaluation Procedures*, prepared by Battelle Columbus Division for Center for Chemical Process Safety, New York, 1985.

American Institute of Chemical Engineers, *Guidelines for Safe Storage and Handling of High Toxic Hazard Materials*, prepared by Arthur D. Little Inc., and Richard LeVine for Center for Chemical Process Safety, New York, 1988.

American Institute of Chemical Engineers, *Guidelines for Vapor Release Mitigation*, prepared by Richard W. Prugh and Robert W. Johnson for Center for Chemical Process Safety, New York, 1988.

Department of Energy, "The International Radioactive Transportation Regulations— A Model for National Regulations," prepared by R.B. Pope and R.R. Rawl, Oak Ridge National Laboratory, Oak Ridge, TN, 1990.

Federal Emergency Management Agency, "Guide for Development of State and Local Emergency Operations Plans," Washington, D.C., 1985.

National Fire Protection Association, *Hazardous Materials Response Handbook*, Martin F. Henry, ed., Quincy, MA, 1989.

National Response Team, "Criteria for Review of Hazardous Materials Emergency Plans," National Response Team of the National Oil and Hazardous Substances Contingency Plan, Washington, D.C., 1988.

National Response Team, "Hazardous Materials Emergency Planning Guide," National Response Team of the National Oil and Hazardous Substances Contingency Plan, Washington, D.C., 1987.

U.S. Environmental Protection Agency, "Environmental Protection Agency Accidental Release Questionnaire," OMB #2050-0065, Washington, D.C., 1991.

U.S. Environmental Protection Agency, "Prevention Reference Manual: User's Guide Overview for Controlling Accidental Releases of Air Toxics," EPA/600/8-87/028, prepared by Radian Corporation for Office of Research and Development, Air and Energy Engineering Research Laboratory, Research Triangle Park, NC, 1987.

U.S. Environmental Protection Agency, "Technical Guidance for Hazards Analysis," U.S. Environmental Protection Agency, Federal Emergency Management Agency, and U.S. Department of Transportation, Washington, D.C., 1987.

BIBLIOGRAPHY

AIChE–CCPS (1989), *Guidelines for Chemical Process Quantitative Risk Analysis*, American Institute of Chemical Engineers—Center for Chemical Process Safety, New York.

_____ (1989), *Guidelines for Human Reliability in Process Safety*, American Institute of Chemical Engineers—Center for Chemical Process Safety, New York.

_____ (1989), *Guidelines for Process Equipment Reliability Data, with Data Tables*, American Institute of Chemical Engineers—Center for Chemical Process Safety, New York.

AIHA (1988), Emergency Response Planning Guidelines Series, "Set 1: Chloroacetyl Chloride; Chloropicrin; Crotonaldehyde; Perfluoroisobutylene; Phosphorus Pentoxide," American Industrial Hygiene Association, Akron, OH.

_____ (1988), *Emergency Response Planning Guidelines Series*, "Set 2: Ammonia; Chlorine; Diketene; Formaldehyde; Hydrogen Fluoride," American Industrial Hygiene Association, Akron, OH.

_____ (1989), *Emergency Response Planning Guidelines Series*, "Set 3: Acrolein; Hexachlorobutadiene; Hydrogen Chloride; Oleum, Sulfur Trioxide, and Sulfuric Acid; Phosgene," American Industrial Hygiene Association, Akron, OH.

_____ (1990), *Emergency Response Planning Guidelines Series*, "Set 4: Dimethylamine; Methyl Mercaptan; Monomethylamine; Sulfur Dioxide; Trimethylamine," American Industrial Hygiene Association, Akron, OH.

_____ (1991), *Emergency Response Planning Guidelines Series*, "Set 5: Acrylic Acid; 1,3-Butadiene; Epichlorohydrin; Tetrafluoroethylene; Vinyl Acetate," American Industrial Hygiene Association, Akron, OH.

_____ (1991), *Emergency Response Planning Guidelines Series*, "Set 6: Bromine; Chlorosulfonic Acid; Hydrogen Sulfide; Methyl Iodide; Phenol," American Industrial Hygiene Association, Akron, OH.

_____ (1992), *Emergency Response Planning Guidelines Series*, "Set 7: Allyl Chloride; Carbon Disulfide; Chlorotrifluoroethylene; Isobutyronitrile; Titanium Tetrachloride," American Industrial Hygiene Association, Akron, OH.

Baltic Marine Environment Protection Commission (1991), "Study of the Risk for Accidents and the Related Environmental Hazards from the Transportation of Chemicals by Tankers in the Baltic Sea Area," Baltic Marine Environment Protection Commission, Helsinki.

CMA (1985), "Community Awareness and Emergency Response Program Handbook," Chemical Manufacturers Association, Washington, D.C.

_____ (1985), "Process Safety Management (Control of Acute Hazards)," Chemical Manufacturers Association, Washington, D.C.

_____ (1986), "Community Emergency Response Exercise Program," Chemical Manufacturers Association, Washington, D.C.

_____ (1986), "Site Emergency Response Planning Handbook," Chemical Manufacturers Association, Washington, D.C.

_____ (1987), "An Analysis of Risk Assessment Methodologies for Process Emissions from Chemical Plants," Chemical Manufacturers Association, Washington, D.C.

_____ (1987), "Review of Site Specific Guidance for EPA's Chemical Emergency Preparedness Program, Washington, D.C.

_____ (1989), "Evaluating Process Safety in the Chemical Industry: A Manager's Guide to Quantitative Risk Assessment," Chemical Manufacturers Association, Washington, D.C.

_____ (1990), "A Manager's Guide to Reducing Human Errors," Chemical Manufacturers Association, Washington, D.C.

_____ (1991), "CMA's Manager Guide," Chemical Manufacturers Association, Washington, D.C.

_____ (1991), "Crisis Management Planning for the Chemical Industry Manual and Videotape," Chemical Manufacturers Association, Washington, D.C.

Davis, Daniel S., et al. (1989), *Accidental Releases of Air Toxics: Prevention, Control and Mitigation*, Noyes Data Corporation, Park Ridge, NJ.

DOE (1988), "Accident Analysis of Railway Transportation of Low-Level Radioactive and Hazardous Chemical Wastes," DE89 001389, Oak Ridge National Laboratory, Department of Energy, Oak Ridge, TN.

_____ (1989), "Development of On-Site Accident Criteria for Waste Transfer Casks," DE89 010201, Idaho National Engineering Laboratory, Department of Energy, Idaho Falls, ID.

_____ (1991), "Historical Overview of Domestic Spent Fuel Shipments—Update," DE91 016051, Oak Ridge National Laboratory, Department of Energy, Oak Ridge, TN.

DOT (1990), "Chemical Data Guide for Bulk Shipment by Water," U.S. Coast Guard, U.S. Department of Transportation, Washington, D.C.

_____ (1990), "Hazardous Materials Flows by Rail," DOT-TSC-RSPA-90-1, Office of Hazardous Materials Transportation, Department of Transportation, Washington, D.C.

Dow Chemical Company (1987), "Fire & Explosion Index Hazard Classification Guide," 6th ed., Dow Chemical Company, Midland, MI.

_____ (1988), "Chemical Exposure Index," Dow Chemical Company, Midland, MI.

Fawcett, H.H.. and W.S. Wood, eds. (1982). *Safety and Accident Prevention in Chemical*

Operations, 2nd ed., John Wiley & Sons, New York.

Fawcett, H.H. (1988), *Hazardous and Toxic Materials: Safe Handling and Disposal*, John Wiley & Sons, New York.

Kelly, Robert B. (1989), *Industrial Emergency Preparedness*, Van Nostrand Reinhold, New York.

Lees, F.P. (1980), *Loss Prevention in the Process Industries*, Butterworths, London.

Malasky, S.W. (1983), *System Safety: Technology and Application*, 2nd ed., Garland STPM Press, New York.

National Tank Truck Carriers (1989), *Hazardous Commodity Handbook*, 9th ed., National Tank Truck Carriers, Alexandria, VA.

National Safety Council (1983), "Accident Investigation: A New Approach, National Safety Council, Chicago, IL.

NRC (1974), "Guide and Checklist for the Development and Evaluation of State and Local Government Radiological Emergency Response Plans in Support of Fixed Nuclear Facilities," U.S. Nuclear Regulatory Commission, Washington, D.C.

_____ (1981), *Fault Tree Handbook*, prepared by N.H. Roberts, W.E. Wesley and D.F. Hassel, U.S. Nuclear Regulatory Commission, Washington D.C.

OMB (1988), "Using Community Right to Know: A Guide to a New Federal Law," OMB Watch, Office of Management and Budget, Washington, D.C., 1988.

Owen, Ronald D. (1986), "Process Hazard Review," Technical Report 02.1277, International Business Machines, General Products Division, San Jose, CA.

Rohm and Haas Co. (1988), "Reducing the Risk of Spillage in the Transportation of Chemical Wastes by Truck," prepared by Michigan University for Rohm and Haas Co., Bristol, PN.

Roland, H.E., and B. Moriarty (1983), *System Safety Engineering and Management*, John Wiley & Sons, New York.

Student, Pat, ed. (1981), *Emergency Handling of Hazardous Materials in Surface Transportation*, Association of American Railroads, Bureau of Explosives, Washington, D.C.

Swain, A.D., and H.E. Guttman (1983), *Handbook of Human Reliability Analysis with Emphasis on Nuclear Power Plant Applications*, U.S. Nuclear Regulatory Commission, Washington, D.C.

Tierney, K.J. (1980), "A Primer for Preparedness for Acute Chemical Emergencies," Book and Monograph 14, Disaster Research Center, University of Delaware, Newark, DL.

Transportation Research Board (1988), "Transportation of Hazardous Materials," prepared by F.F Saccomanno et al., for Transportation Research Board, Washington, D.C.

U.N. (1990), "Recommendations on the Transport of Dangerous Goods," 6th ed., United Nations, Published by Labelmasters, Chicago, IL.

_____ (1990), "Recommendations on the Transport of Dangerous Goods, Test and Criteria," 2nd ed., United Nations, published by Labelmasters, Chicago, IL.

U.S. Congress (1986), "Transportation of Hazardous Materials," OTA-SET-304, Office of Technology Assessment, Congress of the United States, Washington, D.C.

U.S. EPA (1987), "Manual for Preventing Spills of Hazardous Substances at Fixed Facilities," prepared by Environmental Monitoring Services, Inc., for U.S. Environmental Protection Agency, Washington, D.C.

———— (1987), "User's Guide Overview for Controlling Accidental Releases of Air Toxics," prepared by Radian Corporation for the U.S. Environmental Protection Agency, Research Triangle Park, N.C.

———— (1988), "Probability and Control Cost Effectiveness for Accidental Toxic Chemical Releases," prepared by Radian Corporation for the U.S. Environmental Protection Agency, Washington, D.C.

———— (1988), "Review of Emergency Systems," Office of Solid Waste and Emergency Response, U.S. Environmental Protection Agency, Washington, D.C.

Unterberg, W., et al. (1988), *How to Prevent Spills of Hazardous Substances*, Noyes Data Corporation, Park Ridge, NJ.

18

INCIDENT HANDLING TECHNIQUES

Any unplanned release of hazardous materials or hazardous waste must be handled by trained responders, as set forth by OSHA in 29 CFR §1910.120. The National Fire Protection Association (NFPA) has provided guidance for an organized approach to emergency response in NFPA 471 and 472, and these standards are referenced by OSHA as an excellent method of response to a hazardous materials release. Additionally, NFPA has published a textbook, *Hazardous Materials Response Handbook* (see references), which provides the complete text of NFPA 471 and 472, as well as useful information about personal protective equipment and other emergency response considerations.

The foundation of emergency response, as outlined by the NFPA, is the incident command system, now used universally for incident management. The incident command system, which is detailed in NFPA 1561 (see references), provides structure, coordination, and effectiveness in emergency situations, thus enhancing the safety of all responding personnel. The following sections outline the major aspects of this command system and give additional information on other aspects of incident management, including how to zone an area and techniques for incident mitigation.

INCIDENT COMMAND SYSTEM

The incident command system requires emergency responders to perform in assigned roles during the incident. The incident commander is at the head of the command system, and all decisions pertaining to the incident are made by this person, with advice and counsel from qualified experts. This ensures a strong central command, which is necessary to keep the incident response organized and safe. The number and types of personnel needed to respond to an incident will depend on the size and complexity of the incident. Necessary personnel and their functions for proper management of an incident are documented by NFPA and others. Examples of the types of personnel that typically are needed during an incident and types of assignments or duties expected of these personnel are listed in Table 18.1.

In addition to these resources, other "behind the scenes" personnel will be needed to carry out other duties, including:

- Bringing water and food to the area for people involved in the incident
- Acquiring additional emergency equipment such as shovels, blankets, rain gear, and other items that are needed
- Acquiring decontamination equipment and other equipment
- Calling families of the responders to inform them of status of spouse's return home
- Performing other errands as the situation warrants

TABLE 18.1   Personnel that Might be Needed During a Hazardous Materials Incident

Personnel	Responsibilities
Incident commander	Establish and manage the incident response plan; allocate resources, assign activities, manage information, and ensure response is completed per plan
Sector officers	Manage geographical areas, including hazardous materials response teams in that area, as needed; provide specific functions such as serving as safety officer
Command staff	Assist incident commander and/or sector officers; gather data such as chemical information, weather data, and status of other operations
Police or security	Manage the location and activities of the general public
Hazardous materials response teams	Directly manage and mitigate the hazardous materials release under the direction of a specified team leader
Communications personnel	Perform central communication function for incident including requests of additional emergency resources; notify proper agencies of the release; answer media questions
Technical information specialist	Provide expertise directly to the incident commander in terms of chemical information and exposure hazards

Source: Information from NFPA (1992) and Noll, Hildebrand and Yvorra (1988).

Levels of training for hazardous materials (HAZMAT) incident responders is defined in 29 CFR §1910.120. Requirements for each of the five OSHA-defined levels of response are outlined in Table 18.2. Requirements for personnel such as medical professionals and communications personnel are not cited in the OSHA regulations, but training should be adequate to successfully perform duties.

TABLE 18.2   OSHA Requirements for HAZMAT Training

First Responder Awareness Level (e.g., security or police)

- Must demonstrate a basic understanding of what hazardous materials are and the risks associated with them in an incident. This includes understanding the potential outcomes associated with an emergency created when hazardous materials are present.

- Must be able to recognize the presence of hazardous materials in an emergency and, if possible, identify the hazardous materials.

- Must demonstrate an understanding of the role of the first responder awareness individual in the site emergency plan. Must also be able to realize the need for additional resources and make appropriate notifications to the communication center.

First Responder Operations Level (e.g., responders stationed in the decontamination zone or individuals who contain the leak at a distance using diking or other techniques).

- Must demonstrate competency in requirements of first responder awareness level.

- Must demonstrate knowledge of the basic hazard and risk assessment techniques, and must have a basic understanding of hazardous materials terms.

- Must be able to select and use proper personal protective equipment (PPE).

- Must know how to perform basic control, containment and/or confinement operations within the capabilities of the resources and PPE of the unit. Must know how to implement basic decontamination procedures, as well as understand relevant operating procedures and termination procedures.

Hazardous Materials Technician (e.g., responders mitigating the incident)

- Must demonstrate competency in requirements of first responder operations level.

- Must know how to implement the site emergency response plan.

- Must know how to identify, classify, and verify known and unknown materials by using field survey instruments.

- Must be able to function within an assigned role in the incident command system.

- Must know how to select and use specialized chemical PPE provided to the hazardous materials technician.

TABLE 18.2 *Continued*

- Must understand hazard and risk assessment techniques.

- Must be able to perform advanced control, containment, and confinement operations within the capabilities of the resources and PPE available to the unit, and must understand decontamination procedures.

- Must understand termination procedures.

- Must understand basic chemical and toxicological terminology and behavior.

Hazardous Materials Specialist (e.g. safety officer, chemical specialist, or other technical information specialist)

- Must demonstrate competency in requirements of hazardous materials technician.

- Must know how to implement the local emergency response plan, as well as understand the state emergency response plan.

- Must understand in-depth hazard and risk techniques.

- Must be able to determine and implement decontamination procedures.

- Must have the ability to develop a site safety and control plan.

- Must understand chemical, radiological, and toxicological terminology and behavior.

On Scene Incident Commander

- Must demonstrate competency in requirements of first responder awareness level and first responder operations level.

- Must know and be able to implement the site incident command system.

- Must know how to implement the site emergency response plan and the local emergency response plan.

- Must know of the state emergency response plan and of the federal regional response team.

- Must know and understand the hazards and risks associated with employees working in chemical protective clothing.

- Must know and understand the importance of decontamination procedures.

Source: Adapted from 29 CFR §1910.120(q)(6).

COMMAND POST AND ZONING DURING THE INCIDENT

The incident command post is the first area to be established during a hazardous materials incident, and the position of this post is determined by wind factors and materials released. The command post should be visible and recognizable, with a colored light, flag, or other symbol used to mark its location. Once the command post is set, other zones can be set up by appropriate responders. Figure 18.1 illustrates the appropriate zones that should be established prior to an response to the emergency. As can be seen in the figure, the zones include the support zone, the limited access zone, and the restricted or hazard zone. The designated activities in each of these zone are as follows:

- Support zone—The command post is within this zone and communication equipment for dispatching and calling outside agencies is maintained within the command post. Staging of equipment and resources occurs in this zone and key technical personnel such as the safety officer, medical support, and the communications specialist are stationed in this zone. Backup resources to the responders also reside in this zone until needed, and decontaminated equipment and personnel exit through this zone. Nonessential personnel such as news media should remain outside this zone and the communications specialist should exit the support zone to communicate with these personnel.

- Limited access zone—Within this zone are personnel who are integral to managing the incident. Persons who are awaiting entry into or returning from the restricted zone are staged in this zone. Decontamination activities including personnel and equipment decontamination occur in this zone. There is one point of entry into the restricted zone and one point of entry into and out of the limited access zone that are established early and enforced. Protective equipment is required in this zone, and the HAZMAT team leader and/or other control officer maintains communication equipment.

- Restricted zone—Personnel entering this zone—often termed the hazard zone or hot zone—are those who are responding to and mitigating the incident. Two-way communication equipment is carried by the responding personnel. Full protective equipment (usually SCBA and fully encapsulating suit) is worn in this zone. An example of this type of protective equipment (donned during a training session) is shown in Figure 18.2.

Use of precisely marked zones ensures that activities can be carried out safely and in a methodical way. In addition, the area directly outside the support zone should be protected by security or the police, and only a limited number of persons who are prespecified—such as the media or local or state officials—should be allowed near the support zone for briefings by designated support personnel.

TECHNIQUES FOR INCIDENT MITIGATION

Techniques for incident mitigation are determined by the materials involved in the incident. Mitigation activities during the incident include:

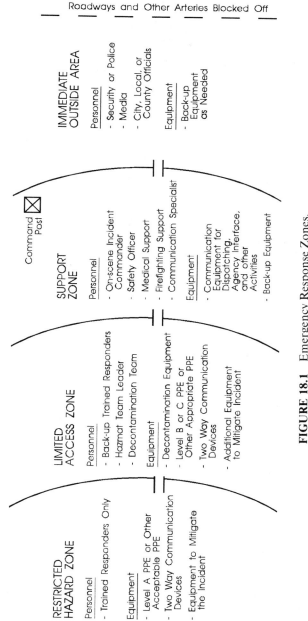

FIGURE 18.1 Emergency Response Zones.

FIGURE 18.2

- Stopping the release—This can be done by turning off flow valves, plugging or patching the hole, or capping the release mechanism. In the case of fire, it must be suppressed first by means of water or foam suppressants before other activities can take place.
- Stopping the spread of material—This can be done by closing storm drain valves, by diking or berming the area, or by diverting the flow to a confined area. An example of diking activities, performed during training, is shown in Figure 18.3.
- Cleanup of the released material in a timely fashion so as to limit seepage into waterways or groundwater—This can be done using adsorbents, neutralizers, earth moving equipment, and other cleanup techniques.

Factors that are considered in selecting the proper incident management and mitigation techniques include:

- Chemical hazard (if known)—This includes physical properties such as specific gravity, flash point, vapor pressure, corrosivity reactivity, and toxicity.
- Physical situation—This includes aspects such as container type, other chemicals in area, release type and rate, and LEL readings or the existence of fire.
- Worker/community hazard—This includes things such as distance of the incident to community and potential impact via air or stream, the

FIGURE 18.3

need for evacuation of workers and/or community members, the need for response from outside agencies such as state or national emergency response teams.

In all cases, incident management techniques should be selected so as to disallow or reduce, to the extent possible, contamination of area soils and waterways from run-off associated with the incident.

CONCLUSION

Incidents and accidents can be mitigated by good planning techniques and well-trained personnel. NFPA has developed an incident command system for responding to fires and other emergencies with an organized approach that has aided in safe handling of emergencies. This approach has been referenced by OSHA as excellent for chemical emergency response. In addition, OSHA has defined training and certification requirements for five levels of HAZMAT responders. It is the goal of both NFPA and OSHA to provide a structure for emergency management so that chemical emergencies can be handled both expediently and safely.

REFERENCES

International Fire Service Training Association, *Hazardous Materials for First Responders*, Fire Protection Publications, Oklahoma State University, Stillwater, OK, 1988.

Noll, Gregory G., Michael S. Hildebrand, and James G. Yvorra, *Hazardous Materials: Managing the Incident*, Fire Protection Publications, Oklahoma State University, Stillwater, OK, 1988.

National Fire Protection Association, *Hazardous Materials Response Handbook*, 2nd ed., Gary Tokle, ed., Quincy, MA, 1992.
National Fire Protection Association, "Recommended Practice for Responding to Hazardous Materials Incidents," NFPA 471, Quincy, MA, 1992.

National Fire Protection Association, "Standard for Professional Competence of Responders to Hazardous Materials," NFPA 472, Quincy, MA, 1989.

National Fire Protection Association, "Fire Department Incident Management System," NFPA 1561, Quincy, MA, 1990.

BIBLIOGRAPHY

CMA (1987), "Teamwork—Safe Handling of a Hazardous Materials Incident Videotape," Chemical Manufacturer's Association, Washington, D.C.

_____ (1990), "CHEMNET Brochure," Chemical Manufacturer's Association, Washington, D.C.

_____ (1992), "National Chemical Response and Information Center (NCRIC) Brochure," Chemical Manufacturer's Association, Washington, D.C.

_____ (1992), "CHEMTREC Brochure," Chemical Manufacturer's Association, Washington, D.C.

_____ (1992), "Chemical Referral Center Brochure," Chemical Manufacturer's Association, Washington, D.C.

DOE (1991), "OSHA Training Requirements for Hazardous Waste Operations," DE92 004780, DOE/EH-0227P, Office of Environment, Safety and Health, U.S. Department of Energy, Washington, D.C.

Fire Protection Publications (1983), *Incident Command System*, Fire Protection Publications, Oklahoma State University, Stillwater, OK.

Theodore, Louis, Joseph P. Reynolds, and Frank B. Taylor (1989), *Accident and Emergency Management*, John Wiley & Sons, New York.

U.S. EPA (1985), "Field Standard Operating Procedures for Establishing Work Zones F.S.O.P. 6," OSWER Directive 9285.2-04, Office of Emergency and Remedial Response, U.S. Environmental Protection Agency, Washington, D.C.

_____ (1988), "EPA Standard Operating Safety Guides," OSWER Directive 9285.1-01C, Office of Emergency and Remedial Response, Environmental Response Team, U.S. Environmental Protection Agency, Washington, D.C.

_____ (1989), "EPA Health and Safety Audit Guidelines," EPA/540/G-89/010, OSWER Directive 9285.8-02, Office of Solid Waste and Emergency Response, Emergency Response Division, U.S. Environmental Protection Agency, Washington, D.C.

APPENDIX A

LIST OF ACRONYMS

AA   Atomic Absorption

ACGIH   American Conference of Governmental Industrial Hygienists

ACL   Alternate Concentration Limit

AEC   Atomic Energy Commission

AIHA   American Industrial Hygiene Association

AIChE   American Institute of Chemical Engineers

AIChE-CCPS   American Institute of Chemical Engineers-Center for Chemical Process Safety

ALARA   As Low As Reasonably Achievable

ANPR   Advanced Notice of Proposed Rulemaking

ANSI   American National Standards Institute

API   American Petroleum Institute

ASHRAE   American Society of Heating, Refrigeration, and Air Conditioning Engineers

ASTM   American Society for Testing and Materials

ASME   American Society of Mechanical Engineers

BACT   Best Available Control Technology

BOD   Biochemical Oxygen Demand

BTX   Benzene, Toluene, and Xylene

CAA   Clean Air Act

CAAA   Clean Air Act Amendments

CERCLA   Comprehensive Environmental Response, Compensation, and Liability Act

CFCs   Chlorofluorocarbons

CFR   Code of Federal Regulations

CGI   Combustible Gas Indicator

CWA   Clean Water Act

DOE   Department of Energy

DOL   Department of Labor

DOT   Department of Transportation

EPA   Environmental Protection Agency

FID   Flame Ionization Detector

FML   Flexible Membrane Liner

FTIR   Fourier Transform Infrared Spectrometer

GC   Gas Chromatograph

GC/MS   Gas Chromatograph/Mass Spectrometer

HAZMAT   Hazardous Materials

HLW   High-Level Waste (Radioactive)

HSWA   Hazardous and Solid Waste Amendments

IARC   International Agency for Research on Cancer

ICS   Incident Command System

IDLH   Immediately Dangerous to Life or Health

IR   Infrared

ISC   Industrial Source Complex

LAER   Lowest Achievable Emission Rate

LEL   Lower Explosive Limit

LEPC   Local Emergency Planning Committee

LLW   Low-Level Waste (Radioactive)

MACT   Maximum Achievable Control Technology

MQL   Method Quantitation Limit

MSDS   Material Safety Data Sheet

MSHA   Mine Safety and Health Administration

MS   Mass Spectrometer

NAAQS   National Ambient Air Quality Standards

NACE   National Association of Corrosion Engineers

NARM   Naturally Occurring and Accelerator-Produced Radioactive Materials

NEPA   National Environmental Policy Act

NESHAP   National Emission Standards for Hazardous Air Pollutants

NFPA   National Fire Protection Association

NIOSH   National Institute for Occupational Safety and Health

NPDES   National Pollution Discharge Elimination System

NRC   National Response Center

NRC   Nuclear Regulatory Commission

NRT   National Response Team

NSPS   New Source Performance Standards

NTIS   National Technical Information System

NVLAP   National Voluntary Laboratory Accreditation Program

ODC   Ozone Depleting Chemical

OSHA   Occupational Safety and Health Administration

OSH Act   Occupational Safety and Health Act

OVA   Organic Vapor Analyzer

PCBs   Polychlorinated Biphenyls

PEL   Permissible Exposure Limit

PI   Plasticity Index

PID   Photoionization Detector

PMN   Premanufacture Notice

PPE   Personal Protective Equipment

POTW   Publicly Owned Treatment Works

PQL   Practical Quantitation Limit

RACT   Reasonably Available Control Technology

RCRA   Resource Conservation and Recovery Act

SARA   Superfund Amendments and Reauthorization Act

SCBA   Self-Contained Breathing Apparatus

SDWA   Safe Drinking Water Act

SE   Synchronous Excitation (Fluorescence Spectroscopy)

SERC   State Emergency Response Commission

SFE   Supercritical Fluid Extraction

SPC   Statistical Process Control

SPCC   Spill Prevention, Control and Countermeasure

SWE   Single Wavelength Excitation (Fluorescence Spectroscopy)

TCLP   Toxicity Characteristic Leaching Procedure

TLV-TWA   Threshold Limit Value-Time Weighted Average

TQM   Total Quality Management

TRU   Transuranic Waste (Radioactive)

TSCA   Toxic Substance Control Act

TWA   Time Weighted Average

UEL   Upper Explosive Limit
UIC   Underground Injection Control
UL   Underwriters Laboratories
USGS   United States Geological Survey
USEPA   United States Environmental Protection Agency
UV   Ultraviolet
WRAP   Waste Reduction Assessments Program
WREAFS   Waste Reduction Evaluations at Federal Sites
WRITE   Waste Reduction Innovative Technology Evaluation

APPENDIX B

SELECTED STANDARDS OF THE AMERICAN NATIONAL STANDARDS INSTITUTE

STANDARDS RELATED TO RADIATION MONITORING

ANSI N13.1-1969(R1982)—Guide to Sampling Airborne Radioactive Materials in Nuclear Facilities

ANSI N13.2-1969(R1982)—Administrative Practices in Radiation Monitoring (A Guide for Management)

ANSI N13.3-1969(R1981)—Dosimetry for Criticality Accidents

ANSI N13.4-1971(R1983)—Specification of Portable X-or Gamma-Radiation Survey Instruments

ANSI N13.5-1972(R1989)—Performance and Specification for Direct Reading and Indirect Reading Pocket Dosimeters for X-and Gamma-Radiation

ANSI N13.6-1966(R1989)—Practice for Occupational Radiation Exposure Records System

ANSI N13.7-1983(R1989)—Criteria for Photographic Film Dosimeter Performance

ANSI N13.11-1983—Personnel Dosimeter Performance

ANSI N13.14-1983—Internal Dosimetry Standards for Tritium

ANSI N13.15-1985—Dosimetry Systems, Performance of Personnel Thermoluminescence

ANSI N13.27-1981—Dosimeters and Alarm Ratemeters, Performance Requirements for Pocket-Sized Alarm

ANSI N13.30 (Draft Standard)—Performance Criteria for Radiobioassay

ANSI N42.6-1980(R1991)—Interrelationship of Quartz-Fiber Electrometer Type Exposure Meters and Companion Meter Chargers

ANSI N42.14-1978(R1985)—Calibration and Usage of Germanium Detectors for Measurement of Gamma-Ray Emission of Radionuclides

ANSI N42.15-1990—Performance Verification of Liquid-Scintillation Counting Systems

ANSI N 42.17B-1989—Performance Specifications for Health Physics Instrumentation—Occupational Airborne Radioactivity Monitoring Instrumentation

ANSI N42.18-1988(R1991)—Specification and Performance of On-site Instrumentation for Continuously Monitoring Radioactivity in Effluents

ANSI N317-1980(R1991)—Performance Criteria for Instrumentation Used for Inplant Plutonium Monitoring

ANSI N320-1979(R1985)—Performance Specifications for Reactor Emergency Radiological Monitoring Instrumentation

ANSI N322-1977(R1991)—Inspection and Test Specification for Direct and Indirect Reading Quartz Fiber Pocket Dosimeters

ANSI N545-1975(R1982)—Performance Testing, and Procedural Specifications for Thermoluminescence Dosimetry: Environmental Applications

ANSI/IEEE 309-1970(R1991)—Test Procedures for Geiger-Muller Counters

ANSI/IEEE 325-1986—Test Procedures for Germanium Gamma-Ray Detectors

ANSI/IEEE 398-1972(R1991)—Standard Test Procedures for Photo-Multipliers for Scintillation Counting and Glossary for Scintillation Counting Field

STANDARDS RELATED TO TRANSPORTATION OF RADIOACTIVE MATERIALS

ANSI N2.1-1989—Warning Symbols—Radiation Symbol

ANSI N14-Guide—Administrative Guide for the Liability Insurance Aspects of Shipping Nuclear Materials

ANSI N14.1-1990—Nuclear Materials—Uranium Hexafluoride—Packaging for Transport

ANSI N14.5-1987—Radioactive Materials—Leakage Tests on Packages for Shipment

ANSI N14.6-1986—Radioactive Materials—Special Lifting Devices for Shipping containers Weighing 10 000 Pounds (4500 kg) or More

ANSI N14.19-1986—Ancillary Features of irradiated Fuel Shipping Casks

ANSI N14.24-1985—Domestic Barge Transport of Highway Route Controlled Quantities of Radioactive Materials

ANSI N14.27-1986—Carrier and Shipper Responsibilities and Emergency Response Procedures for Highway Transportation Accidents Involving Truckload Quantities of Radioactive Materials

OTHER STANDARDS RELATED TO RADIOACTIVE MATERIALS

ANSI/ASME N509-1989—Nuclear Power Plant Air-Cleaning Units and Components

ANSI/ASME N510-1989—Testing of Nuclear Air-Cleaning Systems

ANSI/ASME N626-1990—Qualifications and Duties for Authorized Nuclear Inspection Agencies and Personnel

ANSI/ASME N626.3-1988—Qualifications and Duties of Specialized Professional Engineers

ANSI/ASME NQA-1-1989—Quality Assurance Program Requirements for Nuclear Facilities

ANSI/ASME NQA-2-1989—Quality Assurance Requirements for Nuclear Power Plants

ANSI/ASME NQA-3-1989—Quality Assurance Requirement for the Collection of Scientific and Technical Information for Site characterization of High-Level Nuclear Waste Repositories

ANSI/ASME OM-1990—Operation and Maintenance of Nuclear Power Plants

STANDARDS RELATED TO PERSONAL PROTECTIVE EQUIPMENT

ANSI/ISEA 101-1985—Size and Labeling Requirements for Men's Limited-Use and Disposable Protective Coveralls

ANSI/NFPA 1971-1986—Protective Clothing for Structural Fire Fighting

ANSI Z87.1-1989—Practice for Occupational and Educational Eye and Face Protection

ANSI Z88.2-1980(R1992)—Practices for Respiratory Protection

ANSI Z88.5-1981—Practices for Respiratory Protection for the Fire Service

ANSI Z88.6-1984—Physical Qualification for Respirator Use

ANSI Z89.1-1986—Protective Headwear for Industrial Workers

ANSI Z358.1-1990—Emergency Eyewash and Shower Equipment

STANDARDS RELATED TO VENTILATION

ANSI Z9.2-1979—Fundamentals Governing the Design and Operation of Local Exhaust Systems

ANSI Z9.4-1985—Ventilation and Safe Practices of Abrasive Blasting Operations

ANSI/ASHRE 41.2-1987—Standard Methods for Laboratory Air Flow Measurement

ANSI/ASHRE 41.3-1989—Method for Pressure Measurement

ANSI/ASHRE 41.6-1982—Method of Measurement of Moist Air Properties

ANSI/ASHRE 55-1981—Thermal Environmental Conditions for Human Occupancy

ANSI/ASHRE 62-1989—Ventilation for Acceptable Indoor Air Quality

ANSI/ASHRE 62a-1991—Ventilation for Acceptable Indoor Air Quality

ANSI/ASHRE 110-1985—Method of Testing Performance of Laboratory Fume Hoods

ANSI/ASHRE 111-1988—Practices for Measurement, Testing, Adjusting, and Balancing of Building Heating, Ventilation, Air-Conditioning, and Refrigeration Systems

ANSI/ASHRE 113-1990—Method of Testing for Room Air Diffusion

ANSI/ASHRE 114-1986—Energy Management Control Systems Instrumentation

ANSI/ASHRE/IEEE 90A-1-1988—New Building Design, Energy Conservation

ANSI/ASHRE/IES 100.4 (1984)—Energy Conservation in Existing Facilities—Industrial

STANDARDS RELATED TO TANK AND PIPING SYSTEMS

ANSI/ASME B31G-1991—Manual for Determining the Remaining Strength of Corroded Pipelines

ANSI/ASME B31.3-1990—Chemical Plant and Petroleum Refinery Piping

ANSO/ASME B31.4-1989—Liquid Transportation Systems for Hydrocarbons, Liquid Petroleum Gas, Anhydrous Ammonia, and Alcohol

ANSI/ASME B31.8-1989—Gas Transmission and Distribution Piping Systems

ANSI/ASME B31.11-1989—Slurry Transportation Piping Systems

STANDARDS RELATED TO WASTE TREATMENT AND POLLUTION PREVENTION EQUIPMENT

ANSI/ASME PTC 33-1978(R1985)—Performance Test Code for Large Incinerators

ANSI/ASME PTC 33a-1980(R1987)—Performance Test Code for Large Incinerators

ANSI/ASME PTC 40-1991—Flue Gas Desulfurization Units

ANSI/ASME QRO-1-1989—Qualification and Certification of Resource Recovery Facility Operators

ANSI/ASME SPPE-1-1991—Quality Assurance and Certification of Safety and Pollution Prevention Equipment Used in Offshore Oil and Gas Operations

ANSI/ASME SPPE-2-1991—Quality Assurance and Certification of Safety and Pollution Prevention Equipment Used in Offshore Oil and Gas Operations

APPENDIX C

SELECTED STANDARDS OF THE AMERICAN SOCIETY OF TESTING AND MATERIALS

STANDARDS RELATED TO AIR MONITORING

ASTM D 1356-73(1991)—Standard Definitions of Terms Relating to Atmospheric Sampling and Analysis

ASTM D 1357-82(1989)—Standard Practice for Planning the Sampling of the Ambient Atmosphere

ASTM D 1605-60(1990)—Standard Practices for Sampling Atmospheres for Analysis of Gases or Vapors

ASTM D 1704-78—Test Method for Particulate Matter in the Atmosphere (Optical Density of Filtered Deposit)

ASTM D 2009-65(1990)—Recommended Practice for Collection by Filtration and Determination of Mass, Number, and Optical Sizing of Atmospheric Particulates

ASTM D-2913-87—Standard Test Method for Mercaptan Content of the Atmosphere

ASTM D 2914-1990—Test Method for Sulfur Dioxide Content of the Atmosphere (West-Gaeke Method)

ASTM D 3162-88—Test Method for Carbon Monoxide in the Atmosphere (Continuous Measurement by Nondispersive Infrared Spectrometry)

ASTM D 3249-90—Recommended Practice for General Ambient Air Analyzer Procedures

ASTM D 3416-88—Test Method for Total Hydrocarbons, Methane, and Carbon Monoxide (Gas Chromatographic Method) in the Atmosphere

ASTM D 3608-89—Test Method for Nitrogen Oxides (Combined) Content in the Atmosphere by the Greiss-Saltzman Reaction

ASTM D 3686-89—Standard Practice for Sampling Atmospheres to Collect Organic Compound Vapors (Activated Charcoal Tube Adsorption Method)

ASTM D 3687-89—Standard Practice for Analysis of Organic Compound Vapors Collected by Activated Charcoal Tube Adsorption Method (using Gas/Liquid Chromatography)

ASTM D 3824-88—Standard Test Method for Continuous Measurement of Oxides of Nitrogen in the Ambient or Workplace Atmosphere by the Chemiluminescent Method

ASTM D 4096-89—Practice for Application of Hi-Vol (High-Volume) Sampler Method for Collection and Mass Determination of Airborne Particulate Matter

ASTM D 5149-90—Test Method for Ozone in the Atmosphere: Continuous Measure by Ethylene Chemiluminescence

ASTM D 4240-83(1989)—Standard Test Method for Airborne Asbestos Concentration in Workplace Atmosphere

ASTM D 4490-90—Standard Practice for Measuring the Concentration of Toxic Gases or Vapors using Detector Tubes

ASTM D 4532-85(1990)—Standard Test Method for Respirable Dust in Workplace Atmospheres

ASTM D 4597-87—Standard Practice for Sampling Workplace Atmospheres to Collect Organic Gases or Vapors with Activated Charcoal Diffusional Samplers

ASTM D 4598-87—Standard Practice for Sampling Workplace Atmospheres to Collect Organic Gases or Vapors with Liquid Sorbent Diffusional Samplers

ASTM D 4599-86—Standard Practice for Measuring the Concentration of Toxic Gases or Vapors using Length-of-Stain Dosimeter

ASTM D 4844-88—Guide for Air Monitoring at Waste Management Facilities for Worker Protection

ASTM D 4861-91—Standard Practice for Sampling and Analysis of Pesticides and Polychlorinated Biphenyls in Indoor Atmospheres

ASTM D 4947-89—Standard Test Method for Chlordane and Heptachlor Residues in Indoor Air

ASTM D 5015-89—Test Method for pH of Atmospheric Wet Deposition Samples by Electrometric Determination

ASTM E 1370-90—Guide to Air Sampling Strategies for Worker and Workplace Protection

ASTM F 328-1980(1989)—Practice for Determining Counting and Sizing Accuracy of an Airborne Particle Counter Using Near-Monodisperse Spherical Particulate Materials

ASTM G 91-86—Test Method for Monitoring Atmospheric SO_2 Using Sulfation Plate Technique

STANDARDS RELATED TO TESTING OF PROTECTIVE FABRICS

ASTM D 751-89—Method of Testing Coated Fabrics

ASTM D 1003-92—Test Method for Haze and Luminous Transmittance of Transparent Fabrics

ASTM D 1043-92—Test Method for Stiffness Properties of Plastics as a Function of Temperature by Means of a Torsion Test

ASTM D 2582-92—Test Method for Puncture Propagation Tear Resistance of Plastic Film and Thin Sheeting

ASTM D 4157-92—Test Method for Abrasion Resistance of Textile Fabrics (Oscillatory Cylinder Method)

ASTM F 739-91—Test Method for Resistance of Protective Clothing Materials to Penetration by Liquids

ASTM F 903-90—Test Method for Resistance of Protective Clothing Materials to Penetration by Liquids

ASTM F 1001-89—Guide for Selection of Test Chemicals to Evaluate Protective Clothing Materials

ASTM F 1052-87(1991)—Practice for Pressure Testing of Gas-Tight Totally-Encapsulating Chemical Protective Suits

STANDARDS RELATED TO RADIATION PROTECTION

ASTM E 1167-87—Guide for a Radiation Protection Program for Decommissioning Operations

ASTM E 1168-87—Guide for Radiological Protection Training for Nuclear Facility Workers

STANDARDS RELATED TO CATHODIC PROTECTION

ASTM G 8-90—Test Method for Cathodic Disbonding of Pipeline Coatings

ASTM G 19-83—Standard Test Method for Disbonding Characteristics of Pipeline Coatings by Direct Soil Burial

ASTM G 42-90—Method for Cathodic Disbonding of Pipeline Coatings Subjected to Elevated or Cyclic Temperatures

ASTM G 80-88—Test Method for Specific Cathodic Disbonding of Pipeline Coatings

STANDARDS RELATED TO TANK TESTING

ASTM A 275/A 275M-90—Method for Magnetic Particle Examination of Steel Forgings

ASTM A 754-79(1990)—Test Method for Coating Thickness by X-ray fluorescence

ASTM C 868-85(1990)—Standard Test Method for Chemical Resistance of Protective Linings

ASTM C 982-88—Guide for Selecting Components for Generic Energy Dispersive X-Ray Fluorescence (XRF) Systems for Nuclear Related Material Analysis

ASTM C 1118-89—Guide for Selecting Components for Wavelength-Dispersive X-Ray Fluorescence (XRF) Systems

ASTM D 471-79(1991)—Test Method for Rubber Property—Effect of Liquids

ASTM D 3491-85—Standard Methods of Testing Vulcanizable Rubber Tank and Pipe Lining

ASTM E 125-63(1985)—Reference Photographs for Magnetic Particle Indications on Ferrous Castings

ASTM E 165-91—Practice for Liquid Penetrant Inspection Method

ASTM E 432-91—Guide for the Selection of a Leak Testing Method

ASTM E 433-71(1985)—Reference Photographs for Liquid Penetrant Inspection

ASTM E 709-80(1985)—Practice for Magnetic Particle Examination

ASTM E 750-88—Practice for Measuring Operating Characteristics of Acoustic Emission Instrumentation

ASTM E 976-84(1988)—Guide for Determining the Reproducibility of Acoustic Emission Sensor Response

ASTM E 1002-86—Method for Testing for Leaks Using Ultrasonics Ultrasonics

ASTM E 1003-84(1990)—Method for Hydrostatic Leak Testing

ASTM E 1066-85(1990)—Method for Ammonia Colorimetric Leak Testing

ASTM E 1067-89—Practice for Acoustic Emission Testing of Fiberglass Reinforced Plastic Resin (FRP) Tanks/Vessels

ASTM E 1139-87—Practice for Continuous Monitoring of Acoustic Emission from Metal Pressure Boundaries

ASTM E 1172-87—Practice for Describing and Specifying a Wavelength-Dispersive X-Ray Spectrometer

ASTM E 1208-91—Test Method for Fluorescent Liquid Penetrant Examination Using the Lipophilic Post-Emulsification Process

ASTM E 1209-91—Test Method for Fluorescent Penetrant Examination Using the Water-Washable Process

ASTM E 1210-91—Test Method for Fluorescent Penetrant Examination Using the Hydrophilic Post-Emulsification Process

ASTM E 1211-87—Practice for Leak Detection and Location Using Surface-Mounted Acoustic Emission Sensors

ASTM E 1219-91—Test Method for Fluorescent Penetrant Examination Using the Solvent-Removable Process

ASTM E 1220-91—Test Method for Visible Penetrant Examination Using the Solvent-Removable Process

ASTM G 1-90—Recommended Practice for Preparing, Cleaning, and Evaluating Corrosion Test Specimens

ASTM G 4-84—Guide for Conducting Corrosion Coupon Tests in Plant Equipment

ASTM G 5-87—Reference Test Method for Making Potentiostatic and Potentiodynamic Anodic Polarization Measurements

ASTM G 15-90—Terminology Relating to Corrosion and Corrosion Testing

ASTM G 41-90—Practice for Determining Cracking Susceptibility of Metals Exposed Under Stress to a Hot Salt Environment

ASTM G 46-76(1986)—Recommended Practice for Examination and Evaluation of Pitting Corrosion

ASTM G 48-76(1980)—Testing for Pitting and Crevice Corrosion Resistance of Stainless Steels and Related Alloys by Use of Ferric Chloride

ASTM G 49-85(1990)—Recommended Practice for Preparation and Use of Direct Tension Stress Corrosion Test Specimen

ASTM G 50-76(1984)—Recommended Practice for Conducting Atmospheric Corrosion Tests on Metals

ASTM G 51-77(1984)—Test Method for pH of Soil for Use in Corrosion Testing

ASTM G 62-87—Test Method for Holiday Detection in Pipeline Coatings

ASTM G 71-81(1986)—Practice for Conducting and Evaluating Galvanic Corrosion Tests in Electrolytes

ASTM G 78-89—Guide for Crevice Corrosion Testing of Iron Base and Nickel Base Stainless Alloys in Seawater and Other Chloride-Containing Aqueous Environments

ASTM G 82-83(1989)—Guide for Development and Use of a Galvanic Series for Predicting Galvanic Corrosion Performance

ASTM G 90-85—Practice for Performing Accelerated Outdoor Weathering of Nonmetallic Materials Using Concentrated Natural Sunlight

ASTM G 96-90—Guide for On-Line Monitoring of Corrosion in Plant Equipment (Electrical and Electrochemical Methods)

ASTM G 97-89—Test Method for Laboratory Evaluation of Magnesium Sacrificial Anode Test Specimens for Underground Applications

ASTM G 101-89—Guide for Estimating the Atmospheric Corrosion Resistance of Low-Alloy Steels

ASTM G 104-89—Test Method for Assessing Galvanic Corrosion Caused by the Atmosphere

OTHER STANDARDS RELATED TO TANK SYSTEMS

ASTM D 2487-85—Standard Test Method for Classification of Soils for Engineering Purposes

ASTM D 3299-88—Specification for Filament-Wound Glass Fiber Reinforced Thermoset Resin Chemical-Resistant Tanks

ASTM D 4021-86—Standard Specification for Glass-Fiber-Reinforced Polyester Underground Petroleum Storage Tanks

ASTM D 4097-88—Specification for Contact Molded Glass-Fiber-Reinforced Thermoset Resin Chemical-Resistant Tanks

ASTM D 4865-90—Guide for Generation and Dissipation of Static Electricity in Petroleum Fuel Systems

ASTM F 670-87—Specification for 5 and 10 Gallon Lube Oil Dispensing Tanks

STANDARDS RELATED TO TESTING OF LANDFILL LINERS

ASTM D 2434-68(1974)—Test Method for Permeability of Granular Soils (Constant Head)

ASTM D 4491-89—Test Methods for Water Permeability of Geotextiles by Permittivity

ASTM D 4716-87—Test Method for Constant Head Hydraulic Transmissivity (In-Plane Flow) of Geotextiles and Geotextile Related Products

ASTM D 4751-87—Test Method for Determining Apparent Opening Size of a Geotextile

ASTM E 96-80—Test Methods for Water Vapor Transmission

STANDARDS RELATED TO GROUNDWATER MONITORING

ASTM C 150-89—Specification for Portland Cement

ASTM D 1785-89—Specification for Poly (Vinyl Chloride)(PVC) Plastic Pipe, Schedules 40, 80, and 120

ASTM D 2113-83(1987)—Method for Diamond Core Drilling for Site Investigation

ASTM F 480-90—Specification for Thermoplastic Water Well Casing Pipe and Couplings Made in Standard Dimension Ratio (SDR)

STANDARDS RELATED TO SOILS INVESTIGATION

ASTM D 420-87—Recommended Practice for Investigating and Sampling Soil and Rock for Engineering Purposes

ASTM D 421-85—Practice for Dry Preparation of Soil Samples for Particle-Size Analysis and Determination of Soil Constants

ASTM D 422-63(1990)—Method for Particle-Size Analysis of Soils

ASTM D 1452-80(1990)—Practice for Soil Investigation and Sampling by Auger Borings

ASTM D 1586-84—Method for Penetration Test and Split-Barrel Sampling of Soils

ASTM D 1587-83—Method for Thin-Walled Tube Sampling of Soils

ASTM D 2216-90—Test Method for Laboratory Determination of Water (Moisture) Content of Soil and Rock

ASTM D 2217-85—Practice for Wet Preparation of Soil Samples for Particle-Size Analysis and Determination of Soil Constants

ASTM D 2487-90—Classification of Soils for Engineering Purposes

ASTM D 2488-90—Practice for Description and Identification of Soils (Visual-Manual Procedure)

ASTM D 4318-84—Test Method for Liquid Limit, Plastic Limit, and Plasticity Index of Soils

ASTM D 4452-85(1990)—Methods for X-Ray Radiography of Soil Samples

ASTM D 5079-90—Practices for Preserving and Transporting Rock Core Samples

APPENDIX D

SELECTED STANDARDS OF THE NATIONAL FIRE PROTECTION ASSOCIATION

STANDARDS RELATED TO FIRE PROTECTION SYSTEMS

NFPA 11 (1988)—Low Expansion Foam and Combined Agent Systems

NFPA 11A (1988)—Medium and High Expansion Foam Systems

NFPA 11C (1990)—Mobile Foam Apparatus

NFPA 12 (1993)—Carbon Dioxide Extinguishing Systems

NFPA 12A (1992)—Halon 1301 Fire Extinguishing Systems

NFPA 12B (1990)—Halon 1211 Fire Extinguishing Systems

NFPA 13 (1991)—Installation of Sprinkler Systems

NFPA 13A (1987)—Inspection, Testing, and Maintenance of Sprinkler Systems

NFPA 15 (1990)—Water Spray Fixed Systems

NFPA 16 (1991)—Deluge Foam-Water Sprinkler and Spray Systems

NFPA 16A (1988)—Installation of Closed-Head Foam-Water Sprinkler Systems

NFPA 17 (1990)—Dry Chemical Extinguishing Systems

NFPA 25 (1992)—Inspection, Testing, and Maintenance of Water Based Fire Protection Systems

STANDARDS RELATED TO IDENTIFICATION OF FIRE HAZARDS OF CHEMICALS AND MATERIALS

NFPA 49 (1991)—Hazardous Chemicals Data

NFPA 321 (1991)—Basic Classification of Flammable and Combustible Liquids

NFPA 325M (1991)—Fire Hazard Properties of Flammable Liquids, Gases, and Volatile Solids

NFPA 491M (1991)—Hazardous Chemical Reactions

NFPA 495 (1992)—Explosives Materials Code

NFPA 704 (1990)—Identification of Fire Hazards of Materials

STANDARDS RELATED TO STORAGE OF HAZARDOUS MATERIALS

NFPA 30 (1990)—Flammable and Combustible Liquids Code

NFPA 30B (1990)—Manufacture and Storage of Aerosol Products

NFPA 40E (1993)—Storage of Pyroxylin Plastics

NFPA 43A (1990)—Storage of Liquid and Solid Oxidizers

NFPA 43B (1993)—Storage of Organic Peroxide Formulations

NFPA 43C (1986)—Storage of Gaseous Oxidizing Materials

NFPA 43D (1986)—Storage of Pesticides in Portable Containers

NFPA 55 (1993)—Standard for Storage, Use and Handling of Compressed and Liquified Gases in Portable Cylinders

NFPA 58 (1992)—Standard for Storage and Handling of Liquefied Petroleum Gases

NFPA 59 (1992)—Storage and Handling of Liquefied Petroleum Gases at Utility Gas Plants

NFPA 59A (1990)—Production, Storage, and Handling of Liqu3fied Natural Gas (LNG)

NFPA 231C (1991)—Rack Storage of Materials

NFPA 327 (1987)—Cleaning or Safeguarding Small Tanks and Containers

NFPA 329 (1992)—Underground Leakage of Flammable and Combustible Liquids

NFPA 490 (1993)—Storage of Ammonium

STANDARDS RELATED TO PERSONAL PROTECTIVE EQUIPMENT

NFPA 1404 (1989)—Fire Department Self-Contained Breathing Apparatus Program

NFPA 1500 (1992)—Fire Department Safety and Health Programs

NFPA 1971 (1991)—Protective Clothing for Structural Fire Fighting

NFPA 1972 (1992)—Helmets for Structural Fire Fighting

NFPA 1973 (1988)—Gloves for Structural Fire Fighting

NFPA 1974 (1992)—Protective Footwear for Structural Fire Fighting

NFPA 1976 (1992)—Protective Clothing for Proximity Fire Fighting

NFPA 1981 (1992)—Open Circuit Self-Contained Breathing Apparatus for Fire Fighters

NFPA 1991 (1990)—Vapor-Protective Suits for Hazardous Chemical Emergencies

NFPA 1992 (1990)—Liquid Splash-Protective Suits for Hazardous Chemical Emergencies

NFPA 1993 (1990)—Support Function Protective Garments for Hazardous Chemical Operations

NFPA 1999 (1992)—Protective Clothing for Emergency Medical Operations

STANDARDS RELATED TO MANUFACTURING AND HANDLING OF HAZARDOUS MATERIALS

NFPA 35 (1987)—Manufacture of Organic Coatings

NFPA 91 (1992)—Exhaust Systems for Air Conveying of Materials

NFPA 480 (1987)—Storage, Handling, and Processing of Magnesium

NFPA 481 (1987)—Production, Processing, Handling, and Storage of Titanium

NFPA 482 (1987)—Production, Processing, Handling, and Storage of Zirconium

NFPA 650 (1990)—Pneumatic Conveying Systems for Handling Combustible Materials

STANDARDS RELATED TO FIRE PROTECTION OF SPECIFIC INDUSTRIES OR PROCESSES

NFPA 32 (1990)—Dry Cleaning Plants

NFPA 33 (1989)—Spray Application Using Flammable and Combustible Materials

NFPA 34 (1989)—Dipping and Coating Processes Using Flammable or Combustible Liquids

NFPA 36 (1993)—Solvent Extraction Plants

NFPA 45 (1991)—Fire Protection for Laboratories Using Chemicals

NFPA 120 (1988)—Coal Preparation Plants

NFPA 654 (1988)—Prevention of Fire and Dust Explosions in the Chemical, Dye, Pharmaceutical, and Plastics Industries

NFPA 801 (1991)—Recommended Fire Protection Practice for Facilities Handling Radioactive Materials

NFPA 802 (1993)—Nuclear Research and Production Reactors

NFPA 803 (1993)—Standard for Fire Protection for Light Water Nuclear Power Plants

NFPA 820 (1992)—Fire Protection for Wastewater Treatment and Collection Facilities

STANDARDS RELATED TO VENTILATION AND EXHAUST SYSTEMS

NFPA 90A (1993)—Installation of Air Conditioning and Ventilating Systems

NFPA 90B (1993)—Installation of Warm Air Heating and Air Conditioning Systems

NFPA 204M (1991)—Smoke and Heat Venting

STANDARDS RELATED TO FIRE AND LIFE SAFETY

NFPA 69 (1992)—Explosion Prevention Systems

NFPA 101 (1991)—Life Safety Code®

NFPA 101M (1992)—Alternate Approaches to Life Safety

NFPA 110 (1993)—Emergency and Standby Power Systems

NFPA 111 (1993)—Stored Energy Emergency and Standby Power Systems

NFPA 170 (1991)—Standard Fire Safety Symbols

NFPA 220 (1992)—Standard Types of Building Construction

NFPA 241 (1989)—Safeguarding Building Construction and Demolition Operations

NFPA 251 (1990)—Fire Tests of Building Construction and Materials

NFPA 252 (1990)—Fire Tests of Door Assemblies

NFPA 253 (1990)—Test for Critical Radiant Flux of Floor Covering Systems Using a Radiant Heat Energy Source

NFPA 255 (1990)—Method of Test of Surface Burning Characteristics of Building Materials

NFPA 256 (1993)—Methods of Fire Test of Roof Coverings

NFPA 257 (1990)—Fire Tests of Window Assemblies

NFPA 780 (1992)—Lightning Protection

STANDARDS RELATED TO TRANSPORTATION OF HAZARDOUS MATERIALS

NFPA 306 (1993)—Control of Gas Hazards on Vessels

NFPA 385 (1990)—Tank Vehicles for Flammable and Combustible Liquids

NFPA 386 (1990)—Portable Shipping Tanks for Flammable and Combustible Liquids

NFPA 1124 (1988)—Manufacture, Transportation, and Storage of Fireworks

STANDARDS RELATED TO INCIDENT MANAGEMENT

NFPA 471 (1992)—Responding to Hazardous Materials Incidents

NFPA 472 (1992)—Professional Competence of Responders to Hazardous Materials Incidents

NFPA 473 (1992)—Competencies for EMS Personnel Responding to Hazardous Materials Incidents

NFPA 495 (1990)—Explosive Materials Code

NFPA 902M (1990)—Fire Reporting Field Incident Manual

NFPA 903 (1992)—Fire Reporting Property Survey Manual

NFPA 904 (1992)—Incident Follow-up Report Manual

NFPA 906M (1988)—Guide for Fire Incident Field Notes

NFPA 921 (1992)—Guide for Fire and Explosion Investigations

NFPA 1033 (1993)—Professional Qualifications for Fire Investigator

NFPA 1405 (1990)—Guide for Landbased Fire Fighters Who Respond to Marine Vessel Fires

NFPA 1561 (1990)—Standard on Fire Department Incident Management System

APPENDIX E

SELECTED STANDARDS, RECOMMENDED PRACTICES, AND PUBLICATIONS OF THE AMERICAN PETROLEUM INSTITUTE

DOCUMENTS RELATED TO FIRE PROTECTION

API Recommended Practice 2001 (1984)—Fire Protection in Refineries, 6th ed.

API Publication 2008 (1984)—Safe Operation of Inland Bulk Plants, 4th ed.

API Publication 2009 (1988)—Safe Welding and Cutting Practices in Refineries, Gasoline Plants, and Petrochemical Plants, 5th ed.

API Publication 2021 (1991)—Guide for Fighting Fires In and Around Petroleum Storage Tanks, 3rd ed.

API Publication 2027 (1988)—Ignition Hazards Involved in Abrasive Blasting of Atmospheric Hydrocarbon Tanks In Service, 2nd ed.

API Publication 2028 (1991)—Flame Arresters in Piping Systems, 2nd ed.

API Publication 2201 (1985)—Procedures for Welding or Hot Tapping on Equipment Containing Flammables, 3rd ed.

API Publication 2210 (1986)—Flame Arrester for Vents of Tanks Storing Petroleum Products, 2nd ed.

API Publication 2300 (1985)—Evaluation of Fire-Fighting Foams as Fire Protection for Alcohol Containing Fuels

DOCUMENTS RELATED TO TANK SYSTEMS

API Recommended Practice 2X (1988)—Recommended Practice for Ultrasonic Examination of Offshore Structural Fabrication and Guidelines for Qualification of Ultrasonic Technicians, 2nd ed.

API Publication 301 (1991)—Aboveground Storage Tank Survey

API Publication 306 (1991)—An Engineering Assessment of Volumetric Methods of Leak Detection in Aboveground Storage Tanks

API Publication 307 (1991)—An Engineering Assessment of Acoustic Methods of Leak Detection In Aboveground Tanks

API Standard 510 (1989)—Pressure Vessel Inspection Code, 6th ed.

API Recommended Practice 520 (1990)—Sizing, Selection, and Installation of Pressure Relieving Systems in Refineries, Part 1, "Sizing and Selection," 5th ed.

API Recommended Practice 520 (1988)—Sizing, Selection, and Installation of Pressure Relieving Systems in Refineries, Part 2, "Installation," 3rd ed.

API Recommended Practice 521 (1990)—Guide for Pressure-Relieving and Depressuring Systems, 3rd ed.

API Standard 620 (1990)—Design and Construction of Large, Welded, Low-Pressure Storage Tanks, 8th ed.

API Standard 650 (1988)—Welded Steel Tanks for Oil Storage, 8th ed.

API Recommended Practice 651 (1991)—Cathodic Protection of Aboveground Petroleum Storage Tanks

API Recommended Practice 652 (1991)—Lining of Aboveground Petroleum Storage Tank Bottoms

API Standard 653 (1991)—Tank Inspection, Repair, Alteration, and Reconstruction

API Recommended Practice 1110 (1981)—Recommended Practice for the Pressure Testing of Liquid Petroleum Pipelines

API Recommended Practice 1615 (1987)—Installation of Underground Petroleum Storage Systems, 4th ed.

API Recommended Practice 1621 (1987)—Recommended Practice for Bulk Liquid Stock Control at Retail Outlets

API Recommended Practice 1631 (1992)—Interior Lining of Underground Storage Tanks, 3rd ed.

API Publication 1632 (1987)—Cathodic Protection of Underground Petroleum Storage Tanks and Piping Systems, 2nd ed.

API Publication 2015 (1991)—Cleaning Petroleum Storage Tanks, 4th ed.

API Recommended Practice 2350 (1987)—Overfill Protection for Petroleum Storage Tanks

DOCUMENTS RELATED TO HAZARDOUS MATERIALS TRANSPORTATION

API Publication 1003 (1986)—Precautions Against Electrostatic Ignition During Loading of Tank Truck Motor Vehicles, 3rd ed.

API Recommended Practice 1112 (1984)—Recommended Practice for Developing a Highway Emergency Response Plan for Incidents Involving Hazardous Materials

API Recommended Practice 1122 (1991)—Emergency Preparedness and Response for Hazardous Liquids Pipeline

API Recommended Practice 1125 (1991)—Overfill Control Systems for Tank Barges

DOCUMENTS RELATED TO AIR QUALITY

API Publication 4311 (1979)—NO_x Emissions from Petroleum Industry Operations

API Publication 4322 (1980)—Fugitive Hydrocarbon Emissions from Petroleum Production Operations, Vols. 1 and 2

API Publication 4387 (1983)—Model Performance Evaluation for Offshore Releases

API Publication 4392 (1985)—Test Protocol for Automotive Evaporative Emissions

API Publication 4403 (1985)—Uncertainties Associated with Modeling Regional Haze in the Southwest

API Publication 4409 (1985)—Development and Application of a Simple Method for Evaluating Air Quality Models

API Publication 4413 (1985)—Evaluation of Proposed Downwash Modifications to the Industrial Source Complex Model

API Publication 4421 (1986)—Plume Rise Research for Refinery Facilities

DOCUMENTS RELATED TO GROUNDWATER AND SOILS

API Publication 1628 (1989)—A Guide to the Assessment and Remediation of Underground Petroleum Releases, 2nd ed.

API Publication 4410 (1985)—Subsurface Venting of Hydrocarbons from an Underground Aquifer

API Publication 4416 (1986)—Guide to State Groundwater Programs and Standards

API Publication 4442 (1986)—Cost Model for Selected Technologies for Removal of Gasoline Components from Groundwater

API Publication 4448 (1987)—Field Study of Enhanced Subsurface Biodegradation of Hydrocarbons Using Hydrogen Peroxide as an Oxygen Source

API Publication 4449 (1987)—Manual of Sampling and Analytical Methods for Petroleum Hydrocarbons in Groundwater and Soil

API Publication 4454 (1987)—Capability of EPA Methods 624 and 625 to Measure Appendix IX Compounds

API Publication 4476 (1989)—Hydrogeologic Data Base for Groundwater Modeling

API Publication 4497 (1991)—Cost-Effective, Alternative Treatment Technologies for Reducing the Concentrations of Ethers and Alcohols in Groundwater

API Publication 4499 (1989)—Evaluation of Analytical Methods for Measuring Appendix IX Constituents in Groundwater

API Publication 4510 (1991)—Technological Limits of Groundwater Remediation: A Statistical Evaluation Method

API Publication 4525 (1990)—A Compilation of Field-Collected Cost and Treatment Effectiveness Data for the Removal of Dissolved Gasoline Components from Groundwater

DOCUMENTS RELATED TO PROCESS SAFETY

API Recommended Practice 750 (1990)—Management of Process Hazards

API Publication 2219 (1986)—Safe Operating Guidelines for Vacuum Trucks in Petroleum Service

API Recommended Practice 2220 (1991)—Improving Owner and Contractor Safety Performance

APPENDIX F

SELECTED STANDARDS OF THE UNDERWRITERS LABORATORIES

STANDARDS RELATED TO FIRE PROTECTION

UL 8 (1990)—Foam Fire Extinguishers

UL 154 (1990)—Carbon-Dioxide Fire Extinguishers

UL 162 (1989)—Foam Equipment and Liquid Concentrates

UL 193 (1988)—Alarm Valves for Fire-Protection Service

UL 199 (1990)—Automatic Sprinklers for Fire-Protection Service

UL 268 (1989)—Smoke Detectors for Fire Protective Signaling Systems

UL 268A (1983)—Smoke Detectors for Duct Application

UL 299 (1990)—Dry Chemical fire Extinguishers

UL 346 (1988)—Waterflow Indicators for Fire Protective Signaling Systems

UL 401 (1986)—Portable Spray Hose Nozzles for Fire Protection Service

UL 448 (1984)—Pumps for Fire Protection Service

UL 497B (1987)—Protectors for Data Communication and Fire Alarm Circuits

UL 521 (1988)—Heat Detectors for Fire Protective Signaling Systems

UL 539 (1991)—Single and Multiple Station Heat Detectors

UL 626 (1990)—2½ Gallon Stored-Pressure, Water-Type Fire Extinguishers

UL 711 (1990)—Rating and Fire Testing of Fire Extinguishers

UL 753 (1989)—Alarm Accessories for Automatic Water-Supply Control Valves for Fire Protection Service

UL 864 (1991)—Control Units for Fire-Protective Signaling Systems

UL 1058 (1989)—Halogenated Agent Extinguishing System Units

UL 1093 (1990)—Halogenated Agent Fire Extinguishers

UL 1254 (1992)—Pre-Engineered Dry Chemical Extinguishing System Units

UL 1767 (1990)—Early-Suppression Fast-Response Sprinklers

UL 2006 (1991)—Halon 1211 Recovery/Recharge Equipment

STANDARDS RELATED TO VENTILATION SYSTEMS

UL 181 (1990)—Factory-Made Air Ducts and Air Connectors

UL 181A (1991)—Closure Systems for Use with Rigid Air Ducts and Air Connectors

UL 474 (1987)—Dehumidifiers

UL 555 (1990)—Fire Dampers

UL 555S (1983)—Leakage Rated Dampers for Use in Smoke Control Systems

UL 586 (1990)—High-Efficiency, Particulate Air Filter Units

UL 867 (1989)—Electrostatic Air Cleaners

UL 900 (1987)—Test Performance of Air Filter Units

UL 998 (1987)—Humidifiers

UL 1046 (1979)—Grease Filters for Exhaust Ducts

UL 1784 (1990)—Air Leakage Tests of Door Assemblies

STANDARDS RELATED TO MATERIAL IDENTIFICATION

UL 340 (1979)—Test for Comparative Flammability of Liquids

UL 746A (1990)—Polymeric Materials—Short Term Property Evaluations

UL 746B (1979)—Polymeric Materials—Long Term Property Evaluations

STANDARDS RELATED TO STORAGE OF HAZARDOUS MATERIALS

UL 1275 (1985)—Flammable Liquid Storage Cabinets

UL 1314 (1989)—Special-Purpose Containers

UL 1853 (1991)—Nonreusable Plastic Containers for Flammable and Combustible Liquids

STANDARDS RELATED TO TANK SYSTEMS

UL 58 (1986)—Steel Underground Tanks for Flammable and Combustible Liquids

UL 142 (1987)—Steel Aboveground Tanks for Flammable and Combustible Liquids

UL 180 (1991)—Liquid-Level Indicating Gauges and Tank-Filling Signals for Petroleum Products

UL 353 (1989)—Limit Controls

UL 443 (1990)—Steel Auxiliary Tanks for Oil-Burner Fuel

UL 525 (1984)—Flame Arresters for Use on Vents of Storage Tanks for Petroleum Oil and Gasoline

UL 1238 (1975)—Control Equipment for Use with Flammable Liquid Dispensing Devices

UL 1316 (1983)—Glass-Fiber-Reinforced Plastic Underground Storage Tanks for Petroleum Products

UL 1746 (1989)—External Corrosion Protection Systems for Steel Underground Storage Tanks

STANDARDS RELATED TO FIRE AND LIFE SAFETY

UL 96 (1985)—Lightning Protection Components

UL 96A (1982)—Installation Requirement for Lightning Protection Systems

UL 263 (1992)—Fire Tests of Building Construction and Materials

UL 410 (1992)—Slip Resistance of Floor Surface Materials

UL 924 (1990)—Emergency Lighting and Power Equipment

UL 969 (1989)—Marking and Labeling Systems

UL 1715 (1989)—Fire Test of Interior Finish Material

INDEX